Computation and Communication Technologies

Computation and Communication Technologies

—

Edited by
Kumar T. Senthil, B. Mathivanan

DE GRUYTER

Editors

Dr. Senthil Kumar T
Amrita School of Engineering
India
senkumar604@gmail.com

Dr. B Mathivanan
Sri Ramakrishna Engg. College
India
mathivananbala6@gmail.com

ISBN 978-3-11-045007-1
e-ISBN (PDF) 978-3-11-045010-1
Set-ISBN 978-3-11-045293-8

Library of Congress Cataloging-in-Publication Data
A CIP catalog record for this book has been applied for at the Library of Congress.

Bibliographic information published by the Deutsche Nationalbibliothek
The Deutsche Nationalbibliothek lists this publication in the Deutsche Nationalbibliografie;
detailed bibliographic data are available on the Internet at http://dnb.dnb.de.

© 2016 Walter de Gruyter GmbH, Berlin/Boston
Printing and binding: CPI books GmbH, Leck
♾ Printed on acid-free paper
Printed in Germany

www.degruyter.com

Preface

I am happy to write the preface for the IDES joint International conferences. The joint conferences have attracted 39 papers for registration for the conference. The topics are Computer Science, Electrical and Electronics. There has been good response for the conference from the different parts of the world. However keeping in concern the quality of the conference and the current need to reach quality to the audience, the organizers has made sincere effort to keep the quality of the conference high.

The conference has attracted papers in areas of computer science in thrust areas of machine learning, networking, cloud computing. There has been papers accepted giving focus to modelling in applications like facial recognition systems. Certain papers have good quality as it shows several months of extensive research being done in the same area. I wish the conference organizers a grand success and suggest them to conduct more quality conferences in the future for the research community. I suggest the participants to present the papers in good spirit and take discussions with the research experts as part of the conference.

Dr. T. Senthil Kumar
Amrita School of Engineering, India

Editorial Board

Editors-in-Chief

Dr. Senthil Kumar T (Amrita School of Engineering, India)
Dr. B Mathivanan (Sri Ramakrishna Engg. College, India)

Committees

Honorary Chair
Dr. Pawel Hitczenko (Drexel University, USA)
Dr. Jiguo Yu (Qufu Normal University, China)

Technical Chair
Dr. Narayan C Debnath (Winona State University, USA)
Dr. Xuelong Li (Birkbeck College, University of London, U.K.)

General Chair
Dr. Janahanlal Stephen (Matha College of Technology, India)
Dr. Yogesh Chaba (Guru Jambeswara University, India)

Organizing Chair
Prof. K U Abraham (Holykings College of Engg & Tech, India)

Publication Chair
Dr. Rajiv Pandey (Amity University, India)

Publicity Chair
Prof. Ford Lumban Gaol (University of Indonesia)
Dr. Amlan Chakrabarti (University of Culcutta, India)
Prof. Prafulla Kumar Behera, PhD(Utkal University, India)
Dr. O. P. Rishi (BITS Ranchi, India)
Dr. A Louise Perkins (University of Sounth Mississippi, USA)
Dr. R D Sudhakar Samueli (S J College of Engg., India)

International Committee
Dr. Yuan-Ting Zhang, Hong Kong
Dr. Koichi Shimizu, Japan
Dr. Dimitris Lymperopoulo, Greek
Dr. Frank Emmert-Streib, Queen's University Belfast, UK
Dr. Liviu MOLDOVAN, Romania
Dr. Bogdan Patrut, Romania
Dr. Valentina Emilia Balas, Romania
Dr. Constantinos Pattichis, University of Cyprus
Dr. Thomas Falck, Netherlands

Programming Committee
Dr. Laura M. Roa Romero, University of Valencia, Spain
Dr. Abdel-Badeeh M. Salem, Ain Shames University, Egypt
Dr. Luminita Duta, Valahia University of Targoviste, Romania
Dr. Andriana Prentza, University of Piraeus, Greek
Dr. Elena Nechita, University Vasile Alecsandri of Bacau, Romania
Dr. Alexander Bekiarski, Technical University of Sofia, Bulgaria
Dr. Viorel Negru, West University of Timisoara, Romania
Dr. Ilias Maglogiannis, University of Central Greece

Technical Co-Chair
Dr Pingkun Yan (Philip Research North America)
Prof Yuan Yuan (Aston University, UK)
Prof. Dacheng Tao (NTU, Singapore)
Prof. Bing Xiao (Chinese Academy of Sciences)
Dr Yuancheng Huang (Chinese Academy of Sciences)
Dr shiming Ge (Nokia Research)
Dr Richang Hong (National University of Singapore)
Dr Jingen Liu (UCF, US)
Dr Yuping Shen (UCF, US)
Dr Qieshi Zhang (Wasade University)
Dr Jun Zhang (Waseda University)
Dr Xiaoqiang Lu (Chinese Academy of Sciences)
Dr Yi Tang (Chinese Academy of Sciences)
Dr Bin Wang (Xidian University)
Dr Xiumei Wang (Xidian University)
Dr Biqin Song (Hubei University)
Prof. Wuxia Zhang (Chinese Academy of Sciences)
Prof. Meijuan Yang (Chinese Academy of Sciences)
Prof. Xiaofei He (Zhejiang University)
Dr Dell Zhang (University of London)
Prof Yanwei Pang (Tianjin University, China)
Dr Huiyu Zhou (Queens Univ, N. Ireland)
Dr Xianbin Cao (Beihang University)
Prof Liangpei Zhang (Wuhan University, China)
Prof Jialie Shen (Singapore Management University, Singapore)
Dr Qingxiang Wu (Ulster University, N. Ireland)
Dr Xiao Bai (Bath University, UK)
Dr Jing Li (Sheffield University, U.K.)

Prof Kaiqi Huang (Chinese Academy of Sciences)
Prof Xiaowei Shao (Tokyo University, Japan)
Prof Mingli Song (Hong Kong Polytechnic University, Hong Kong)
Dr Tianhao Zhang (UPENN, US)

Table of Contents

1. Iris Recognition using Support Vector Machine —— 1–10
    Adeetya Sawant and Sudha Gupta

2. Nonlinear Unknown Input Sliding Mode Observer for Chaotic Synchronization
    and Message Recovery —— 11–26
    Vivek Sharma, Sharma B B and Nath R

3. An Efficient Lightweight Proactive Source Routing Protocol for Handling Overhead,
    Packet Delivery, Throughput in MANETS —— 27–47
    Rengarajan A, Rajasekaran S and Kumaran P

4. Classification of Mobile Videos on Sports Genre —— 48–56
    Neeharika K and Ravi Kishan S

5. Genetic Algorithm with Dual Mutation Probabilities for TCSC based
    ATC Enhancement —— 57–72
    Naresh Kumar Yadav

6. WSN Lifetime Enhancement through Residual energy balancing
    by Avoiding Redundant Data Transmission —— 73–83
    Mohanaradhya and Sumithra Devi K A

7. Neural Network based Facial Expression Recognition System —— 84–98
    Rutooja D Gholap and Saylee Gharge

8. Indoor Surveillance Robot with Live Video Transmission —— 99–108
    Vinay Kumar S B, Naveen Kumar S, Monica B V and Ravi U S

9. In-Page Semantic Ranking of Snippets for WebPages —— 109–120
    Leena Giri G, Praveen Gowda I V, Manjula S H, Venugopal K R and Patnaik L M

10. Information Diffusion in Online Social Network: Techniques, Applications and
    Challenges —— 121–134
    Kumaran P and Chitrakala S

11. Harmony Search Algorithm for PAPR Reduction —— 135–143
    Shruti S Sharma and Achala Deshmukh

12. Closed Loop Control System based Hybrid Energy System Model —— 144–156
Allam Venkata Surendra Babu, Pradeepa S and Siva Subba Rao Patange

13. Topology Preserving Skeletonization Techniques for Grayscale Images —— 157–170
Sri Krishna A, Gangadhar K, Neelima N and Ratna Sahithi K

14. Control Office Application on Tracking Movement in Indian Railways —— 171–185
Apurvakeerthi M and Ranga Rao J

15. New Change Management Implementation Methods for
IT Operational Process —— 186–194
Ramesh K and Sanjeevkumar K M

16. Enhanced Multiple-Microcantilevers based Microsensor for Ozone Sensing —— 195–203
Jayachandiran J and Nedumaran D

17. Secure Data Management – Secret Sharing Principles Applied to Data or
Password Protection —— 204–216
Adithya H K Upadhya, Avinash K and Chandrasekaran K

18. Enhancing the Uncertainty of Hardware Efficient Substitution Box based on
Linear Cryptanalysis —— 217–227
Jithendra K B and Shahana T K

19. Comparative Study of Software Defined Network Controller Platforms —— 228–244
Nishtha and Manu Sood

20. Home Automation using Android Application —— 245–253
Arya G S, Oormila R Varma, Sooryalakshmi S, Vani Hariharan and Siji Rani S

21. Missing Value Estimation in DNA Microarrays using Linear Regression And
Fuzzy Approach —— 254–268
Sujay Saha, Praveen Kumar Singh and Kashi Nath Dey

22. Detection of Different Types of Diabetic Retinopathy and
Age Related Macular Degeneration —— 269–279
Amrita Roy Chowdhury, Rituparna Saha and Sreeparna Banerjee

23. Design and Implementation of Intrusion Detection System using
Data Mining Techniques and Artificial Neural Networks —— 280–292
Inadyuti Dutt and Samarjeet Borah

24. Development of Sliding mode and Feedback Linearization Controllers for
the Single Fluid Heat Transfer System —— 293–305
Pranavanand S and Raghuram A

25. Implementation of SVPWM Controlled Three Phase Inverter for
use as a Dynamic Voltage Restorer —— 306–317
Sridevi H R, Aruna Prabha B S, Ravikumar H M and Meena P

26. Enhancing the Uncertainty of Hardware Efficient Substitution Box based on
Differential Cryptanalysis —— 318–329
Jithendra K B and Shahana T K

27. Load Balancing of MongoDB with Tag Aware Sharding —— 330–343
Piyush Shegokar, Manoj V Thomas and Chandrasekaran K

28. Implementation of Multilevel Z-Source Inverter using Solar Photovoltaic Energy system
for Hybrid Electric Vehicles —— 344–359
Arulmozhiyal R, Thirumalini P and Murali M

29. Use of Rogowski Coil to Overcome X/R Ratio Effect on Distance Relay Reach —— 360–373
Sarwade A N, Katti P K and Ghodekar J G

30. Stability Analysis of Discrete Time MIMO Nonlinear System by
Backstepping Technique —— 374–388
Mani Mishra, Aasheesh Shukla and Vinay kumar Deolia

31. Comparative Study of Comparator Topologies: A Review —— 389–398
Prerna shukla and Gaurav kumar sharma

32. Implementation of Temperature Process Control using Conventional and
Parameter Optimization Technique —— 399–406
Poongodi P, Prema N and Madhu Sudhanan R

33. A Novel Compact Dual-band slot Microstrip Antenna for Internal
Wireless Communication —— 407–414
Rajkumar S and Ramkumar Prabhu M

34. Comparing Data of Left and Right Hemisphere of Brain Recorded
    using EEGLAB —— 415–428
    Annushree Bablani and Prakriti Trivedi

35. Non-Word Error Correction for Luganda —— 429–448
    Robert Ssali Balagadde and Parvataneni Premchand

36. A New Approach for Texture based Script Identification at Block Level
    using Quad- Tree Decomposition —— 449–461
    Pawan Kumar Singh, Supratim Das, Ram Sarkar and Mita Nasipuri

37. Ligand based Virtual Screening using Graph Wavelet Alignment Kernel based on
    Boiling Point and Electronegativity —— 462–472
    Preeja Babu and Dhanya Sethumadhavan

38. Word Segmentation from Unconstrained Handwritten Bangla Document Images
    using Distance Transform —— 473–484
    Pawan Kumar Singh, Shubham Sinha, Sagnik Pal Chowdhury, Ram Sarkar and Mita Nasipuri

39. Electromagnetically Coupled Microstrip Patch Antennas with Defective Ground Structure
    for High Frequency Sensing Applications —— 485–492
    Sreenath Kashyap S, Vedvyas Dwivedi and Kosta Y P

Adeetya Sawant [1] and Sudha Gupta [2]

Iris Recognition using Support Vector Machine

Abstract: Biometrics involves identification of an individual from certain secernating features of human beings. These include speech, retina, iris, fingerprints, facial features, script and hand geometry. Amongst these, iris is unique, has highly discerning pattern, does not change with age and is easy to acquire. This work concentrates on iris recognition system that consists of an automatic segmentation system based on Hough transform, which separates out the circular iris and pupil region, occlusion of eyelids and eyelashes, and reflections. The extracted iris region was then normalized to constant dimensions to eliminate imaging inconsistencies. Further, 1D Log-Gabor filter was extracted and quantized to four levels to encode the epigenetic pattern of the iris into a bit-wise biometric template. SVM has been used for classification purposes which is a kernel-based supervised method. RBF kernel proves to be more prominent over Polynomial kernel.

Keywords: Iris Recognition; Hough transform; Normalisation; Support Vector Machine; Kernel function

1 Introduction

The field of biometrics contributes significantly to systems wherein security and personal authentication are of utmost importance. The need of a robust biometric technology arises due to the increase in number of frauds and security breaches. The traditional technologies currently used for personal identification and data privacy, such as passwords (something you know and can be easily forgotten), cards or keys (something you have and can be copied), are proving unreliable [8]. Thus, biometric technology outwits the conventional methods of identification and verification since it involves recognition of an individual by incorporating directly the peculiar characteristics of human beings which

1 K. J. Somaiya College of Engineering, Mumbai
adeetya.sawant@gmail.com
2 K. J. Somaiya College of Engineering, Mumbai
sudhagupta@somaiya.edu

cannot be borrowed, stolen or forged by any means. Biometric-based systems provide trustworthy solutions to safeguard confidential transactions and personal information. The demand for biometrics can be found in federal, state and local governments, in the military, and in commercial applications. The iris is an externally visible, yet protected organ whose unique epigenetic pattern remains unchanged throughout adult life [5].In the training process, features which are extracted are stored in the database, while the matching process correlates the extracted features with stored features. Both training and matching process consists of image acquisition, iris localization [10] using Hough transform followed by Canny edge detection [3], iris normalization [1], and feature extraction with the help of Gabor wavelet [7]. As suggested in [9][6], Support Vector Machine performs more efficiently for biometric classification. The vision behind this work is:

- To obtain features which best distinguishes the data.
- To evaluate Support Vector Machine for Iris recognition.
- To find out the optimum kernel parameter and cost function to obtain acceptable performance.

The organization of this paper is as follows. First, section II describes the related work on image pre-processing for iris recognition. Feature extraction technique is described in Section III. Matching approach using Support Vector Machine has been discussed in Section IV. Experimental approaches and results are analyzed in section V. Finally, section VI concludes this paper.

2 Image Preprocessing

Image Preprocessing is performed to increase recognition accuracy [5]. This mainly includes segmentation and normalization. Along with Iris, eye region is comprised of certain regions such as pupil, sclera, eyelids, eyelashes, eyebrows and reflections, which are redundant. Segmentation aims to eliminate such inconsistencies. The steps for recognition can be summarized in Figure. 1.

3 Image Acquisition

Samples of iris are taken from CASIA database [5]. Good segmentation is provided by this database, since iris pupil boundary and sclera boundary is clearly distinguished, samples from this database is taken for iris recognition specifically. Iris radius range varies from 90 to 150 pixels, while that of pupil radius ranges from 28 to 75 pixels.

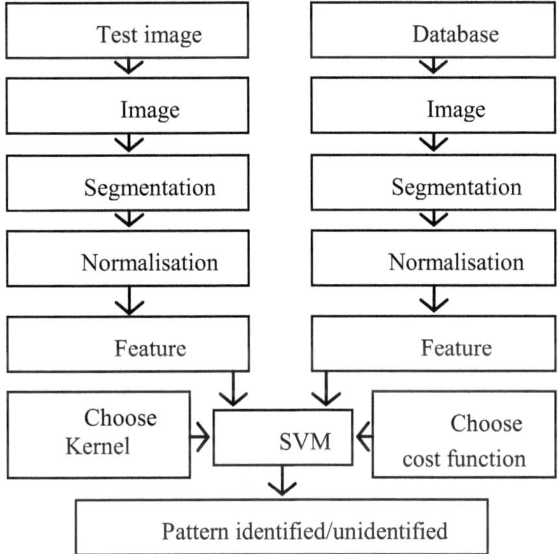

Figure. 1: Flow diagram of the system

3.1 Iris Localisation

The technique used to identify the iris and pupil boundaries is Circular Hough transform [3]. At first Edge map is generated by employing Canny edge detection. Gradients were based in the vertical direction for the outer iris/sclera boundary. Vertical and horizontal gradients were weighted equally for the inner iris/pupil boundary.

Initially the Hough transform for the iris/sclera boundary was computed followed by the Hough transform for the iris/pupil boundary within the iris region. This ensures an efficient and accurate process for circle detection, since detecting the pupil within the iris region would be easier instead of the whole eye region. Once the above process is complete, six parameters are recorded, the radius, and x and y center coordinates for both circles.

3.2 Eyelid Occlusion

The linear Hough transform is applied to the upper and lower eyelids and they were isolated by first fitting a line [5]. Then, a horizontal line is fitted, such that it intersects with the first line at the edge of iris nearest to the pupil. Eyelid regions are isolated to a maximum by the horizontal line. Only horizontal gradient information is retained from the edge map created by using Canny edge detection. The eyelid detection system was found to be effective, and fairly managed to detach most occluding eyelid regions.

3.3 Normalisation

The next stage after productive segmentation of the iris region from an eye image is to transform the iris region so that it has fixed dimensions in order to make comparisons possible [10]. The disparities in dimensions between eye images result from stretching of the iris due to pupil dilation caused by varying levels of illumination. Other sources of inconsistency include rotation of the eye within the eye socket, fluctuating imaging distance, turning of the camera and head tilt. The result of the normalization process enables iris regions to have fix dimensions, so that two images of the same iris under contrast circumstances will have characteristics with same location in feature space. For normalization of iris regions a technique based on Daugman's rubber sheet model is employed as shown in Figure. 2.

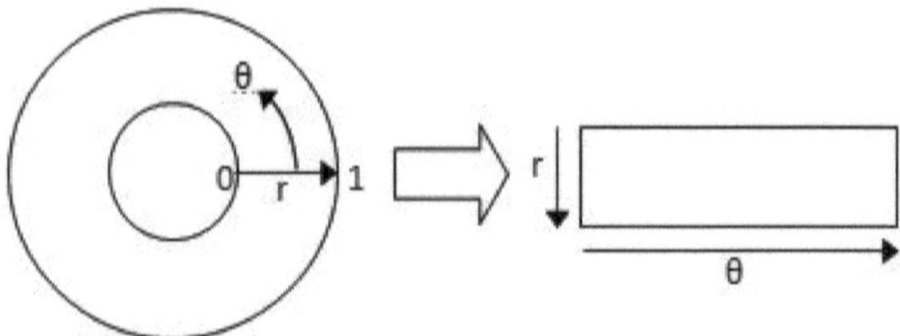

Figure. 2: Daugman's rubber sheet model

The remapping of the iris region from *(x, y)* Cartesian coordinates to the normalized non-concentric polar representation is modeled as

$$I(x(r,\theta)y(r,\theta)) \rightarrow I(r,\theta) \tag{1}$$

With

$$x(r,\theta) \rightarrow (1-r)x_n(\theta) + rx_l(\theta) \tag{2}$$
$$y(r,\theta) \rightarrow (1-r)y_n(\theta) + ry_l(\theta) \tag{3}$$

Where $I(x, y)$ is the iris region image, (x, y) are the original Cartesian coordinates, (r, θ) are the corresponding normalised polar coordinates, and x_p, x_p and x_l, y_l are the coordinates of the pupil and iris boundaries along the θ direction.

4 Feature Extraction

Accurate recognition of an individual can be achieved by extracting the most distinguishing information contained in an iris pattern. Only the noteworthy features of the iris are encoded so that comparisons between templates can be made. Feature encoding is implemented by convolving the normalised iris pattern with 1D Log-Gabor wavelets [5]. The 2D normalised pattern is broken up into a number of 1D signals. Then these 1D signals are convolved with 1D Gabor wavelets.

$$G(f) = \exp\left(\frac{-\left(\log\left(\frac{f}{f_0}\right)\right)^2}{2(\log(\frac{\sigma}{f_0}))^2}\right) \tag{4}$$

Where f_0 and σ are parameters of the filters, f_0 will give centre frequency and σ gives bandwidth of filter.

The Feature matrix consists of three features:
- Difference between radius of pupil and sclera.
- Distance between centre coordinates of pupil and sclera.
- Iris patterns (using 1-D Log Gabor wavelet).

5 Classification Algorithm

Support Vector Machine has been used as a classifier. SVM is supervised kernel based statistical learning theory [5]. A SVM kernel maps the low dimensional points of the input space (feature matrix) to the higher dimensional feature space.

The input data points of various classes are separated by means of a hyper-plane which is optimally determined from feature space. This enables the SVM to recognize and classify the patterns. An input to SVM model and SVM is trained in such a way that the direct decision function maximizes the generalization ability for the classification. Consider that M, m dimensional training inputs xi, $(i = 1, \cdots, M)$ belong to class 1 or class 2 and then the associated labels be $yi= 1$ for class 1 and $yi= -1$ for class 2. If this data is linearly separable, the decision function is determined by:

$$D(x) = w^T x_i + b \tag{5}$$

Where w is an m dimensional vector, b is a bias term, and i varies from 1 to M.

$$w^T x_i + b \begin{cases} \leq 1, & for \quad y_i = -1 \\ \geq 1, & for \quad y_i = \ \ 1 \end{cases} \tag{6}$$

This equation can be precisely termed as:

$$y_i(w^T x_i + b) \geq 1, \ for \ i = 1,2, \dots M \tag{7}$$

The convex quadratic optimization problem for w, b and ξ can be solved to obtain optimally separated hyper plane by the following equation.

$$minQ(w, b, \xi) = \frac{1}{2}\|w^2\| + C \sum_{i=1}^{M} \xi_i \tag{8}$$

subject to:

$$y_i(w^T x_i + b) \geq 1 - \xi, \xi \geq 0, \ for \ i = 1,2, \dots M \tag{9}$$

where, $\xi = (\xi 1, \cdots, \xi M)^T$ is non-negative slack variable ($\xi \geq 0$) and C is the margin parameter which is used to ascertain the trade-off between the maximization of the margin and the minimization of the classification error. Linear separability can be enhanced by employing a kernel function which maps the input feature set to its higher dimensional space known as feature space. In SVM, selection of kernel depends upon the nature of pattern to be classified. Once the kernel is selected, the values of kernel parameter and the margin parameter C can be determined [4]. The value of kernel parameter decides the complexity and the generalization capability of the network, which has an influence on the smoothness of SVM response and it affects the number of support vectors. Hence, to build an optimised classifier, the optimised value of margin parameter and kernel parameter must be determined. This is called as model selection. In this work cross-validation procedure has been used to avoid over fitting problem for model selection [2]. In cross-validation a complete training set is divided into v subsets of fixed size, then sequentially one subset is tested using the classifier trained on the remaining $(v - 1)$ subsets. In this way each instance of the whole training set is predicted once during SVM modeling. Output matrix from 1-D Log Gabor filter is used as input feature vector to SVM for training, testing and classification. SVM modeling steps are as follows:

– Data acquisition: Feature matrix obtained from 1-D Log Gabor wavelet.
– Kernel selection: The different kernels are applied to the input which shows that the RBF kernel gives maximum training accuracy. Hence, RBF kernel has been selected.

Model selection: Using cross-validation, find out the optimal values for kernel parameter and cost function.

6 Experimental Results

Selected raw data from CASIA database shown in Figure. 3 being optimally preprocessed is subjected to normalization.

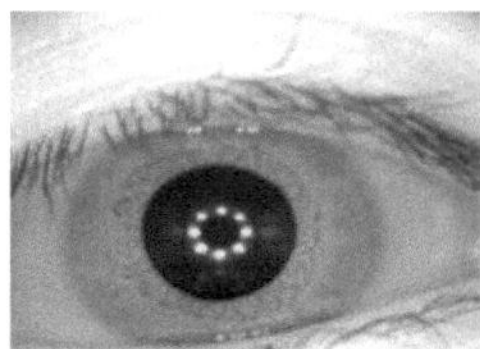

Figure. 3: Original image from CASIA database

Normalization remaps iris image from Cartesian form to non-concentric polar form. Before doing this non-relevant data such as eyelid, eyelashes and pupil has been removed from original image which is shown in Figure. 4

Figure. 4: Occlusion of eyelid, eyelashes and pupil region

Normalized image shown in Figure. 5 has a constant dimension of 250x30 on which then Gabor filter is applied to get feature matrix.

Figure. 5: Normalised image from Original image

Feature matrix of size 54x6, where 54 denotes number of training samples and 6 are number of features, is given as an input to SVM which is use as a classifier. Results show that Radial Basis Function (RBF) kernel gives more accuracy than other (Polynomial) kernel.

The training and testing accuracies for Polynomial and RBF kernel are given in Table 1 and Table 2.

Table 1. Accuracy for Polynomial kernel, γ=0.25

Cost Function C	Training Accuracy %	Testing Accuracy %
0.002	88.86	72.72
0.02	90.74	72.72
0.2	64.81	64.64
0.8	87.03	54.54
1	87.03	54.54
4	87.03	45.45

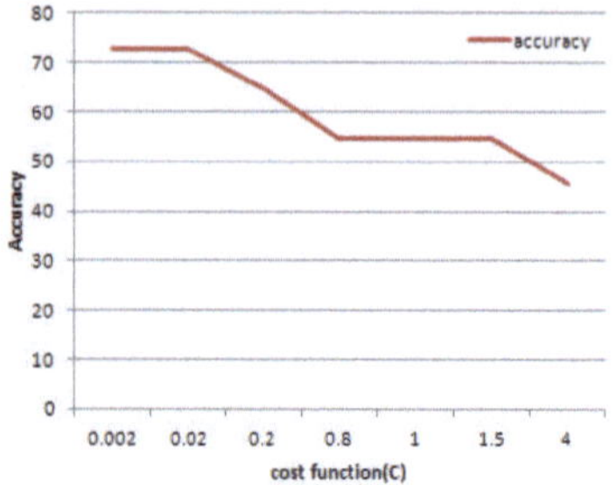

Figure. 6: Accuracy for Polynomial Kernel

Table 2. Accuracy for RBF kernel, C=0.02

Kernel Parameter γ	Training Accuracy %	Testing Accuracy %
0	94.44	81.81
0.5	94.44	81.81
0.2	94.44	81.81
0.25	94.44	90.90
0.35	94.44	90.90

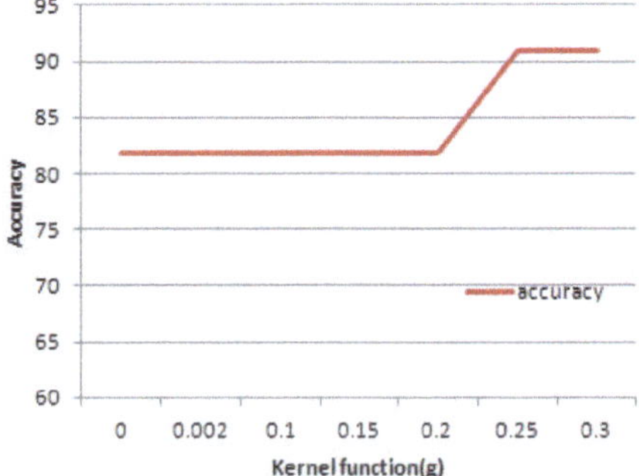

Figure. 7: Accuracy for RBF Kernel

In Figure. 6 Polynomial kernel is chosen where kernel parameter is taken as 0.2 and value of C is varied to get accuracy. It is shown that accuracy is 72.72% maximum for C ≤ 0.02.

Whereas in Figure. 7 RBF kernel is used where C = 0.02 and value of kernel parameter is varied to get optimum accuracy. It is shown that accuracy is 90.90% maximum for ≥ 0.25.

Conclusion

Iris recognition system has been successfully implemented using SVM algorithm and gives highly accurate results for human identification and verification for mentioned feature set. The proposed work can perform better if segmentation algorithm gives eyelid and eyelashes detection more efficiently. Iris liveness detection and fake iris can be a issue that should be handled carefully. Hence for relative high performance multi-modal biometric recognition can be proposed. From the results obtained, we can conclude that RBF kernel of SVM gives maximum testing accuracy. The strategy adopted in this work can prove beneficial for different real world biometric assessment systems.

References

1   Ali, A. N. (2013, November), "Simple features generation method for SVM based iris classification", In 2013 IEEE International Conference on Control System, Computing and Engineering (ICCSCE), pp. 238-242

2   Burges, C. J. (1998), "A tutorial on support vector machines for pattern recognition", Data mining and knowledge discovery, 2(2), pp. 121-167.

3   Elgamal, M., & Al-Biqami, N. (2013), "An efficient feature extraction method for iris recognition based on wavelet transformation", Int. J. Comput. Inf. Technol, 2(03), 521-527.

4   Gonzalez, R. C., Woods, R. E., & Eddins, S. L. (2004), "Digital image processing using MATLAB", Pearson Education India, pp. 407-421.

5   Gupta, S., Doshi, V., Jain, A., & Iyer, S. (2010), "Iris recognition system using biometric template matching technology", International Journal of Computer Applications, 1(2), pp. 1-4.

6   Gupta, S., Kambli, R., Wagh, S., & Kazi, F. (2015), "Support-Vector-Machine-Based Proactive Cascade Prediction in Smart Grid Using Probabilistic Framework", IEEE Transactions on Industrial Electronics, 62(4), pp. 2478-2486.

7   Lee, T. S. (1996), "Image representation using 2D Gabor wavelets", IEEE Transactions on Pattern Analysis and Machine Intelligence, 18(10), pp. 959-971.

8   Mabrukar, S. S., Sonawane, N. S., & Bagban, J. A. (2013), "Biometric system using Iris pattern recognition", International Journal of Innovative Technology and Exploring Engineering, ISSN, 2278-3075.

9   Son, B., Won, H., Kee, G., & Lee, Y. (2004, October), "Discriminant iris feature and support vector machines for iris recognition", In 2004 International Conference on Image Processing (ICIP'04), Vol. 2, pp. 865-868.

10 Zafar, M. F., Zaheer, Z., & Khurshid, J. (2013, January), "Novel iris segmentation and recognition system for human identification", In 2013 10th International Bhurban Conference on Applied Sciences and Technology (IBCAST), pp. 128-131.

Vivek Sharma [1], B. B. Sharma [2] and R. Nath [3]

Nonlinear Unknown Input Sliding Mode Observer for Chaotic Synchronization and Message Recovery

Abstract: Unknown Input Observers (UIOs) have been designed for both linear and nonlinear systems, to estimate the states when excited by some unknown input. These observers find their extensive application in Fault Detection and Isolation (FDI), considering fault to be an unknown input. UIO designed using Sliding Mode approach has the additional advantage of easy reconstruction of unknown input using equivalent output injection. For this reason sliding mode observers (SMO) have also been used in secure communication to recover unknown message signal. Sliding Mode approach is known to be robust against matched uncertainties but its performance degrades in the presence of unmatched uncertainties. In present work, Nonlinear Unknown Input Sliding Mode Observer (NUISMO) is proposed to achieve synchronization and recover message while considering unmatched uncertainty in the chaotic transmitter. First, LMI (Linear Matrix Inequality) condition is derived to decouple the unmatched uncertainty and then sliding mode technique is used for synchronization and message recovery. Simulation results are presented on third order Chua circuit and a Rossler system to show the efficacy of proposed approach.

Keywords: Unknown Input Observer (UIO), Linear Matrix Inequality (LMI), Sliding Mode Observer (SMO), Chaos Synchronization and message recovery.

1 Introduction

Design of observers for linear and nonlinear systems has been a benchmark problem in control engineering. Sliding mode control and observer design

1 Dept. of Electrical Engineering, National Institute of Technology, Hamirpur, India
Vivekmay12@gmail.com
2 Dept. of Electrical Engineering, National Institute of Technology, Hamirpur, India
bbs.iit@gmail.com
3 Dept. of Electrical Engineering, National Institute of Technology, Hamirpur, India
r.nath1964@gmail.com

technique is used extensively in systems having uncertainties associated with their dynamics. In sliding mode control the system is constrained on sliding surface using discontinuous control and operates with the reduced order dynamics while in sliding mode. Moreover, it is known for its ability to be robust against matched disturbances and uncertainties. Sliding mode control became popular since the pioneer work of Utkin (Vadim, 1977) in Soviet Union. Sliding mode control is considered problematic for many control applications because of its discontinuous nature and chattering, but in observer design it is easily implemented in software (Spurgeon K. S., 2008). One of the early contribution in sliding mode observer design came from (Slotine, Hedrick, & Misawa, 1987). (Walcott & Zak, 1987) gave structural conditions which linear systems must qualify for the existence of sliding mode observer. (Edwards & Spurgeon, 1998) proposed coordinate transformation to represent linear system in canonical form and further gave necessary and sufficient conditions for the existence of sliding mode observer based on this canonical form. Unknown input observers are important for systems subject to disturbances or with inaccessible inputs, and are of use in many applications such as fault detection and isolation (FDI) (Venkatasubramanian, 2003) or parameter identification. (De Persis, 2001) used differential geometry approach to the problem of FDI for nonlinear systems. (Patton, 1997) addressed robustness issue in FDI using unknown input observers. Unknown input observer for linear system have been proposed by (Darouach, Zasadzinski, Xu, & al, 1994) and (Chen, Patton, & Zhang, 1996). The concept of decoupled unknown input used by (Chen, Patton, & Zhang, 1996) was extended to nonlinear systems by (Chen & Saif, 2006) and (Imsland, Johansen, Grip, & Fossen, 2007). Similar to equivalent control, (Edwards, Spurgeon, & Patton, 2000) developed the methodology of equivalent output injection to reconstruct the actuator fault signal. Later, the work was extended to include sensor fault (Tan & Edwards, 2002) and fault reconstruction in the presence of uncertainties (Yan & Edwards, 2007). While the error trajectories are in sliding mode, because of the designed dynamics on sliding mode error converges to zero. When error has significantly converged to zero, unknown input because of fault can be recovered from equivalent output injection.

Such Unknown input observers for fault detection and isolation have also been extended to recover message signal in secure communication using chaotic transmitter and receiver (Observer), considering message signal to be unknown input (Bowong & Tewa, 2008). Chaotic systems are the subclass of nonlinear systems which show high sensitivity to their initial conditions and exhibit oscillatory behavior but these oscillations are non-periodic in nature (Boccaletti & al., 2002). In the pioneering work of (Pecora & Carroll, 1990), it was proposed that chaos synchronization can be used for secure communication. Since then, chaos synchronization for secure communication has been explored to a great extent (Kocarev, Halle, Eckert, Chua, & Parlitz, 1992), (Fallahi & Leung, 2010), (Theesar & Balasubramaniam, 2014), (Sharma & Kar, 2009), (Sharma & Kar, 2011) using various methodologies. (Fallahi & Leung, 2010) proposed chaotic multiplication modulation, and for message recovery Extended Kalman filter (EKF) based receiver was designed. (Theesar & Balasubramaniam, 2014) used sampled data feedback controller is for the synchronization of chaotic Lur'e system. Along with other techniques, sliding mode has been used for its ability to be robust against matched uncertainties but its performance degrades in case of unmatched uncertainties. To make it robust against unmatched disturbances, strategy has been to use hybrid approach (Spurgeon & Davies, 1993), like design of optimal sliding surface using min-max control (Poznyak, Shtessel, & Gallegos, 2003), or LMI based approach in which sliding mode dynamics becomes invariant to mismatched uncertainties (Choi, 2003).

Keeping in view the present state of art in the field of nonlinear unknown input observer (NUIO) and their applications, nonlinear unknown input sliding mode observer (NUISMO) based synchronization scheme is presented in the presence of unmatched uncertainty. Methodology adopted here assumes bounded nature of nonlinearities which is common in many nonlinear systems, especially chaotic systems. Using the scheme, secure communication using chaotic transmitter – receiver configuration, in the presence of unmatched uncertainties is proposed. Here LMI condition is derived for the existence of NUIO, which can decouple the unmatched uncertainties and further the error dynamics become independent of unmatched uncertainties. Along with that, existence conditions for the feasibility of LMI solution are also proposed. Synchronization between system and observer is achieved under the influence of coupling based on sliding mode approach. Effectiveness of the proposed sliding mode coupling is proved using Lyapunove theory. Effective message recovery is shown using equivalent output injection. Moreover, Lipschitz condition on nonlinearity is relaxed, which enhances the scope of the proposed approach. The simulation results presented in the end justifies the effectiveness of proposed theory.

The paper is organized as follows: Problem formulation for the proposed class of nonlinear system is given in Section 2. In Section 3 Main results of the paper are presented. First, LMI formulation along with methodology to decouple the error dynamics from unmatched uncertainty is presented. Further, design of sliding mode controller to stabilize the error dynamics is presented.

Later, methodology to recover message is also given. Section 4 shows the simulation results to justify the proposed claim.

2 Problem Formulation

In this section, observer frame work required for the proposed class of nonlinear systems is presented. Consider the master chaotic system subject to unmatched uncertainties and the unknown input. Unknown input is a message signal and is to be recovered at the receiver side.

$$\dot{x} = Ax + Bf(x) + Bu + g(y, u_1) + Dv(t)$$
$$y = Cx \tag{1}$$

where $x \in \Re^n$ is the state vector, $y \in \Re^p$ is the output vector, $u \in \Re^m$ is the message signal, $v \in \Re^k$ is the unmatched uncertainty, $f(x): \Re^n \to \Re^m$ is the nonlinear vector function. $A \in \Re^{n \times n}$ is the linear part of the system dynamics. $B \in \Re^{n \times m}$ associates the nonlinear part of the dynamics, $g(\cdot) \in \Re^n$, u_1 is known part of input, $D \in \Re^{n \times k}$ is uncertainty distribution matrix and $C \in \Re^{p \times n}$ is the output matrix.

Assumption 1: It is assumed that there exists $K > 0$ for the operating range of x such that $\sup|f(x)| < K$.

Remark 1: The above assumption is realistic in many nonlinear systems especially chaotic systems where system states are bounded over the entire range of operation.

Assumption 1: D is a full column rank matrix.

With above assumptions nonlinear unknown input observer for the given system (1) is proposed as

$$\dot{z} = Nz + Ly + MBf(\hat{x}) + Mg(y, u_1) + MBv$$
$$\hat{x} = z - Ey \tag{2}$$

where $z \in \Re^n$, $\hat{x} \in \Re^n$ is the state estimate, N, L, M and E are matrices of appropriate dimension. v is the coupling to be designed while selecting N, L, M and E suitably.

3 Main Results

State estimation error for the system (1) and the observer (2) is defined as

$$e = \hat{x} - x = z - (I + EC)x \tag{3}$$

For this system – observer pair, error dynamics can be written as

$$\begin{aligned}
\dot{e} &= \dot{z} - (I + EC)\dot{x} \\
&= Ne + [N(I + EC) + LC - (I + EC)A]x - (I + EC)Bu \\
&\quad + MBf(\hat{x}) - (I + EC)Bf(x) - (I + EC)Dv + MBv \tag{4}
\end{aligned}$$

For (2) to be an observer it is required that the estimated state $\hat{x}$ converges to the system state x. That means error dynamics (4) should be convergent. Moreover, for this observer to be robust against unmatched uncertainties, error dynamics should be somehow independent of unmatched uncertainties. In view of these requirements, in error dynamics (4) it is required that

$$N < 0 \tag{5}$$
$$N(I + EC) + LC - (I + EC)A = 0 \tag{6}$$
$$(I + EC)D = 0 \tag{7}$$

i.e. N has to be stable matrix along with the conditions given in (6) and (7). To meet up these requirements it is proposed that the observer matrices be selected as

$$N \qquad (I + EC)A - KC \tag{8}$$

$$L \qquad K(I + CE) - (I + EC) \tag{9}$$

$$M \qquad (I + EC) \tag{10}$$

Where E and K are design parameters. Using the matrix conditions for N in (8) and L in (9), requirement (6) can be satisfied. So the problem boils down to finding E and K such that (5) and (7) are satisfied.

To decouple the unmatched uncertainty, design parameter E is to be found such that $(I + EC)D = 0$. Now for $ECD = -D$ to have solution for E following rank conditions should be satisfied:

$$Rank\begin{bmatrix} CD \\ D \end{bmatrix} = Rank\begin{bmatrix} CD \end{bmatrix}$$

Since D is full column rank so CD should also be full column rank. If CD is full column rank, all possible solutions of E using generalized inverse will be

$$E \qquad -D(CD)^{+} + Y(I - (CD)(CD)^{+})$$

$$U + YV \qquad \text{(Say)} \tag{11}$$

Where

$$CD^+ \qquad [(CD)^T (CD)]^{-1}(CD)^T$$

$$U \qquad -D(CD)^+$$

$$V \qquad I - (CD)(CD)^+$$

Y is the design parameter. Now the problem is restructured to find Y and K which will satisfy (5).

Lemma 1: If $N < 0$ then there exists $P > 0$ such that $N^T P + PN < 0$

Remark 2: The above result is straight forward application of Lyapunov equation to stability results (Khalil, 2002).

Theorem 1: If there exist matrices P_Y, P_K and P such that

$$\begin{bmatrix} W^T + W^T & 0 \\ 0 & -P \end{bmatrix} \qquad 0 \tag{12}$$

Where

$$W \qquad P((I + UC)A) + P_Y VCA - P_K C \tag{13}$$

$$P_Y \qquad PY$$

$$P_K \qquad PK$$

Then the observer states under the effect of coupling

$$v \qquad -sign(e^T P_e MB)[U_{max} + \beta] - f(\hat{x}) \tag{14}$$

converges asymptotically to system states, where U_{max} is upper bound on $\left[\|u\| + \|f(x)\|\right]$, β is some positive constant.

Proof: It is stated above that using the definitions in (8) – (10) it can be shown that $[N(I + EC) + LC - (I + EC)A] = 0$. Hence error dynamics (4) reduces to

$$\dot{e} = \begin{aligned} &Ne - (I + EC)Bu + MBf(\hat{x}) - \\ &(I + EC)Bf(x) - (I + EC)Dv + MBv \end{aligned}$$

$$\dot{e} = \begin{aligned} &Ne - MBu + MBf(\hat{x}) - MBf(x) - \\ &(I + EC)Dv + MBv \end{aligned}$$

Now for $N < 0$, from lemma 1 and using the value of N from (8), one can obtain

$$[A^T(I+EC)^T - C^T K^T]P + P[(I+EC)A - KC] < 0$$

Now replacing the value of E from (11), which comes from the solution of $(I+EC)D = 0$, the above inequality modifies to

$$[A^T(I+(U+YV)C)^T - C^T K^T]P + P[(I+(U+YV)C)A - KC] < 0$$

Since P, K and Y are unknown above equation is not an LMI in variables P, K and Y. To resolve the issue new variables $P_Y = PY$, and $P_K = PK$ are introduced. By using the definition of W as in (13), the above expression reduces to

$$W^T + W < 0$$

The above expression along with condition $P > 0$ gives (12). The LMI (12) is to be solved for feasibility. If Solution of LMI exists it will subsequently provide the values of design parameters K, Y and $P > 0$. Using solution of the LMI i.e. feasible values of Y and K in (11) and (8 – 10), one can obtain observer matrices satisfying the requisite conditions. It also meets out the requirement of decoupling the unmatched uncertainties and $N < 0$. The error dynamics takes the form

$$\dot{e} = Ne + MB(f(\hat{x}) - f(x) - u) + MBv \qquad (15)$$

In the next stage controller v based on sliding mode control is designed to first stabilize the error dynamics and then recover message signal.

Sliding Mode Controller Design: To stabilize the error dynamics (15) coupling v is proposed as

$$v = \frac{v_f + v_s}{}$$

Where
$$v_f = -f(\hat{x})$$

$$v_s = -sign(e^T P_e MB)[U_{max} + \beta]$$

This reduces the error dynamics (15) to

$$\dot{e} = Ne + MB(-f(x) - u) + MBv_s \qquad (16)$$

To prove the effectiveness of controller to stabilize (15) consider Lyapunov function as $V = (e^T P_e e)/2$ where $P_e > 0$. The rate of change of V w.r.t. time can be written as

$$\dot{V} = (\dot{e}^T P_e e + e^T P_e \dot{e})/2$$

Using the error dynamics (16) in above expression, we get

$$\dot{V} = \begin{aligned}(e^T N^T P_e e + (-u - f(x) + v_s)^T B^T M^T P_e e + \\ e^T P_e N e + e^T P_e M B (-u - f(x) + v_s)) / 2\end{aligned}$$

Since N is found such that it is negative definite, hence $P_e > 0$ is found such that

$$N^T P_e + P_e N \;<\; 0 \qquad (17)$$

$$\Rightarrow \dot{V} \leq e^T P_e M B (-u - f(x) + v_s) \qquad (18)$$

Choosing $v_s = -sign(e^T P_e M B)[U_{max} + \beta]$ makes $\dot{V} < 0$ where U_{max} is upper bound on $\big[\|u\| + \|f(x)\|\big]$, β is some positive constant. If $f(x)$ is known at the receiver the controller may be modified as

$$v = -sign(e^T P_e M B)[U_{max} + \beta] - f(\hat{x}) - f(x) \qquad (19)$$

U_{max} becomes upper bound on $\|u\|$.

Discussion on Coupling: Conventionally in sliding mode control, first sliding surface is designed on which system trajectories show the desired behavior and then discontinuous control is designed to restrain the system on the sliding surface. Here, error dynamics is proved to be convergent using Lyapunov function V and $\dot{V} < 0$. Sliding surface (S) because of discontinuous control becomes $S = e^T P_e M B$.

Remark 3: The observer for the given system will exist under the following existence conditions:

1. $Rank[CD] = Rank[D] = k$

2. $((I + EC)A, C)$ pair is detectable. From (5) where $N < 0$ and the definition of N in (8).

Remark 4: It is to be noted that the proposed approach addresses the coupling design problem in the scenario of unknown input and unmatched uncertainty satisfactorily. Though solution may come out to be conservative in nature (Dimassi & Loria, 2011).

Design Steps: The procedure proposed for designing NUISMO is summarized as follows:

1. Solve LMI (12) for design parameters K, Y and $P > 0$.

2. Use the value of Y to evaluate E from (11).

3. Evaluate the Observer matrices N, L and M from (8-10)

4. To design coupling, find P_e by solving LMI (17)

Message Recovery: From (15) when the error trajectories are in sliding mode and error has considerably converged to zero i.e. system states have been estimated effectively, coupling v approximates the unknown message signal $u(t)$. For message recovery, coupling is approximated as v equivalent (v_{eq}) which is given as follows:

$$v_{eq} = \frac{e^T P_e MB}{\left\| e^T P_e MB \right\| + \delta}\left(U_{max} + \beta\right) - f(\hat{x}) \tag{20}$$

Where δ is a small positive constant. It is to be noted here that smaller the value of δ closer is v_{eq} to v.

4 Numerical Examples

To validate the proposed scheme of designing NUISMO, results of simulation on two systems i.e. Chua circuit (with piecewise linear nonlinearity function $f(x)$) and Rossler system are presented here.

4.1 Chua Circuit

Dynamics of the Chua circuit with uncertainty is given as

$$\dot{x} = Ax + Bf(x) + Bu(t) + Dv(t)$$

$$y = Cx$$

Where

$$f(x) = m_1 x + (m_0 - m_1)\frac{\|x+1\| - \|x-1\|}{2}$$

$$m_0 = -1.143 \qquad m_1 = -0.714$$

$$A = \begin{bmatrix} -c_1 & c_1 & 0 \\ c_2 & -c_2 & c_2 \\ 0 & -c_3 & 0 \end{bmatrix} \qquad B = \begin{bmatrix} -c_1 & 0 & 0 \end{bmatrix}^T$$

$$C = \begin{bmatrix} 0 & 1 & 0 \end{bmatrix} \qquad D = \begin{bmatrix} 1 & 1 & 1 \end{bmatrix}^T$$

Chua circuit exhibits chaotic behavior for $c_1 = 15.6,\ c_2 = 1,\ c_3 = 28$. Comparing to system dynamics defined in (1), here $g(\Box) = 0$. Objective here is to design observer matrices $N, L, M\ \&\ E$ appropriately, so as to estimate the system states effectively irrespective of the unknown message and unknown disturbance. As mentioned in the design algorithm, observer matrices $N, L, M\ \&\ E$ are obtained by solving LMI (12) and further using (11) and (8 – 10)

$$N = \begin{bmatrix} -16.6 & 0 & -1 \\ 0 & -0.5 & 0 \\ -1 & 0 & -1 \end{bmatrix} \qquad MB = \begin{bmatrix} -15.6 & 0 & 0 \end{bmatrix}^T$$

$$L = \begin{bmatrix} -1 & 0 & -29 \end{bmatrix}^T \qquad\qquad E = \begin{bmatrix} -1 & -1 & -1 \end{bmatrix}^T$$

The message signal $u(t)$, which is to be recovered at the observer, and the disturbance signal $v(t)$, for the purpose of simulation are taken as

$$u(t) \qquad\qquad 0.2 square(2\pi\,t)$$

$$v(t) \qquad\qquad 0.1 Sin(2\pi t)$$

To design coupling , parameter P_e is found by solving LMI (17). MATLAB LMI toolbox has been used to solve LMIs. Using the procedure following positive definite matrix P_e is obtained:

$$P_e = \begin{bmatrix} 0.967 & 0 & 0 \\ 0 & 14.767 & 0 \\ 0 & 0 & 8.066 \end{bmatrix} \times 10^2$$

Coupling is designed as per (19) and (20) with $\delta = 0.1$. Depending on the structure of MB the extra information required for the design of coupling is e_1 which is assumed to be available. e_1 is state estimation error for state x_1. Results of the simulation are shown here. Since initial state of the system is not known at the observer, so system and observer start from different initial conditions. Initial condition for the system and the observer are taken as

$$x(0) = \begin{bmatrix} 0.1 & 0 & 0 \end{bmatrix}; \qquad \hat{x}(0) = \begin{bmatrix} 2.2 & -2.9 & -4.9 \end{bmatrix}$$

Fig 1 shows observer starting from different initial condition to that of system, gradually starts tracking the system states, even in the presence of unknown input $u(t)$ and disturbance $v(t)$. Fig 2 shows the asymptotic convergence of state estimation error to zero. Fig 3 shows the message signal

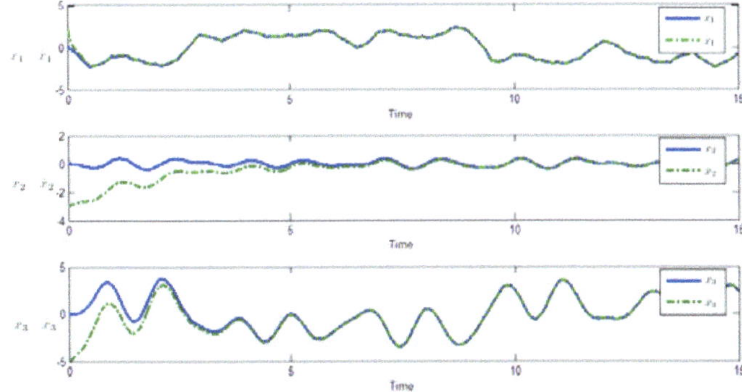

Fig. 1: Convergence of observer and system states $\hat{x}_1$ to x_1, $\hat{x}_2$ to x_2 and $\hat{x}_3$ to x_3

(dotted line) sent encoded in the states of the chaotic transmitter and the message reconstructed (solid line) at the receiver, using the information of two states of the transmitter i.e. x_1 and x_2. Message is recovered effectively only after the sliding motion is set and state estimation error has converged to zero.

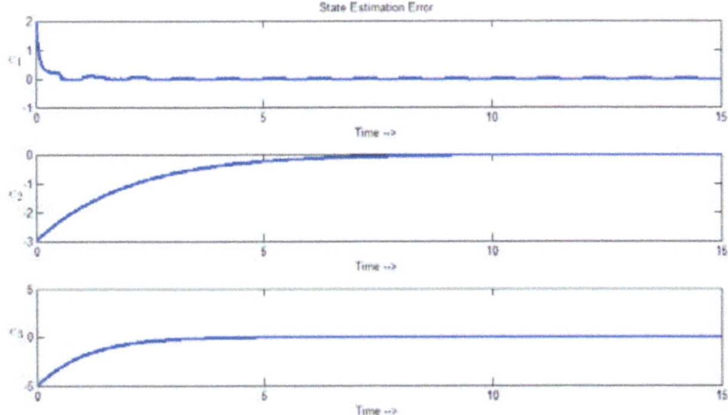

Fig. 2: Convergence of state estimation error e_1, e_2 and e_3 to zero

4.2 Rossler System

To prove the efficacy of proposed approach Rossler system is considered for simulation study, whose dynamics with uncertainty is given as (Stefano Boccaletti, 2002)

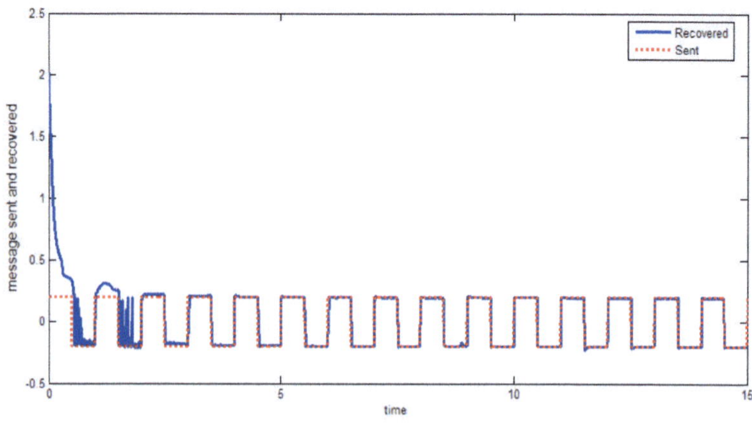

Fig. 3: Message sent from the transmitter (Chua Circuit) and recovered at the observer

$$\dot{x} = Ax + Bf(x) + Bu(t) + g(u_1, y) + Dv(t)$$

Where

$$A = \begin{bmatrix} 0 & -1 & -1 \\ 1 & a & 0 \\ 0 & 0 & -\mu \end{bmatrix} \qquad B = \begin{bmatrix} 0 & 0 & 1 \end{bmatrix}^{T}$$

$$C = \begin{bmatrix} 0 & 1 & 0 \end{bmatrix} \qquad D = \begin{bmatrix} 1 & 1 & 1 \end{bmatrix}^{T} \qquad g(\Box) = f$$

The considered Rossler system shows chaotic behavior for the parameter values as $a = 0.2$, $f = 0.2$, $\mu = 5.7$. Disturbance $v(t)$ and the message signal $u(t)$ are the same as considered in the previous example. solving LMI (12) and following subsequent design procedure observer matrices $N, L, M \& E$ are obtained as

$$N = \begin{bmatrix} -1 & 0 & -1 \\ 0 & -0.5 & 0 \\ -1 & 0 & -5.7 \end{bmatrix} \qquad MB = \begin{bmatrix} 0 & 0 & 1 \end{bmatrix}^{T}$$

$$L = \begin{bmatrix} -3.2 & 0 & -6.9 \end{bmatrix}^{T} \qquad E = \begin{bmatrix} -1 & -1 & -1 \end{bmatrix}^{T}$$

Coupling parameter $P_e > 0$ by solving LMI (17) is obtained as

$$P_e = \begin{bmatrix} 62.01 & 0 & 0 \\ 0 & 96.22 & 0 \\ 0 & 0 & 15.05 \end{bmatrix}$$

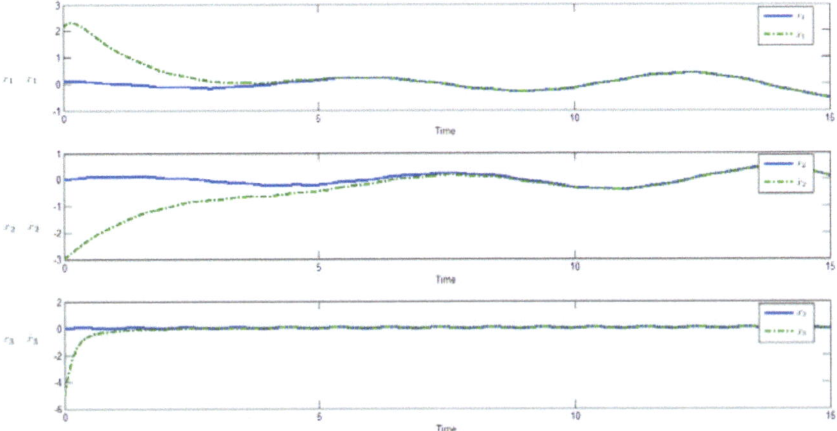

Fig. 4: Convergence of observer and Rossler system states $\hat{x}_1$ to x_1, $\hat{x}_2$ to x_2 and $\hat{x}_3$ to x_3

Coupling is implemented as per equation (20). For simulation study initial condition of observer are taken different from system as

$$x(0) = \begin{bmatrix} 0.1 & 0 & 0 \end{bmatrix}; \qquad \hat{x}(0) = \begin{bmatrix} 2.2 & -2.9 & -4.9 \end{bmatrix}$$

Figure (4) shows the convergence of the system states to the observer states. Figure (5) shows the convergence of state estimation error to zero, and hence the proposed observer with specified observer matrices is an effective observer. Figure (6) shows sufficiently accurate message recovery after 5 sec onwards.

Conclusion

Here problem of designing nonlinear unknown input sliding mode observer (NUISMO) for a class of nonlinear systems with bounded nonlinearities is presented. Moreover the existence conditions for the observer are proposed while relaxing Lipschitz condition on nonlinearities. From the theory and the simulation results presented it may be concluded that, the designed NUISMO is an effective observer to estimate the states of the system, even in the presence of unmatched uncertainties. Moreover, the proposed methodology of using equivalent output injection in sliding mode coupling is effective for message recovery in secure communication, in the presence of unmatched uncertainty.

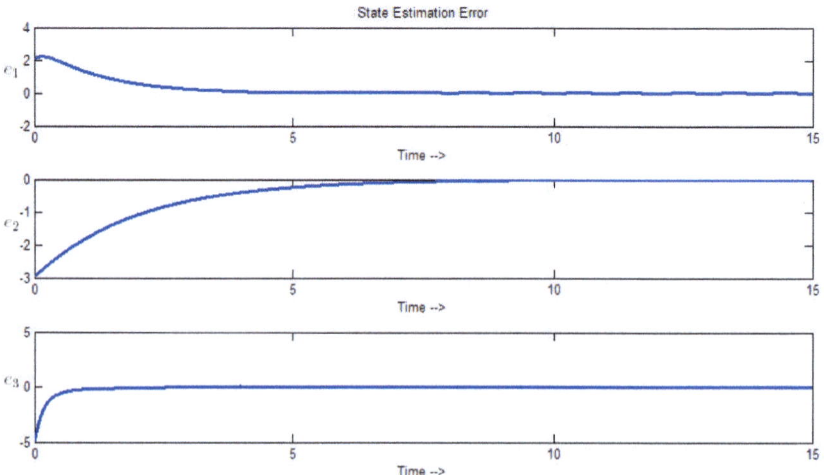

Fig. 5: Convergence of state estimation error e_1, e_2 and e_3 to zero

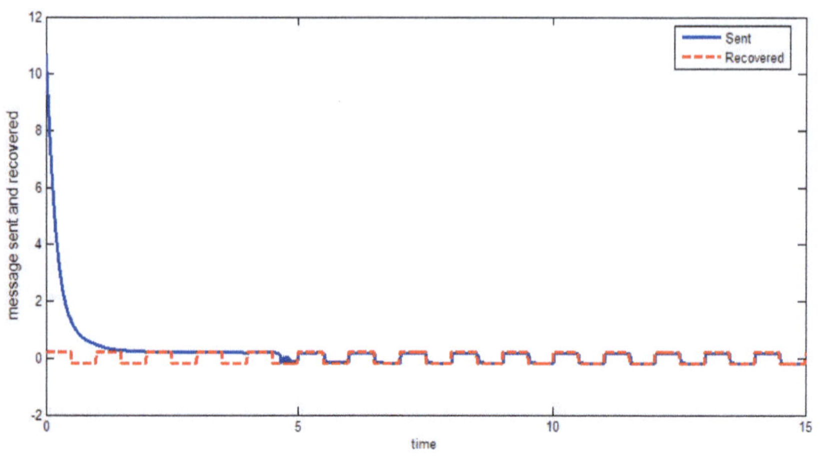

Fig. 6: Message sent from the transmitter (Rossler system) and recovered at the observer

References

1 Boccaletti, S., & al., e. (2002). The synchronization of chaotic systems. Physics reports , 366 (1), 1-101.
2 Bowong, S., & Tewa, J. J. (2008). Unknown inputs adaptive observer for a class of chaotic systems with uncertainties. Mathematical and ComputerModelling , 48 (11), 1826–1839.
3 Chen, J., Patton, R. J., & Zhang, H. Y. (1996). Design of unknown input observers and robust fault detection filters. International Journal of Control , 63 (1), 85–105.
4 Chen, W., & Saif, M. (2006). Unknown input observer design for a class of nonlinear systems: an lmi approach. American Control Conference .
5 Choi, H. H. (2003). An lmi-based switching surface design method for a class of mismatched uncertain systems. IEEE Transactions on Automatic Control , 48 (9), 1634–1638.
6 Darouach, M., Zasadzinski, M., Xu, S. J., & al, e. (1994). Full-order observers for linear systems with unknown inputs. IEEE Transactions on Automatic Control , 39 (3), 606–609.
7 De Persis, C. a. (2001). A geometric approach to nonlinear fault detection and isolation. IEEE Transactions on Automatic Control , 46 (6), 853-865.
8 Dimassi, H., & Loria, A. (2011). Adaptive unknown-input observers-based synchronization of chaotic systems for telecommunication. IEEE Transactions on Circuits and Systems I: , 58 (4), 800–812.
9 Edwards, C., & Spurgeon, S. (1998). Sliding mode control: theory and applications.
10 Edwards, C., Spurgeon, S. K., & Patton, J. R. (2000). Sliding mode observers for fault detection and isolation. Automatica , 36 (4), 541–553.
11 Fallahi, K., & Leung, H. (2010). A chaos secure communication scheme based on multiplication modulation. Communications in Nonlinear Science and Numerical Simulation , 15 (2), 368–383.
12 Imsland, L., Johansen, A. T., Grip, F. H., & Fossen, I. T. (2007). On non-linear unknown input observers–applied to lateral vehicle velocity estimation on banked roads. International Journal of Control , 80 (11), 1741–1750.
13 Khalil, H. K. (2002). Nonlinear Systems.
14 Kocarev, L., Halle, K. S., Eckert, K., Chua, L. O., & Parlitz, U. (1992). Experimental demonstration of secure communications via chaotic synchronization. International Journal of Bifurcation and Chaos , 2 (03), 709–713.
15 Patton, R. J. (1997). Observer-based fault detection and isolation: robustness and applications. Control Engineering Practice , 5 (5), 671-682.
16 Pecora, L. M., & Carroll, T. L. (1990). Synchronization in chaotic systems. Physical review letters , 64 (8), 821.
17 Poznyak, A. S., Shtessel, Y. B., & Gallegos, C. J. (2003). Min-max sliding-mode control for multimodel linear time varying systems. IEEE Transactions on Automatic Control , 48 (12), 2141–2150.
18 Sharma, B. B., & Kar, I. N. (2009). Chaotic synchronization and secure communication using contraction theory. Pattern Recognition and Machine Intelligence Springer , 549–554.
19 Sharma, B. B., & Kar, I. N. (2011). Observer-based synchronization scheme for a class of chaotic systems using contraction theory. Nonlinear Dynamics , 63 (3), 429–445.

20 Slotine, J. J., Hedrick, J. K., & Misawa, E. A. (1987). On sliding observers for nonlinear systems. Journal of Dynamic Systems, Measurement, and Control , 109 (3), 245–252.

21 Spurgeon, K. S. (2008). Sliding mode observers: a survey. International Journal of Systems Science , 39 (8), 751–764.

22 Spurgeon, S. K., & Davies, R. (1993). A nonlinear control strategy for robust sliding mode performance in the presence of unmatched uncertainty. International Journal of Control , 57 (5), 1107–1123.

23 Stefano Boccaletti, J. K. (2002). The synchronization of chaotic systems. Physics reports , 366 (1), 1–101.

24 Tan, C., & Edwards, C. P. (2002). Sliding mode observers for detection and reconstruction of sensor faults. Automatica, , 38 (10), 1815–1821.

25 Theesar, S. J., & Balasubramaniam, P. (2014). Secure communication via synchronization of lure systems using sampled-data controller. Circuits, Systems, and Signal Processing , 33 (1), 37–52.

26 Vadim, U. I. (1977). Survey paper variable structure systems with sliding modes. IEEE Transactions on Automatic control , 22 (2).

27 Venkatasubramanian, V. e. (2003). A review of process fault detection and diagnosis: Part I: Quantitative model-based methods. Computers & chemical engineering , 27 (3), 293-311.

28 Walcott, B., & Zak, S. H. (1987). State observation of nonlinear uncertain dynamical systems. IEEE Transactions on Automatic Control , 32 (2), 166–170.

29 Yan, X. G., & Edwards, C. (2007). Nonlinear robust fault reconstruction and estimation using a sliding mode observer. Automatica , 43 (9), 1605–1614.

Dr. A. Rengarajan [1], S. Rajasekaran [2] and P. Kumaran [3]

An Efficient Lightweight Proactive Source Routing Protocol for Handling Overhead, Packet Delivery, Throughput in MANETS

Abstract: In this paper we proposed an Effective Proactive Routing Protocol (EPSR). EPSR can maintain more network topology information than conventional protocol like Distance Vector (DV) based protocols, Link State (LS)-based routing optimized link state routing and reactive source routing as in LPSR. The paramount function of Mobile Ad Hoc networks is to determine the route between source nodes to destination and forwards the traffic from source node to neighbor node to reach destination when the transmission range exceeds. This can be done proactively using the proactive routing protocol for frequent update of topological information and to avoid the stale that is history of routing information. We propose a brand new routing protocol called efficient proactive routing protocol galvanized by DSDV and OLSR protocols which rely on collective task of routing. The proposed protocol implemented using few techniques called route update, loop detection and multipath route discover algorithm. Using these techniques we have analyzed it can give better performance with help of the parameters includes overhead, throughput, and packet delivery ratio and end- to-end delay.

Keywords: Mobile Ad Hoc Network, DSDV, OLSR, Route Update, Loop Detection

1 Introduction

A Mobile Adhoc network is a wireless communication network, where nodes that are not within direct transmission range of each other will required other nodes to forward data. It can operate without existing infrastructure, supports

1 Professor, CSE, Veltech Multitech Engineering College, Chennai, Tamilnadu, India
rengu_rajan@yahoo.com
2 Junior Research Fellow, Veltech Multitech Engineering College, Chennai, Tamilnadu, India
rajasekaran009@gmail.com
3 Ph.D. Scholar, CSE, Anna University, Chennai, Tamilnadu, India
kumaran.0991@gmail.com

mobile user and falls under the general scope of Multihop wireless networking. Such a networking paradigm was originated from the needs in battle field communication, emergence operation, search and rescue, and disaster relief operations.

2 Literature Survey

Zehua Wang *et al* [1] propose a lightweight proactive source routing (PSR) protocol. PSR can maintain more network topology information than distance vector (DV) routing to facilitate source routing, although it has much smaller overhead than traditional DV-based protocols. PSR is only a fraction of the overhead of these baseline protocols, and PSR yields similar or better data transportation performance than these baseline protocols. In this proposed system they used three different methods to implement the overhead controlled routing protocol. The period of the update cycle is an important parameter in PSR. Furthermore, we go an extra mile to reduce its routing overhead. First, we interleave full dump and differential updates to strike the balance between efficient and robust network operation. Second, we package affected links into forests to avoid duplicating nodes in the data structure. Finally, to further reduce the size of differential update messages, each node tries to minimize the alteration of the routing tree that it maintains as the network changes its structure.

Z.Wang et al [2] propose CORMAN is a network layer solution to the opportunistic data transfer in mobile ad hoc networks. Its node coordination mechanism is largely in line with that of Ex-OR and it is an extension to Ex-OR in order to accommodate node mobility. CORMAN generalizes the opportunistic data forwarding in Ex-OR to suit mobile wireless networks. That is, when a batch of packets are forwarded along the route towards the destination node, if an intermediate node is aware of a new route to the destination, it is able to use this new route to forward the packets that it has already received. That is, a pipeline of data transportation could be achieved by better spatial channel reuse. The design of CORMAN can be further improved to address this explicitly. This may involve timing node back off more precisely and tightly, or even devising a completely different coordination scheme. The potential of cooperative communication in multi-hop wireless networks is yet to be unleashed at higher layers, and CORMAN is only an example.

Cheng Li PSR *et al* [3] said builds a Breadth First Spanning Tree (BFST) in every node of the network. As a proactive routing protocol, we must reduce the overhead of PSR. Ideally, we need to provide each node with abundant routing information using a communication overhead similar to or smaller than that of a proactive DV protocol. High data transportation performance reducing the communication overhead should not penalize the network's capability in data communication. In PSR, nodes maintain and exchange BFSTs periodically. The full dump message containing the entire spanning tree is of the size $O\ (|N|)$, which is in fact much less frequently broadcast than a compact differential updates. While achieving these objectives, PSR yields the same transportation capability as the more expensive protocols like OLSR and DSDV.

Padmavathi.K *et al* [4] purpose of this paper is to provide a better quality of the package delivery rate and the throughput, that is in need of powerful routing protocol standards, which can guarantee delivering of the packages to destinations, and the throughput on a network. We study some parameters of OLSR that forces the inaccuracies in the energy level information of neighboring nodes and show the comparison between ideal and realistic version of OLSR. It operates in a table-driven and proactive manner, i.e., topology information is exchanged between the nodes on periodic basis. Its main objective is to minimize the control traffic by selecting a small number of nodes, known as Multi Point Relays (MPR) for flooding topological information. In route calculation, these MPR nodes are used to form an optimal route from a given node to any destination in the network. This routing protocol is particularly suited for a large and dense network. OLSR generally proposes four types of periodic control messages, namely: Hello messages, Topology Control (TC) messages, Multiple Interface Declaration (MID) messages, Host and Network Association (HNA) messages.

Dhanalakshmi Natarajan *at al* [5] said an optimized link state routing (OLSR) protocol is a proactive routing protocol. An advanced OLSR (AOLSR) protocol is proposed based on a modified Dijkstra's algorithm which enables routing in multiple paths of dense and sparse network topologies. The routing is based on the energy of nodes and links (implied from the lifetime) and the mobility of the nodes. It is a hybrid ad hoc routing protocol because it combines the proactive and reactive features. Two cost functions are introduced to build link-disjoint or node-disjoint paths. Secondary functions, namely path recovery and loop discovery process are involved to manage the topology changes of the network. AOLSR protocol is analyzed and compared with the existing MANET routing protocols namely, dynamic source routing (DSR) and OLSR. Its performance is observed to be satisfactory in terms of average end-to-end delay, packet delivery ratio (PDR), average time in first-in-first-out (FIFO) queue, and throughput.

Gulfishan Firdose Ahmed *at al* [6] said this paper examines routing protocol for mobile ad hoc networks the Destination Sequenced Distance Vector (DSDV) and On Demand protocol that evaluates both protocols based on the packet delivery fraction and average delay while varying number of sources and pause time. In this paper we consider the new approach an Improved-DSDV Protocol is proposed for Ad Hoc networks. Improved -DSDV overcomes the problem of stale routes, and thereby improves the performance of regular DSDV. In our improved DSDV routing protocol, nodes can cooperate together to obtain an objective opinion about other trustworthiness. In this state problem of DSDV routing information will maintain at each node locally in the network. All routing decisions are taken completely distribution fashion. So the local information may be old and invalid, local information updates periodically.

Rahem Abri *at al* [7] proposes a new improvement on DSDV; we introduce a new metric called hop change metric in order to represent the changes in the network topology due to mobility. We determine a threshold value based on this metric in order to decide the full update time dynamically and cost effectively. The proposed approach (La-DSDV) is compared with the original DSDV. The results show that our threshold-based approach improves the packet delivery ratio and the packet drop rate significantly with a reasonable increase in the overhead and the end-to-end delay. A new lightweight threshold-based scheme is proposed in order to improve the low packet delivery ratio of the original DSDV under high mobility. The results show that our approach based on this metric improves the packet delivery ratio and the packet drop rate with a reasonable increase in the overhead and the end-to-end delay. It decides the update time without communicating with other nodes in the network. Since the communication between nodes of battery depletion, it is an important metrics for the nodes that run on battery power in MANETs.

Seon Yeong Han *at al* [8] proposes an adaptive Hello messaging scheme to suppress unnecessary Hello messages without reduced detestability of broken links. Simulation results show that the proposed scheme reduces energy consumption and network overhead without any explicit difference in throughput. An adaptive Hello interval toreduce battery drains through practical suppression of unnecessary Hello messaging. Based on the event interval of a node, the Hello interval can be enlarged without reduced detects ability of a broken link, which decreases network overhead and hidden energy consumption.

Poonkuzhali Ramadoss *at al* [9] said, in this study a new QoS routing protocol is proposed which provides better path selection by avoiding congestion, balancing the load and energy. It minimizes the communication overhead without reducing the network performance. The protocol provides source routing using DFS spanning tree routing in MANET. Thus PSRP is designed which provides nodes with the cost of network structure information for source routing at a communication overhead similar or even less than a distance vector routing protocol. In PSRP nodes maintain and exchange DFSTs periodically. The full dump message which contains the entire spanning tree is of the size O (|n|) which is less frequently broad cast than a compact differential updates. PSRP yields the same transportation capability as the more expensive protocol like OLSR and DSDV.

3 Problem Identification

In PSR, each node maintains a breadth-first search spanning tree of the network rooted at it. This information is periodically exchanged among neighboring nodes for updated network topology information. This allows it to support both source routing and conventional IP forwarding, in which we try to reduce the routing overhead of PSR as much [3]. PSR has only a fraction of overhead of OLSR, DSDV, and DSR but still offers a similar or better data transportation capability compared with these protocols [4], [6]. Many lightweight protocols use "Table Driven" approach. PSR make use of Path Finding Algorithm for routing update and Link Vector Algorithm for reducing the overhead of LS routing algorithm.

Overhead is reduced only fraction when compared to other baseline protocol using TCP. PDR is less than the DSR when using UDP/TCP for data transmission. The overhead is less when compared to other underlying protocol but the data transportation is very low i.e. the packet delivery ratio is less.

3.1 Disadvantages

- Due to overhead there is comparatively less Packet delivery ratio than DSR during data transmission.
- Overhearing the packet during data exchange leads to energy consumption which increases the overhead. Transmission link breakage more thus reduces the PDR and increases the overhead.
- Route discovery and maintenance takes more time to update the information hence delay is more.
- The existing protocols are also vulnerable to the denial-of-service (DoS) attacks, such as RREQ based broadcasting.

- Large amount of bandwidth required to estimate the optimal routes.
- Limited number of control traffic messages

4 Proposed an Effective Proactive Routing Protocol (EPSR)

LPSR having Overhead during the topology changes in the network this can be further reduced using hybrid routing protocol. Proactive reduces the topology updates overhead whereas reactive increases the packet delivery ratio (PDR) thereby the performance and delay can be avoided. Hello Message technique can reduce the delay and overhead by assigning intervals. Using Hybrid Routing Protocol (AOLSR&DSDV) the neighbor node maintenance is easier which further reduces the overhead and increases PDR. An advanced OLSR (AOLSR) protocol, based on a modified version of Dijkstra's algorithm can be proposed. The hello message internal will be added to the hello message packet which identifies the signal strength of neighbor nodes thereby link breakage can be avoided [5], [6], [7]. The hybrid routing protocol focuses on reducing the overhead further and thereby increases the Packet Delivery Ratio (PDR) and throughput. The Tree Based Algorithm will optimize the information to be forwarded to other neighbor nodes.

4.1 Advantages

- It increases the protocols suitability for ad hoc network with the rapid changes of the source and destinations pairs.
- It protocol   does not require that the link is reliable for the control messages since the message are sent periodically and the delivery does not have to be sequential.
- It protocol is well suited for the application which does not allow the long delays in the transmission of the data packets hence reduces overhead.
- Routes maintained only between nodes who need to communicate that reduces overhead of routing maintained.
- Route caching reduces route discovery overhead and a single route may yield many routes to the destination.

- The hello message internal will reduce the time consumption during the topology update and Tree based algorithm optimize the route and reduces the communication overhead.

5 System Architecture

EPSR having Overhead during the topology changes in the network this can be further reduced using hybrid routing protocol. Proactive reduces the topology updates overhead whereas reactive increases the packet delivery ratio (PDR) thereby the performance and delay can be avoided. The route discovery process is initiated by the source who broadcast a new RREQ in the network. The broadcasting is done via flooding. The duplicate RREQs will be discarded by checking the pair of source node address and its broadcast ID present in the RREQ. If a node receives RREQ for the first time, it updates its routing table. This reverse route may later be used to relay the Route Reply (RREP) back to the initiator. Hello Message technique can reduce the delay and overhead by assigning intervals. Using Hybrid Routing Protocol (AOLSR&DSDV) the neighbor node maintenance is easier which further reduces the overhead and increases PDR. An advanced OLSR (AOLSR) protocol, based on a modified version of Dijkstra's algorithm can be proposed. The hello message internal will be added to the hello message packet which identifies the signal strength of neighbor nodes thereby link breakage can be avoided. The hybrid routing protocol focuses on reducing the overhead further and thereby increases the Packet Delivery Ratio (PDR) and throughput. The Tree Based Algorithm will optimize the information to be forwarded to other neighbor nodes. In case of link failure, the node closer to the source of the break initiates RERR and propagates it to all the precursor nodes. When the source node receives the RERR, it will reinitiate the route discovery. The primary functions of AOLSR protocol are topology detection and path estimation [5]. The network topology is sensed to inform the nodes the topology information. It is an optimization of link-state routing. In a classic link-state algorithm, link-state information n is flooded throughout the network. OLSR uses this approach as Well, but the protocol runs in wireless multi-hop scenarios, the message flooding in OLSR is optimized to preserve bandwidth [4]. DSDV has two types of route update packets Full dump for all available routing information and Incremental for Only information changed since the last full dump [10], [11], [12].

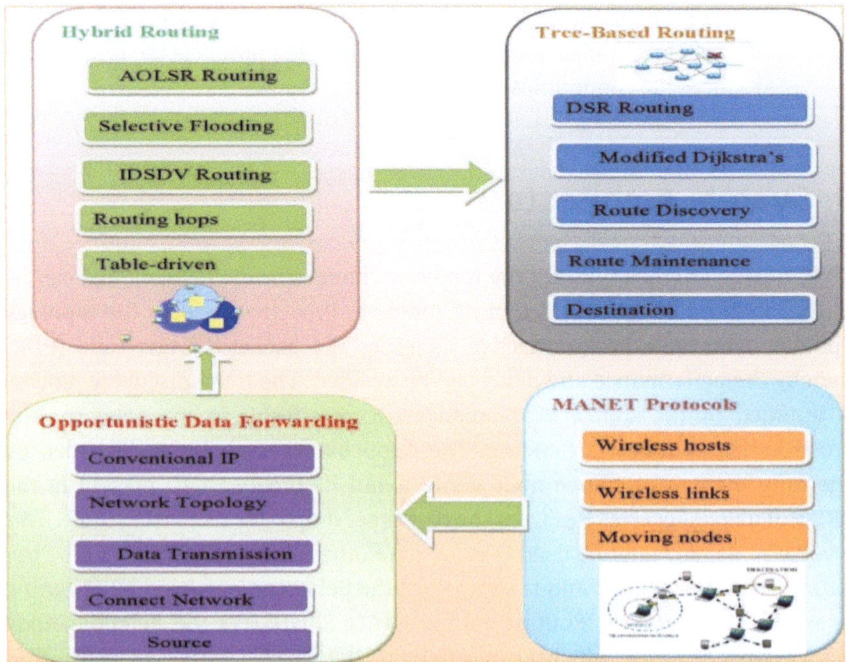

Fig. 1: System Architecture

5.1 Algorithm for Route Discovery

When a packet reaches a router:

- Time delay is measured with all the routes through which the destination can be reached and also connected with the present router, by sending ECHO packet.
- Based upon the fitness value (delay value), the fitter paths/routes are selected.
- The selected paths from same source-destination pair are then sorted and checked on the basis of their freshness.
- The packet is then forwarded based upon the current value of the route.
- Step (2) is repeated until the packet reaches the destination.

6 Module Specification

6.1 Modules1: Mobile Ad Hoc Networks Environment

- Formed by wireless hosts which may be mobile. No pre-existing infrastructure.
- Routes between nodes may potentially contain multiple hops
- Nodes act as routers to forward packets for each other.
- Node mobility may cause the routes change.

Routing protocols in MANETs can be categorized using an array of criteria. The most fundamental among these is the timing of routing information exchange. On one hand; a protocol may require that nodes in the network should maintain valid routes to all destinations periodically [13], [14]. In this case, the protocol is considered *proactive*, which is also known as *table driven*.

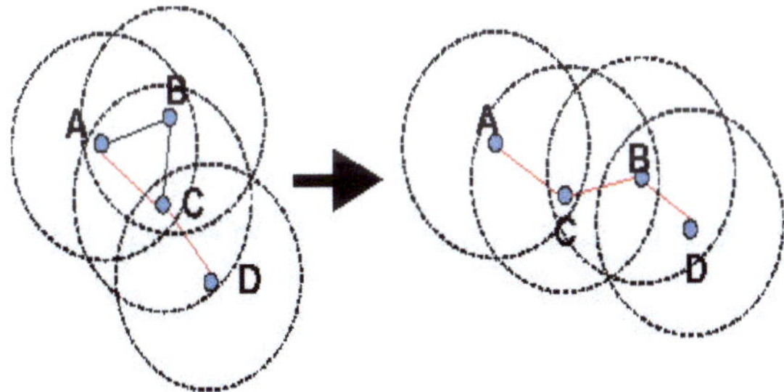

Fig. 2: Mobile Adhoc Network

6.2 Modules2: Opportunistic Data Forwarding Technique

Opportunistic network are a class of mobile ad hoc networks (MANETs) where contacts between mobile nodes occur unpredictably and a complete end-to-end path between source and destination rarely exists at one time **[17]**, **[18]**, **[19]**. There are two important functions provided by the transport layer.

- Ensuring reliable data transmission between source and destination.
- Ensuring that the network does not become congested with traffic.
- These functions are ineffective in opportunistic networks.
- Opportunistic networks require different approaches to those adopted in the more common intermittently connected networks.

6.3 Modules3: Implementing Modified Dijkstra's Algorithm

- The primary functions of AOLSR protocol are topology detection and path estimation. The network topology is sensed to inform the nodes the topology information.
- The path estimation utilizes the modified Dijkstra's algorithm to compute the various paths based on the information from topology detection. A link failure in one path should not affect other routes.
- The source path (route from source to destination including all the hops) is always preserved in the header of the data packets. The data flow diagram of AOLSR protocol.
- Topology detection and path estimation are responsible for the determination of the multiple paths from the source to the destination. Hello message interval will make sure that there is no local link repair.

Destination sequence distance vector (IDSDV)
- Table-driven
- Based on the distributed Bellman-Ford routing algorithm
- Each node maintains a routing table
- ➤ Routing hops to each destination
- ➤ Sequence number

Problem in DSDV is a lot of control traffic in the network and solution Using IDSDV is
- Two types of route update packets
- ➤ Full dump (All available routing information)

Incremental (Only information changed since the last full dump)

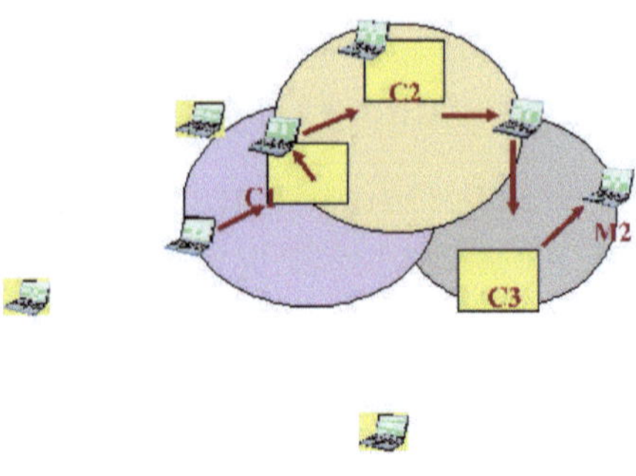

Fig. 3: Destination Sequence Distance Vector

Advanced Optimized Link State Routing (AOLSR)

- AOLSR
  - ➤ Inherits Stability of Link-state protocol
  - ➤ Selective Flooding
  - ➤ Only MPR retransmit control messages and it minimizes the flooding effect and Suitable for large and dense network.
- Advanced Optimized Link State Routing (AOLSR) is a table driven proactive routing protocol for MANET.
- A mobile ad hoc network (MANET) is a self-configuring wireless network of mobile hosts connected through arbitrary topology without the aid of any centralized administration [16].
- The optimization is based on a technique called Multipoint Relaying. This information is exchanged by protocol messages after periodic time.
- AOLSR performs hop-by-hop routing, where each node uses its most recent topology information for routing
- The primary functions of AOLSR protocol are topology detection and path estimation. The network topology is sensed to inform the nodes the topology information.
- The path estimation utilizes the modified Dijkstra's algorithm to compute the various paths based on the information from topology detection. A link failure in one path should not affect other routes [5], [15].

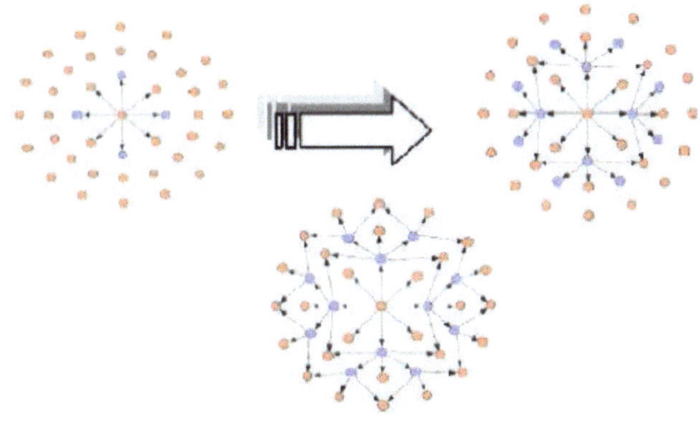

Fig. 4: Multipoint Relays (MPRs)

6.4 Modules4: Hello Messaging Technique

- Dynamic Source Routing
- ➤ On-demand driven
- ➤ Based on the concept of source routing
- ➤ Required to maintain route caches
- Two major phases
- ➤ Route discovery
- ➤ Route maintenance

The Hello Message interval will be added to the Source node when the node searches for its neighbor node thereby it identifies the signal range of each node. If any node having less signal interval when compared to nearest neighbor will be discarded [20].

6.5 Modules5: Tree-Based Routing

- In proposed work, nodes would hold their broadcast after receiving a route update for a period of time. If more updates have been received in this window, all updates are consolidated before triggering one broadcast.
- The period of the update cycle is an important parameter in EPSR. Furthermore, we go an extra mile to reduce its routing overhead. First, we interleave full dump and differential updates to strike the balance between efficient and robust network operation.
- Second, we package affected links into forests to avoid duplicating nodes in the data structure.
- Finally, to further reduce the size of differential update messages, each node tries to minimize the alteration of the routing tree that it maintains as the network changes its structure [9], [18].

7 Simulation and Results

In this scenario the nodes are detected using the HELLO message broadcasting. Those are in red color are detected and it's idle to receive the data packet from the source node. Inside each node there is a number which indicates the signal strength of that node. This signal strength calculated when it receives the Hello Message request i.e. RREQ (Route Request) from the source node. Once all nods

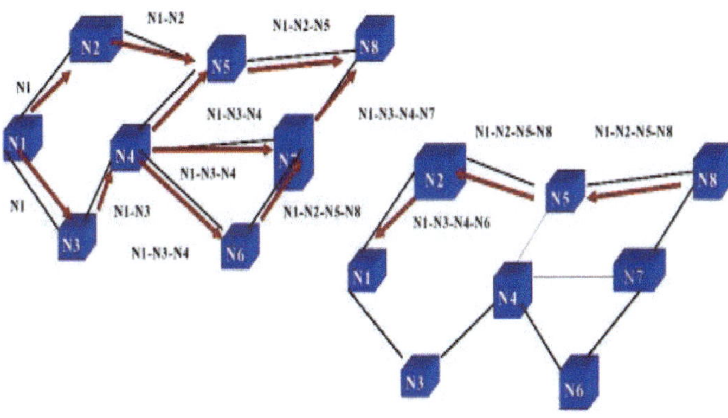

Fig. 5: Dynamic Source Routing Protocol for Hello Messaging Technique

are detected it will enable the multipath route discover algorithm to discover the path which is optimized to reach the destination.

Fig.6 describe about the Overhead of the Proposed EPSR protocol. This evaluation is carried out with 10^3 No. of Nodes and overhead packet with size 10^3 in x-axis and y- axis. This estimation shows Overhead Reduction is high and less packet dropped ratio due to the Hello scheme with signal strength analysis respectively.

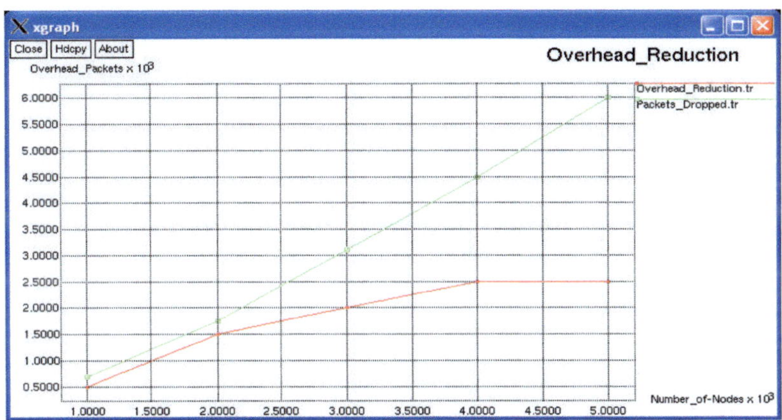

Fig. 6: Overhead vs Number of Nodes

Fig.7 describes about the Throughput of EPSR protocol. This evaluation is carried out with 10^3 No. of Nodes and Throughput in mbps in x-axis and y- axis. This result shows high throughput of sending packets and it outperforms the underlying protocols. The loop detection helps to identify the infinity loops and discards it which further increases the throughput.

Fig.8 depicts the Packet Delivery Ratio (PDR) of proposed EPSR protocol. This evaluation is carried out with 10^3 No. of Nodes and Packet Delivery Ratio in x-axis and y- axis. This estimation shows remarkably high PDR. The EPSR helps in Path recovery, route update and throughput of receiver side packet is exceptionally better.

Fig.9 depicts End- to- End Delay for proposed approach. This evaluation is carried out with 10^3 No. of Nodes and End- to- End Delay with Latency in x-axis and y- axis. This estimation shows decreasing down time when it encounters the transmission of packet from source to destination in estimated optimized path. The End- to- End Delay is reduced due to the Healing of Route and fraction Rerouting using the Multi path Route Discover algorithm.

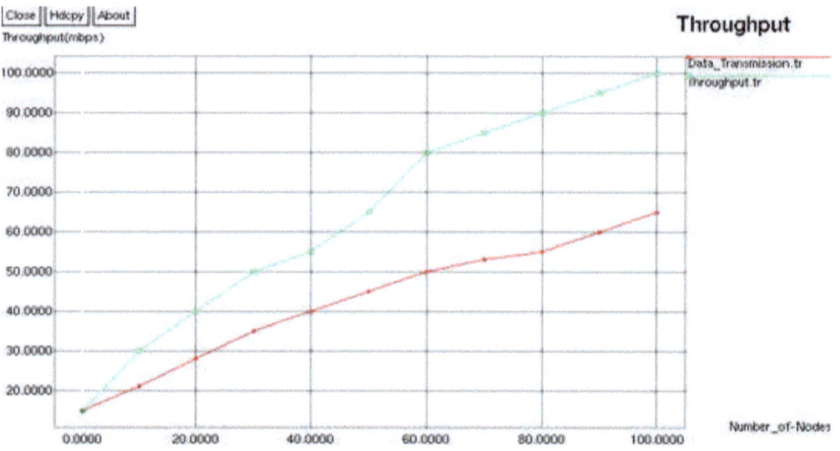

Fig. 7: Throughput vs Number of Nodes

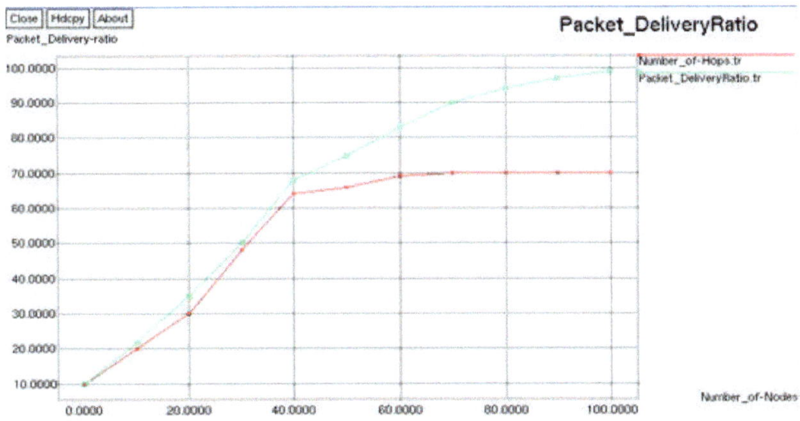

Fig. 8: Packet Delivery Ratio vs Number of Nodes

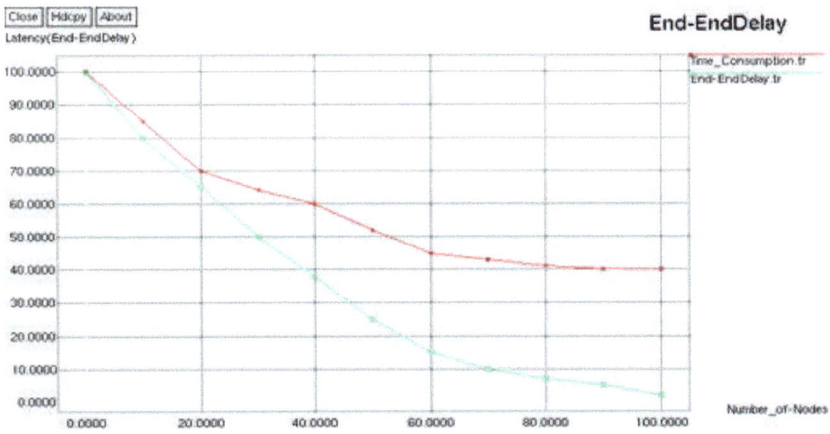

Fig. 9: End- to- End Delay vs. Number of Nodes

8 Snapshots for Simulation Outputs

8.1 MANET Environment

In this snapshot it describes the environment created with wireless nodes (hosts).

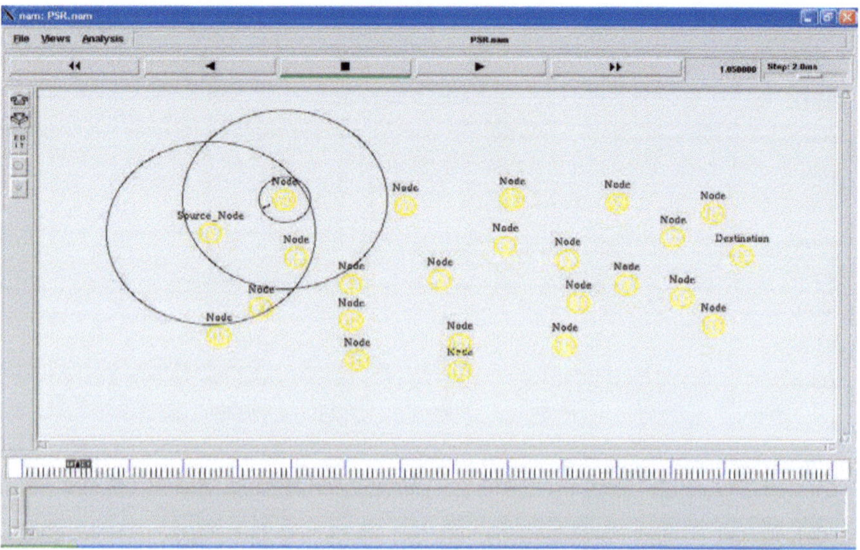

Fig. 10: MANET Environment

8.2 Hello Messaging Scheme

This module senses the neighbor node and reports the status to its source node or requested node.

8.3 Multipath Route Discover

In this module it's trying to identify the optimized path using its Algorithm for Multipath Route Discover and its saves the entire discovered path in each of the nodes.

8.4 Optimized Path

Here it's identified the optimized path and gathering the information of all the nodes and it's trying to update the topological information.

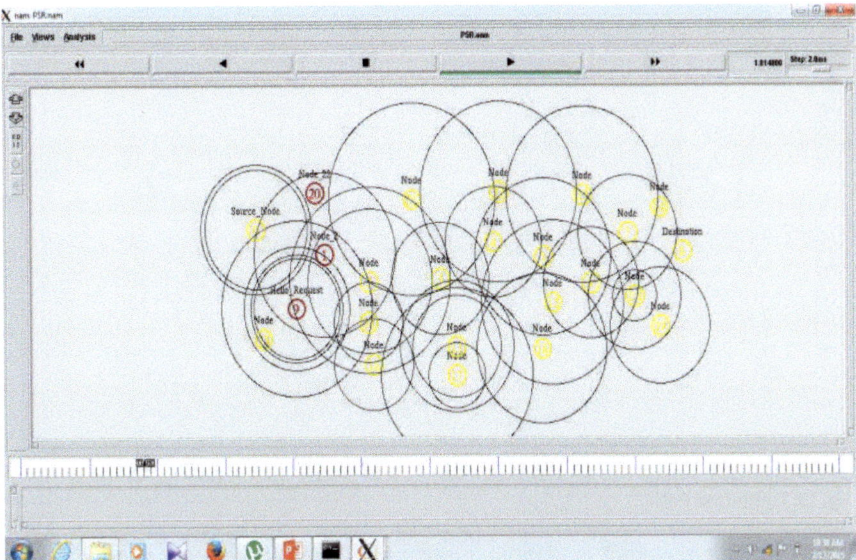

Fig. 11: Hello Messaging Scheme (Neighbor Node Detection)

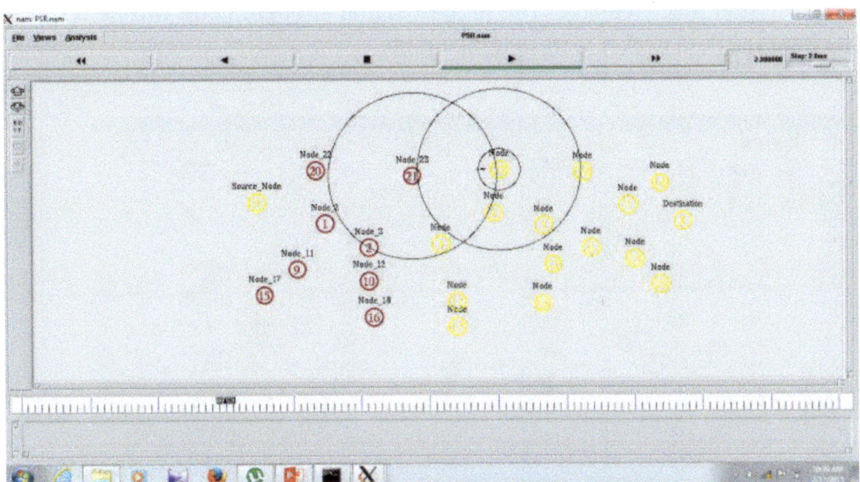

Fig. 12: Using Algorithm for Multipath Route Discover

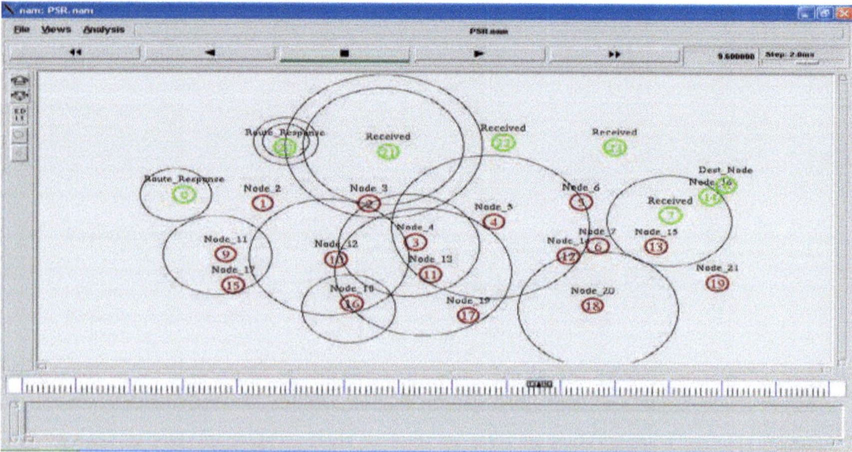

Fig. 13: Optimized Path

8.5 Signal Strength Detection

The data packets are transferred with the help of signal strength to avoid the transmission link breakage. The highest signal strength node will be taken to forward the data packet to its neighbor node.

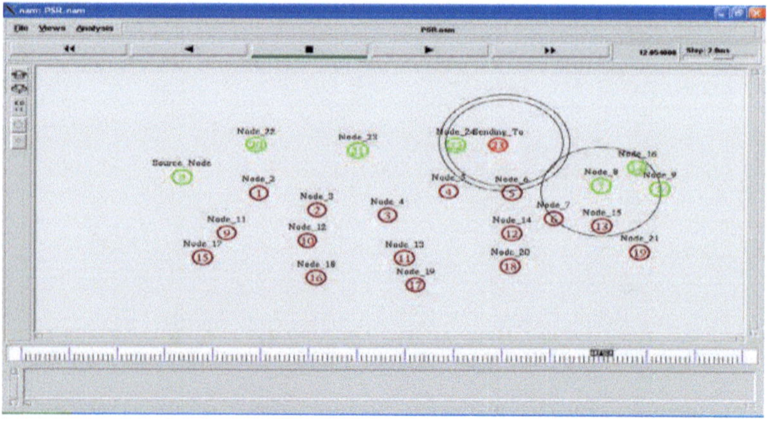

Fig. 14: Signal Strength Detection

8.6 Path Recovery and Rerouting

In this snapshot it's explain how it recovers from transmission link breakage and how it chooses another route to reach its destination. The Rerouting will be done with the help of Multipath Route Discover Algorithm.

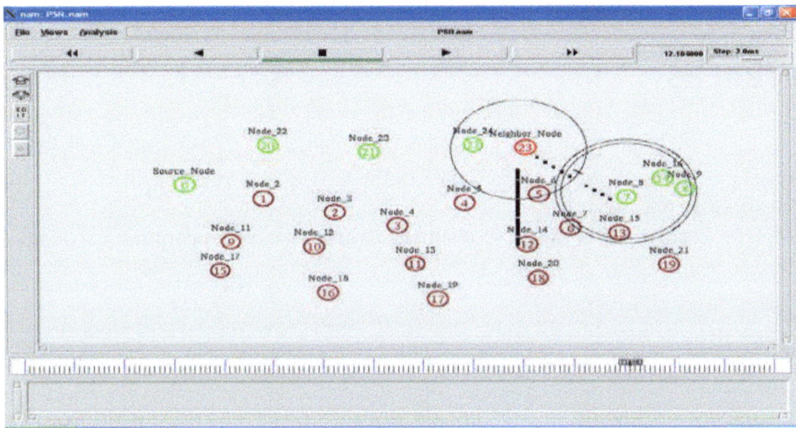

Fig. 15: Path Recovery and Rerouting

Conclusion

This paper has been motivated by the need to support opportunistic data forwarding in MANETs. To generalize the milestone work of ExOR for it to function in such networks, we needed a PSR protocol. Such a protocol should provide more topology information than DVs but must have significantly smaller overhead than LS routing protocols named as EPSR (Efficient Proactive Source Routing Protocol); even the MPR technique in OLSR would not suffice. Thus, we put forward a tree-based routing protocol, i.e., EPSR, which is inspired by modified Dijkstra's Algorithm and threshold based DSDV. Its routing overhead per time unit per node is on the order of the number of the nodes in the network as with IDSDV, but each node has the full-path information to reach all other nodes. For it to have a very small footprint, EPSR's route messaging is designed to be very concise. Hence the overhead can be reduced further and increase the Packet Delivery Ratio (PDR).

Acknowledgement

Authors deliver their graduate to SERB (Young Scientist Scheme, No. SP/FTP/ETA-51/2013) Govt. of India for Financial Assistance and FIST F.NO:SR/FST/College-189/2013.

References

1   Zehua Wang, Yuanzhu Chen, and Cheng Li (2014), "Lightweight Proactive Source Routing Protocol for Mobile Ad Hoc Networks", IEEE TRANSACTIONS ON VEHICULAR TECHNOLOGY, VOL. 63, NO. 2.
2   Z. Wang, Y. Chen, and C. Li (2012), "CORMAN: A novel cooperative opportunistic routing scheme in mobile ad hoc networks", IEEE J. Sel. Areas Commun., vol. 30, no. 2, pp. 289–296.
3   Zehua Wang, Cheng Li, and Yuanzhu Chen (2011), "PSR: Proactive Source Routing in Mobile Ad Hoc Networks", pages: 5818-5823 IEEE global communication.
4   Padmavathi.K, Thivaharan.S (2014), "An Improved OLSR Protocol for Reducing the Routing Overhead in MANETs", International Journal of Innovative Research in Computer and Communication Engineering (An ISO 3297: 2007 Certified Organization) Vol. 2, Issue 4, ISSN (Online): 2320-9801 page no: 3924-3931.
5   Dhanalakshmi Natarajan and Alli P Rajendran (2014), "AOLSR: hybrid ad hoc routing protocol based on a modified Dijkstra's algorithm", EURASIP Journal on Wireless Communications and Networking ,90.
6   Gulfishan Firdose Ahmed, Raju Barskar, and Nepal Barskar (2012), "An Improved DSDV Routing Protocol for Wireless Ad Hoc Networks"
7   Rahem Abri and Sevil Sen (2013) , "A Lightweight Threshold-Based Improvement on DSDV"
8   Seon Yeong Han, and Dongman Lee (2013), "An Adaptive Hello Messaging Scheme for Neighbor Discovery in On-Demand MANET Routing Protocols", IEEE Communications Letters, Vol. 17, No. 5.
9   Poonkuzhali Ramadoss, Sanavullah Mohamed Yakub and Sabari Annaji (2014), "A Preemptive Link State Spanning Tree Source Routing Protocol for Mobile Adhoc Networks", Journal of Computer Science 10 (1): 85-90, ISSN: 1549-3636.
10 Al-Rabayah M and Malaney R (2012), "A new scalable hybrid routing protocol for VANETs", IEEE Trans. Veh. Technol., vol. 61, no. 6, pp. 2625–2635.
11 Ajit R. Bandgar, Sandeep A. Thorat (2013), "An Improved Location-Aware Ant Colony Optimization based routing Algorithm for MANETs", 4th ICCCNT, IEEE – 31661.
12 Biswas s and Morris R (2005), "ExOR: Opportunistic multi-hop routing for wireless networks", in Proc. ACM Conf. SIGCOMM, Philadelphia, PA, USA, pp. 133–144.
13 Chandni, Sharad Chauhan, Kamal Kumar Sharma (2014), "Routing Protocol In MANET – A Survey", Chandni al. IJRRA, Vol. 1, Issue 2, pp. 48-52.
14 Chlamtac I, Conti M, and J.-N. Liu (2003), "Mobile ad hoc networking: Imperatives and challenges, Ad Hoc Networks", vol. 1, no. 1, pp. 13–64.

15 Mrs Kavitha K, Dr.Selvakumar K, Mrs.Nithya T, Ms.Sathyabama S (2013), "Zone Based Multicast Routing Protocol for Mobile Ad-Hoc Network", ISBN: 978-1-4673-5301-4 on IEEE.

16 Larsson P (2001), "Selection diversity forwarding in a multihop packet radio network with fading channel and capture", ACM Mobile Comp. Commun. Rev., vol. 5, no. 4, pp. 47–54.

17 Le Van Minh, Yang Mingchuan, Yang Liang, Guo Qing (2008), "Congestion-aware and hybrid routing protocol for vehicular ad hoc network", 2nd International Conference on Computer Science and Network Technology of IEEE ISBN: 1-4673-2964-4.

18 Rajaraman R (2002), "Topology control and routing in ad hoc networks: A survey", ACM SIGACT News, vol. 33, no. 2, pp. 60–73.

19 ShivaPrakash, J.P.Saini, S.C.Gupta (2011), "Areview of Energy Efficient Routing Protocols for Mobile Ad Hoc WirelessNetworks", InternationalJournalofComputerInformationSystems,Vol.1,No.3

20 Vinod N Biradar, Suresh chimkode, Gouri Patil (2013), "Simulation and Performance Analysis of CORMAN, AODV and DSR Routing Protocol in MANET (IJART)", Volume 2, Issue 8, 148 ISSN 2278-7763.

Authors

Dr.A.Rengarajan received the Ph.d., Degree from Bharath University, Tamilnadu, India in 2011 and the M.E., Degree from Sathyabama University, Tamilnadu, India in 2005 and the B.E., Degree from Madurai Kamaraj University, Tamilnadu, India in 2000. He is a Professor of Computer Science & Engineering in Veltech and he received R & D Project from DST-SERB worth of 12.84 Lakhs. His current research is Network Security. His official Mail id is rengu_rajan@yahoo.com and Mobile number is 9841310814

Rajasekaran S. received the M.E., Degree in Network Engineering from Veltech Multitech Dr.Rangarajan & Dr.Sakunthala Engineering College, Affiliated to Anna University, Tamilnadu, India in 2013 and the B.E., Degree in Computer Science & Engineering from Institute of Road & Transport Technology, Affiliated to Anna University, and Tamilnadu, India in 2011. He is a reviewer of IAJIT journal. He is a JRF of Computer Science & Engineering in Veltech Multitech Research development. His current research is network security and He is doing Funded Project also. His official Mail id is rajasekaran009@gmail.com and Mobile Number is 9751650205

P.Kumaran B.E., M.E., He is currently studying Ph.D scholar in the Department of Computer Science and Engineering in Anna University. His official Mail id is kumaran.0991@gmail.com and Mobile Number is **9585612633**. In addition he got gold medal in P.G and U.G, His Current area of research interest is **Vehicular Adhoc Network**

K. Neeharika[1] and S. Ravi Kishan[2]

Classification of Mobile Videos on Sports Genre

Abstract: Video classification is one of the challenging research environments. Many researchers had done research on video classification using broadcast videos. But few authors had done research on the video classification using user generated videos or mobile videos. This paper explains the classification of mobile videos in particularly sports domain, because still it is a challenging environment due to the diversity of sports rules. We use the single user approach to extract domain knowledge from the recorded video to classify the sport type. In this work video could be separated as two different modalities suchas audio and video. Each modality is separately analyzed and extracts the combination of audio and visual features using HOG descriptors (for visual). The final result of the classification which uses distance measure to obtain result from feature extraction should identify the sport type. The best adaptive technique gives the high accuracy at classification.

Keywords: Video, Sport, Visual, Audio, Histogram of Oriented Gradient (HOG), Principal Component Analysis (PCA), Cross-Correlation.

1 Introduction

The video content which was recorded by mobile users can be termed as Mobile videos. These Mobile videos can also be called as User Generated Videos. The user generated videos are of several categories such as sports, news etc. At any situation user seems to be interesting of a particular event. Every moment user not able to carry the hand held cameras and not available of professional camera setup. But the user is very much fond of recording a video. In such scenario the mobile users can recorded a particular event by using their mobile cameras. The camera enabled mobile users are enormously increasing due to the fast growing culture. Mobile users can share any video in multimedia by

1 M.Tech, Computer Science & Engineering, VR Siddhartha Engineering College, Kanuru, Vijayawada, Andhra Pradesh, India
2 Professor, Computer Science & Engineering, VR Siddhartha Engineering College, Kanuru, Vijayawada, Andhra Pradesh, India

using such services. To share the video user has to search the particular video among different categories of videos. For this purpose automatic video classification system can be developed to classify the videos according to its genre. The main goal of project is to classify the videos particularly in sports domain.

In the past years onwards multimedia content such as the audio and video had been enormous growth and also the sharing of that content through the internet has been increased simultaneously. A typical retrieval scenario is searching an event within a video which considers an interesting feature. In retrieving such events or objects effectively, considering the multimedia content is a primary step. The consideration of multimedia content process can be performed either manually or automatically. Automatic analysis of multimedia data is a very challenging task. Due to diversity of sports rules classification of sports videos is still a challenging research environment. Many researchers were performs classification by considering either audio or visual. Among them for effective classification both audio and visual approaches can be considered for different kinds of sports. This classification follows the single user approach which is shown in figure1. The recorded mobile video can be given as input. Training set is prepared and build the model. When the testing data processed classifier compares with the model and finally classifies the videos. This app-roach is not limited to mobile videos but it can be directly applied to videos captured by other types of recording devices considered in this work, such as some models of hand-held cameras.

2 Related Work

Now a day's users had been showing interest in accessing videos. But it is infeasible to the user to select the video among a lot of available videos. To resolve the inconvenience to the user's, videos could be classifying automatically according to the genres and subgenres. Most of the researchers had done experiments on classification of entire video according to genre. But some researchers had focus on classifying specific type of genre.

The classification can be done automatically according to the features which could be drawn by the 3 main approaches. They are
- Text-based approach
- Audio-based approach
- Visual-based approach

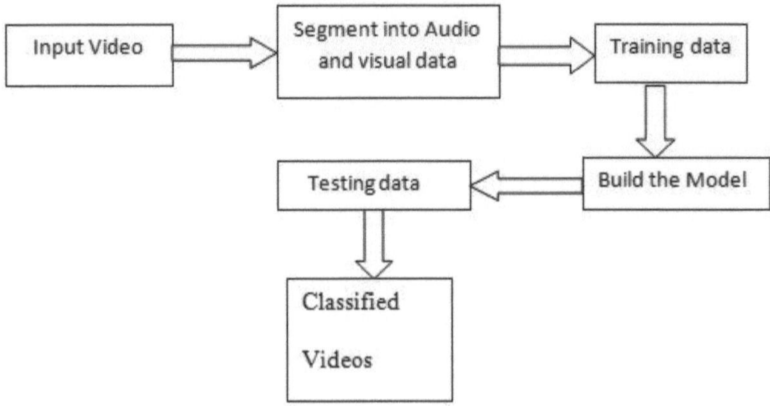

Figure.1: Video Classification using single user approach

In classification we have several standard classifiers such as Bayesian Networks, Support Vector Machine (SVM), Neural Network, Gaussian Mixture Model (GMM), and Hidden Markov Model (HMM).

Jianguo, Wang, Wei Hu, Mingliang Sun, Zhang performs classification on soccer highlight detection using 2D Bayesian network. 2D Bayesian network is a particular Bayesian network which is used to characterize and trained among features by assuming the condition that one variable depends on other two variables.

Many researchers had done work on the highlight detection in the soccer event, but still it is a challenging environment due to the semantic gap between the low level features such as audio visual features and the high level events which explains in figure 2.

The main approach is to detect the semantic keywords from low-level features suchas Goal, shot, foul, free kick, Corner kick based on stack fusion(Stack fusion refers that making decision by considering the unimodal feature which utilizes the supervised learning) and the 2D Bayesian network classifier can be applied on the detected semantic keyword to identify the event.

Advantages: Performance of goal detection shot detection is perfect and easily trained.

Disadvantages: Keywords will not change fast, Shot and free kick can simultaneously occurred due to the false alarms and miss faults.

Yuan , Wei Lai, Tao Mei, Xian-Sheng, Qing Wu, Shipeng , performed classification on automatic video genre categorization using hierarchal SVM. Different genres of videos can be classified using hierarchal SVM by considering

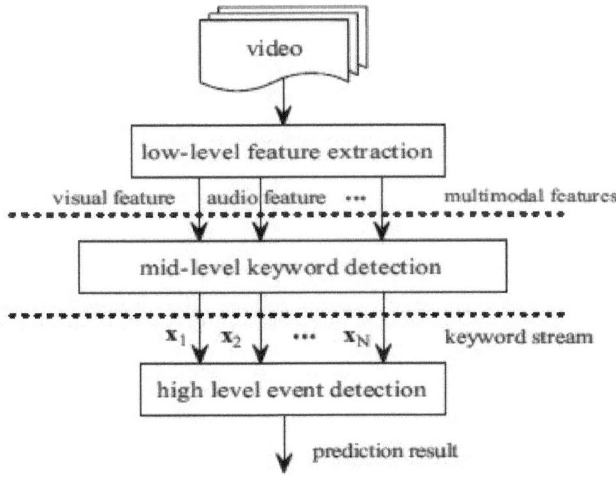

Figure. 2: Representation of video classification using Bayesian network

the computable spatio temporal features. Hierarchal SVM can be represented in the form of binary trees which is shown in figure3 and can be used to deal multiclass problems. In video categorization task by using global SVM binary tree classifier it classifies into 3 stages.

At first sage identify the hierarchal ontology among different genres, at second level categorize into different genres by considering the spatial features suchas average brightness and color entropy and temporal features suchas average shot length, cut percentage, average color difference, camera motion, by using hierarchal SVM classifier, at third level optimization takes place either globally or locally. But global optimization gives best performance than the local optimization.

Advantages: Computation cost is low.

Disadvantages: Increasing the training time, Local optimal binary tree may not be global.

Jinjun, Changsheng, Engsiong performs classification using pseudo 2D HMM. In this paper multilevel framework would be explained and automatically recognize the port event by considering the low level features and apply pseudo 2D HMM classifier.

Consider the video segments from different video clips and apply the 1DHMM classifier to the audio features suchas MFCC and visual features suchas camera motion, dominant color which was extracted from the video segments.

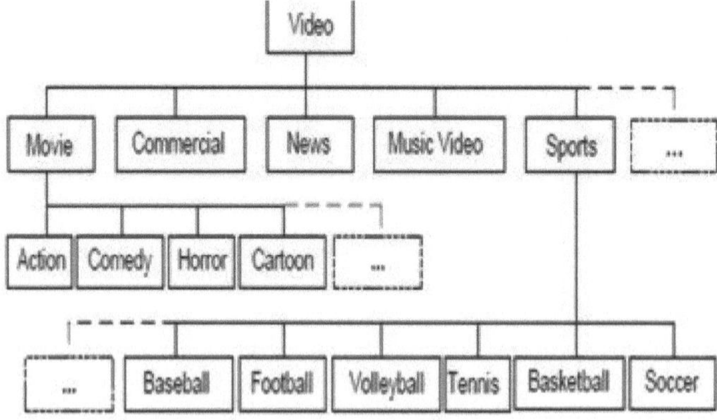

Figure. 3: Representation form of hierarchical SVM

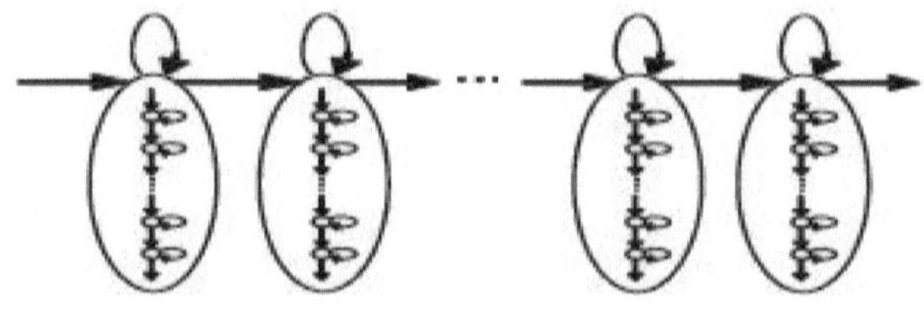

Figure. 4: Structural Representation of Pseudo 2D HMM

The classified segments can be fused and the obtained output can be given as input to the 2DHMM which can be represented and clearly explained in figure 4.If any errors raised that could be recognized in the end of 2DHMM by analyzing the audio.

Advantages: High accuracy and extracts suitable low level features.

2.1 Methodology

The automatic video classification system classifies the video into different kinds of genres. In this paper classification performs particularly on sports genre. There are several kinds of sports. Among them 2 sports had chosen to classify which is basketball, and cricket.

In training phase we have to follow these steps to train the data:

Step 1: Collect the image samples of different sports.

Step 2: Preprocessing: In this step, all the image samples are resized to Particular size.

Step 3: Feature Extraction: Features are extracted using HOG (Histogram of Oriented Gradients).

Step4: Feature Selection: In this step, PCA is used to select the features.

In testing phase the sequence of steps can be explained below:

Step1: Capture the real time video.

Step2: Separate audio and visual data using by using mat lab functions.

Step3: (a) Visual data:

    I.   Divide the video into frames.

    II.   The step 2, 3, 4 implemented in training phase is done in testing phase on the selected frames.

    III.   The data is tested by using distance measure algorithm and the results are analyzed.

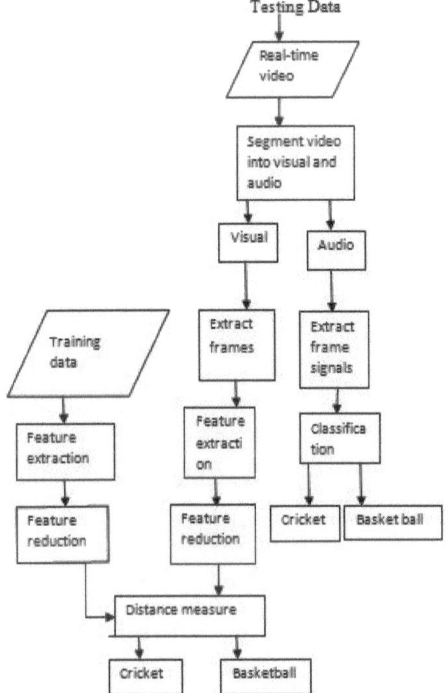

Figure. 3: Proposed method for the automatic video classification

Step3: (b) Audio data:

I. Segment the audio signals into frames.
II. By using matrix algorithm results are analyzed.

3 Results

The real time video which was recorded by mobile phones and uploaded in YouTube of 2 different games such as cricket, basketball can be taken as input.

3.1 Visual Results

To get the visual result 27 training images can be taken from the web and features can be extracted from the trained data. The input video extracts frames and randomly selected 21 testing images are taken. The extracted features can be reduced to 600 features by applying dimensionality reduction technique. By using Euclidean distance measure sport type can be identified very effectively rather than other classification techniques. The main advantage of using this technique is gives the effective and correct results without any misclassifications. The testing image compares with the training image by using distance measure the corresponding matched image can be identifies that particular sport type.

Figure. 4: Visual Results

3.2 Audio Results

The input video segments into two different modalities suchas audio data and visual data by using the mat lab commands.

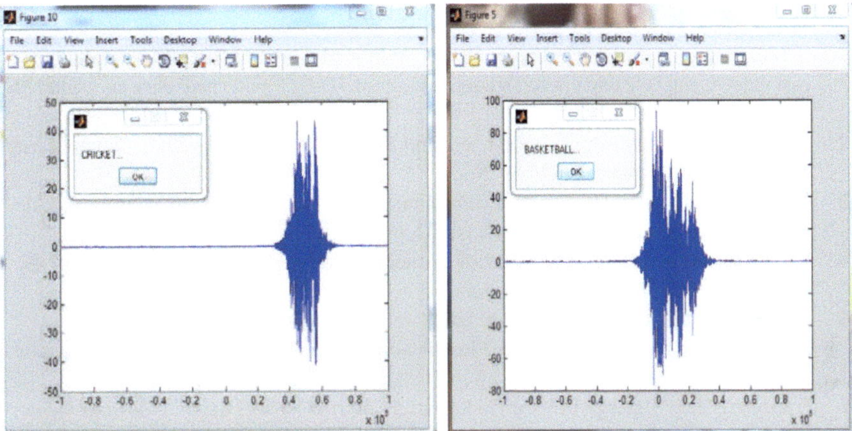

Figure. 5: Audio Results

The audio signals can be divided into different signals. The 10 audio wave signals can be saved in database. By using correlation matrix the input audio samples0 are compared and matched signal can identifies the sport type. The segmented audio signal from the visual data can be divided into different kinds of frame signals. In cricket the keywords suchas SIX, FOUR are identified and compare the audio signals with the database. If the signal is either SIX or FOUR then it matches with the database and the result can be obtained as cricket. If the keywords GOAL, FOUL words can be recognized then it compare with the database and the result obtained as Basket ball.

Conclusion

This project proposes a method for classifying mobile videos on sports genre. It follows single user approach. It is an extended version of existing work. Both visual and audio phases give good results and better accuracy by choosing adaptive techniques. If it extends to multi user approach it gives more good results. Along with multi user approach more work on audio gives the better performance.

References

1   G. Hauptmann, R. Jin, and T. D. Ng ; (2002); Multi-modal information retrieval from broadcast video using OCR and speech recognition; *ACM/IEEE-CS joint conference on Digital libraries*; pp. 160 –161.

2   Anchor Kutchhi Polytechnic, B Mesh ram and B V Patel &, Chembur; (2012); Content Based Video Retrieval Systems; *IJU*; Vol.3, No.2.

3   C.J. Lin, C.W. Hsu; (2002); A comparison of methods for multiclass support vector machines; *IEEE Trans. on Neural Networks*; vol. 13; pp. 415-425.

4   Darin Brezeale and Diane J. Cook; (2007); Automatic Video classification: A Survey of the Literature; *IEEE*; Transaction on System.

5   Francesco Cricri, Jussi Leppänen, Mikko J.Roininen, Sujeet Mate, Stefan Uhlmann; (2014); Sport Type Classification of Mobile Videos; *IEEE Transactions on Multimedia*; Vol. 16; No. 4.

6   Francesco Cricri & Igor D. D. & Curcio Kostadin Dabov & Moncef Gabbouj & Sujeet Mate; (2014); Multimodal extr0action of events and of information about the recording activity in user generated videos; *Springer*; DOI:70:119–158,.

7   F. Sebastiani; (2002); Machine learning in automated text categorization; *ACM Computing Surveys*; vol.34; no. 1; pp. 1–47.

8   H.Tominaga, K. Yokoyama, S. Hattori, S. Takagi; (2003); Sports Video Categorizing Method Using Camera Motion Parameters; *IEEE International Conference on Multimedia and Expo*; vol. 2; pp. 461-464.

9   Jinjun Wang, Changsheng Xu and Engsiong Chng; (2006); Automatic Sports Video Genre Classification using Pseudo-2D-HMM; *IEEE*.

10   Jianguo Li, Tao Wang, Mingliang Sun, Wei Hu, Yimin Zhang; (2006); Soccer Highlight Detection Using Two-Dependence Bayesian Network; *IEEE*.

11   L. Duan, M. Xu, T.S. Chua, Q. Tian; (2006); Amid-level representation framework for semantic sports video analysis; *In ACM Multimedia Conference.*

12   Vapnik; (1998); Statistical Learning Theory.

Naresh Kumar Yadav [1]

Genetic Algorithm with Dual Mutation Probabilities for TCSC – Based ATC Enhancement

Abstract: The growing demand and regular restructuring of transmission system necessitates the need of enhancing the Available Transfer Capability (ATC). Though FACTS devices are found to be promising, they have to be optimally located with adequate compensation level. This paper introduces an improved Genetic Algorithm (GA) to enhance ATC by optimally locating the connections of Thyristor Controlled Series Compensator (TCSC). The proposed GA has Dual Mutation Probabilities (DMP) for mutation operator and hence it is termed here as GA with DMP (GADMP). Here, one of the mutation probabilities entertains static mutation whereas the other probability enables adaptive mutation. As a result, the GADMP exploits the benefits of both the mutation types to enable fast convergence. Experiments are conducted on IEEE 24 RTS, IEEE 30 and IEEE 57 bus systems. The results demonstrate that the GADMP offers better ATC enhancement than the traditional GA (TGA).

Keywords: ATC; FACTS; TCSC; GADMP; power flow

1 Introduction

Since many years, numerous research problems in power systems are based on knowledge – based decision making process for which artificial intelligent techniques remain as promising solutions [1] [2]. The artificial intelligent techniques include bio-inspired optimization algorithms, swarm intelligence, chaos theory, etc. One of such problems that require artificial intelligence is ATC enhancement in power systems under deregulated environment [3] [4]. The ATC enhancement has become mandatory, because the current way of restructuring of the power sector and the rapid power demand have forced to operate the

1 Department of Electrical Engineering, Faculty of Engineering and Technology,
Deenbandhu Chhotu Ram University of Science & Technology, Murthal (Sonepat) - Haryana, India
Email: nareshyadhavdr@gmail.com

power system so close to the stability limits. Increased ATC can ensure secure and reliable transmission with hassle free electricity commerce [26].

The ATC enhancement problem has become constrained due to practical issues in expanding the transmission network and the confining to use only the existing network. Hence, FACTS devices have been seen as emerging and robust solutions since they enhance the ATC through redistribution and regulation of line flows and bus voltages, respectively [5] [6] [7] [12]. However, the FACTS devices can be cost effective only if they are placed in the optimal locations [6] [11]. Moreover, optimal compensation by the connected FACTS devices can enhance the ATC considerably. Hence, the ATC enhancement problem can be formulated as either optimization or decision making problem or both as well. The optimization problems can be further categorized as single objective and multi-objective optimization [4]. Among various methodologies, GA is found to be a prominent solution for optimization problems in power systems [3] [4]. Earlier, GA searched through binary solution encoding principle, whereas the later literature use real encoding [3] [4].

Numerous methodologies have been reported in the literature on optimizing the usage of FACTS devices. Rashidinejad *et al* [4] have worked on hybridizing real GA and fuzzy sets to solve the problem and hence aimed at improving the ATC. T. Nireekshana *et al* [3] have attempted to enhance ATC using FACTS devices such as Static Var Compensator (SVC) and TCSC. They have used real coded GA to determine the optimal location and the compensating reactance for the FACTS devices. Both normal and contingency cases have been considered in their experimentation with IEEE 14 and 24-bus systems and they have studied the performance of both the FACTS devices. H. Farahmand *et al* [8] have proposed a hybrid mutation particle swarm optimization (HMPSO) technique to enhance the ATC. They have dealt with multi-objective optimization problem, which has been solved by the HMPSO. The formulated multi-objective optimization problem has considered optimal installation and allocation of FACTS and their capacities, respectively. The experimental investigation has been done in IEEE 30-bus and 57-bus systems. Ali ES and Abd-Elazim [9] have used swarm intelligence, termed as bacterial swarm optimization for enhancing the ATC. Earlier, Khaburi MA and Haghifam MR [10] have worked on improving the total transfer capability (TTC) since ATC finds its direct determination from TTC. They have modeled in the probabilistic perspective and hence the probabilistic modeling features have been extracted for the betterment of TTC.

Despite numerous works have been reported, the challenges still reside due to the growing demand and continuous restructuring of the power system. Moreover, the introducing cost [4] limits the convergence of the solution,

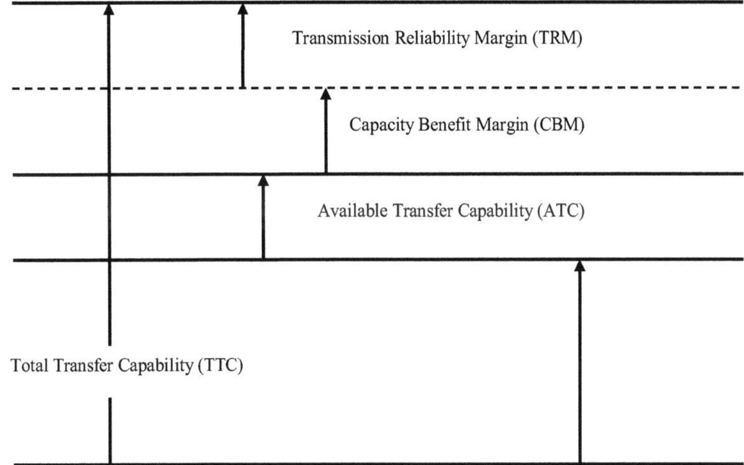

Figure. 1: Definition of ATC

Algorithm: Genetic Algorithm (GA)	
Step 1	Generate arbitrary population
Step 2	Evaluation population
Step 3	Select better solutions
Step 4	Crossover
Step 5	Mutation
Step 6	Evaluation new population
Step 7	Go to Step 3 if termination criterion is not met
Step 8	Return best solution

Figure. 2: Basic steps of GA

whereas the multi-objective scenarios increase the complexity [8]. This paper solves the ATC enhancement problem as an optimization problem using GA with newly introduced DMP. The DMP enables faster convergence than the traditional GA. In the rest of the paper, the preliminaries about the concept are discussed followed by the problem model formulation. Subsequently, the

proposed ATC enhancement procedure is explained and the experimental
results are discussed.

2 Preliminaries

2.1 Available Transfer Capability (ATC)

Under a deregulated electrical market, the generators and customers share a
common power transmission system to transfer the power from the generation
end to the consumer end. Since high competition persists among the multiple
producers, overloading and congestion often occurs here. Due to such
overloading and congestion, violating voltage limits, line flow and stability
limits occur and hence the system remains insecure and unreliable. Hence, ATC
has to be determined for any such system to ensure the system is under secure
and reliable side. This is found to be very significant and hence the worldwide
nations entertain to update the information about ATC in systems such as Open
access same time information system (OASIS) [13].

The general definition of ATC is usually given as *"a measure of the Available
Transfer Capability (ATC) is a measure of the transfer capability remaining in the
physical transmission network for further commercial activity over and above
already committed uses"*. The mathematical representation of ATC is given as

$$ATC = TTC - TRM - \left(ETC + CBM\right) \quad (1)$$

where, TTC, TRM, ETC and CBM refer to Total Transfer Capability, Transmission Reliability Margin, existing transmission commitments and Capacity
Benefit Margin, respectively. The relationship between TTC, TRM, ETC, CBM and
ATC is illustrated in Fig. 1 [13]. The definitions of these parameters are given as
follows.

Definition 1: The TTC is defined as the amount of electric power that can be
transferred over the interconnected transmission network in a reliable manner
while meeting all of a specific set of defined pre- and post-contingency system
conditions.

Definition 2: The TRM is defined as that amount of transmission transfer
capability necessary to ensure that the interconnected transmission network is
secure under a reasonable range of uncertainties in system conditions.

Definition 3: The CBM is defined as that amount of transmission transfer
capability reserved by load serving entities to ensure access to generation from
interconnected systems to meet generation reliability requirements.

2.2 Genetic Algorithm

GA is random global search algorithm that is inspired from the natural evolution. By exploiting the past search results, GA explores new searching points to determine the near-optimal solution. In GA, the chromosome represents a probable solution in the search space. Each chromosome consists of genes, where the number of genes that define the chromosome length refers to dimension of the solution space [18] [19].

The GA is operated by three major operators, namely, selection, crossover and mutation. These operators are often termed as genetic operators. As illustrated in Fig. 2, the GA is initiated by arbitrary population, which is a collection of possible solutions from the solution space. Each chromosome in the population pool is evaluated using fitness function (or objective function). Based on the theory of survival of the fittest, the best solutions (can be said as fittest solutions) are selected by the selection operator, while the other solutions are vanished from the population pool. These selected solutions undergo crossover and mutation to produce new solutions. These solutions are evaluated to find better solutions than those in the population pool.

The entire process is repeated till a defined number of iterations is reached or desired level of solution is obtained. Since its development, GA finds numerous applications like image processing [20], wireless communications [21], vehicular technology [22], software engineering [23], etc.

3 Problem Formulation

Let us consider a typical bus system in which the ATC has to be enhanced using the limited external resources, i.e., using the offered FACTS devices. Since our paper concentrate on TCSC, the problem can be interpreted as maximizing the ATC of the given bus system using the given TCSCs. The number of TCSC can be as per the offering and hence the resources are said as limited. Based on the description, the problem model can be represented as

$$\left[L_t^*, X_t^* \right] = \arg_{[L_t, X_t]} \max ATC; 0 \le t \le N_T - 1 \quad (2)$$

where, L_t and X_t refer to the set of line indices and the compensation level of tth TCSC in the bus system, respectively, whereas the best connections of the set are represented as $\left[L_t^*, X_t^*\right]$. Since the primary objective reside in enhancing the ATC, the formal definition and its description is presented further.

3.1 Determining ATC

Adequate contributions have been made in the literature to determine the ATC. For instance, C.Y. Li and C.W. Liu [14] have proposed a novel technique for calculating ATC by considering the shape of the security boundary. C. Vaithilingam and R.P. Kumudini Devi [15] have used Support Vector Machine (SVM) to estimate ATC.

Despite such new efforts have been put forth to estimate the ATC, the common literary methods are classified as two, namely, distribution factors – based ATC estimation [4] and continuation power flow (CPF) based ATC estimation [16]. Earlier methods exploit AC or DC power flows and hence they are simple and computationally efficient [4], but they do not produce optimal outcomes. Hence CPF methods are preferred that injects at a set of buses with linear variation to determine the maximum possible value of a scalar function [17]. In other words, the CPF linearly increases the controlling parameter and solves the power flow problem at every iteration. The process is repeated till the voltage instability occurs. This paper considers power flow based ATC estimation [3] as given the right part of Fig. 2, which are sequentially listed below.

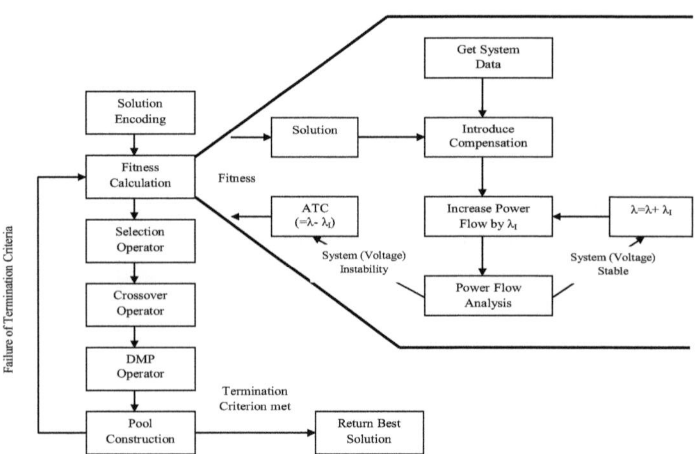

Figure. 3: Architecture of the ATC enhancement using GADMP. Left portion is the GADMP procedure, whereas the Right portion is the ATC estimation process

Sequential steps

1. Initialize and determine basic variables
   a. Read the system line data: From bus, To bus, Line resistance, line reactance, half line charging, off nominal turns ratio, maximum line flows
   b. Read the system bus data: Bus no, Bus type, P_{gen}, Q_{gen}, P_{load}, Q_{load}, P_{min}, P_{max}, V_{sp} shunt capacitance data
   c. Calculate $P_{shed}(i)$, $Q_{shed}(i)$: $i = 1,2,\ldots,n$, where,

   $$P_{shed}(i) = P_{gen}(i) - P_{load}(i) \text{ and } Q_{shed}(i) = Q_{gen}(i) - Q_{load}(i)$$

   d. Construct Y_{bus} using sparsity technique

2. Power flow analysis
   a. Set $iter = 1$ iteration count
   b. Reset $\left|\Delta P_{max}\right|$ and $\left|\Delta Q_{max}\right|$ to zero
   c. Determine $P_{cal}(i) = \sum\limits_{q=1}^{n} |V_i||V_q||Y_{iq}|\cos(\delta_{iq} - \theta_{iq})$ and

   $$Q_{cal}(i) = \sum\limits_{q=1}^{n} |V_i||V_q||Y_{iq}|\sin(\delta_{iq} - \theta_{iq})$$

   d. Determine $P(i) = P_{shed}(i) - P_{cal}(i)$ and $Q(i) = Q_{shed}(i) - Q_{cal}(i)$
   e. Reset P_{slack} and Q_{slack} to zero
   f. Calculate $\left|\Delta P_{max}\right|$ and $\left|\Delta Q_{max}\right|$ form $[\Delta P]$ and $[\Delta Q]$ vectors
   g. Is $\left|\Delta P_{max}\right| \le \epsilon$ and $\left|\Delta Q_{max}\right| \le \epsilon$. If yes, go to Step 7, since the problem is said to be converged

3. Construct Jacobian Matrix
   a. Set $A[i][j] = 0$ for $i = 1$ to $2n + 2$; for $j = 1$ to $2n + 2$
   b. Determine diagonal elements for $i = 1$ to $2n + 2$

   $$H_{pp} = \frac{\partial P_p}{\partial \delta_p} = -Q_p - B_{pp}|V_p|^2 \quad \text{(a)}$$

   $$N_{pp} = \frac{\partial P_p \cdot |V_p|}{\partial V_p} = P_p + G_{pp}|V_p|^2 \quad \text{(b)}$$

   $$M_{pp} = \frac{\partial Q_p}{\partial \delta_p} = P_p - G_{pp}|V_p|^2 \quad \text{(c)}$$

$$L_{pp} = \frac{\partial Q_p |V_p|}{\partial V_p} = Q_p - B_{pp}|V_p|^2 \quad \text{(d)}$$

c. Determine off-diagonal elements

$$H_{pq} = \frac{\partial P_p}{\partial \delta_q} = |V_p||V_q|\left(G_{pq} \sin \delta_{pq} - B_{pq} \cos \delta_{pq}\right) \quad \text{(e)}$$

$$N_{pq} = \frac{\partial P_p |V_q|}{\partial V_q} = |V_p||V_q|\left(G_{pq} \cos \delta_{pq} + B_{pq} \sin \delta_{pq}\right) \quad \text{(f)}$$

$$M_{pq} = \frac{\partial Q_p}{\partial \delta_q} = -N_{pq} \quad \text{(g)}$$

$$L_{pq} = \frac{\partial Q_p |V_q|}{\partial V_q} = H_{pq} \quad \text{(h)}$$

d. Update Jacobian elements

    i. Slack bus as $H_{pp} = 10^{20}$ and $L_{pp} = 10^{20}$

    ii. PV bus as $L_{pp} = 10^{20}$

e. Construct RHS vector as $B[i] = \Delta P[i], B[i+n] = \Delta Q[i]$

4. Gauss-elimination method

    a. Objective function: $[A][\Delta X] = [B]$

    b. Update phase angle and voltage magnitudes $\delta_i = \delta_i + \Delta X_i$ and $V_i = V_i + \{\Delta X_{i+n}\}V_i$ as , respectively

5. Power flow termination criterion evaluation

    a. Increment *iter* by 1

    b. If $iter < iter_{max}$ go to step 2 (b)

6. Determine output data: line flows, bus powers, slack bus power, converged voltages, line flows and powers

7. Check for any overload. If any overload occurs, go to step 10

8. Read the sending bus (seller bus) m and the receiving bus (buyer bus) n

9. Inject positive real power

    a. Update $\Delta tp (= 0.1)$,i.e. λ-factor at seller bus- m and negative injection $\Delta tp (= -0.1)$, i.e. λ-factor at the buyer bus- n and form mismatch vector

    b. Go to step 2 to perform power flow analysis

10. Determine the maximum increasable power over the base-case load at the sink bus and return it as the ATC.

4 GADMP for ATC Enhancement

GADMP resembles with TGA, but it uses a sophisticated mutation operator rather than traditional uniform random mutation. GADMP solves the objective function given in Eq. (2) using the basic steps of GA with DMP. Moreover, GADMP encodes the solution in such a way that the GA can search effectively. Fig. 3 portrays the GADMP as well as the fitness calculation procedure in the right portion.

4.1 Solution encoding

Since the objective of the model is to determine the optimal location and the compensation level, each pair of gene represents an index of line and the compensation to be introduced. Let $S_p : 0 \le p \le N_p - 1$ be the chromosome, which can be represented as $S_p = \{s_0, s_1, ..., s_{N_t-1}\}_p$, where N_p is the population size. Each gene of the chromosome s_t can be referred as tuple as $s_t = \{L, X\}_t$, where L refers to the line where the TCSC has to be connected and X is the compensation level to incurred by the TCSC.

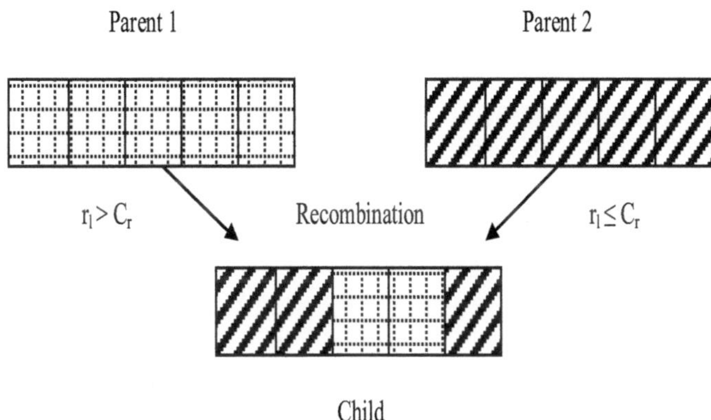

Figure. 4: Standard uniform crossover operation

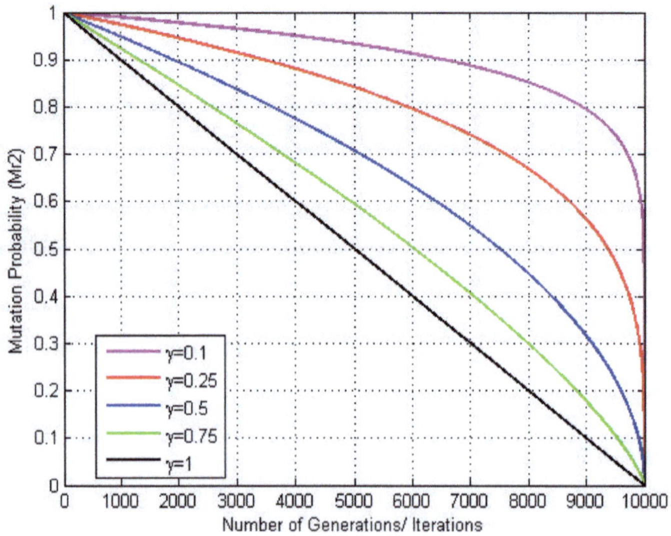

Figure. 5: M_{r2} in various iterations and the effect of γ on it

Remark 1: Since each gene refers to a tuple, the actual length of the chromosome N_L is twice of N_T. Despite the technical interpretation considers the tuple as a gene, the actual gene refers to each element of tuple too.

4.2 Fitness Calculation

The fitness of S_p is calculated using an iterative process involved in the ATC estimation as shown in Fig. 3. The system data referred in Fig. 3 consists of bus data as well as the line data of the test case system. The compensation level for a particular line index given by S_p is introduced in the line data. Initially, the lambda λ is set as small as possible. Let us say it as 0.1, i.e., the initial lambda ($\lambda = \lambda_I$), where λ_I = 0.1. The continuous power flow analysis is made after the compensation reactance is included and the bus voltages are observed. If any voltage limit violations are determined, the process is terminated and the estimated ATC is considered as the fitness of the respective solution, else, the lambda is keep on increased till the system violates the voltage limits. A detailed mathematical description about the ATC estimation is presented in [3].

4.3 Genetic Operators

Selection is first genetic operator that selects $N_p\big/2$ best solutions from the population pool based on the estimated ATC. Despite numerous operators are available, we use rank based selection in which a solution with highest ATC is given first rank and vice versa. The top ranked $N_p\big/2$ chromosomes are selected and subjected to crossover operation. Renowned uniform crossover operator is used in this paper. The crossover operator gets two chromosomes as parent chromosomes to generate a child chromosome by recombining them at rate of C_r as illustrated in Fig. 4. The child chromosome is subjected to proposed DMP.

DMP Operator: Literature has revealed numerous mutation operators and their significance on aiding GA to converge near the best solution [24] [25]. The DMP operator proposed in this paper uses two mutation probabilities M_{r1} and M_{r2}. M_{r1} is a static mutation probability, i.e., set as constant throughout the GA executions, whereas M_{r2} is set as adaptive towards the number of generations, i.e. it is varying for every iteration. The adaptive nature of M_{r2} is illustrated in Fig. 5 in which the M_{r2} degrades asymptotically when the coefficient γ tends to zero, while it exhibits linear decay when γ tends to one. The new offspring produced by the proposed DMP is given as

$$C_p^{new}(l) = \begin{cases} \hat{C}_p(l); if\ r_1 > M_{r1} \\ C_p(l); otherwise \end{cases} \quad (3)$$

$$\hat{C}_p(l) = M_{r2}\left[C^{min}(l) + \left(C^{max}(l) - C^{min}(l) \right) r_2 \right] \quad (4)$$

$$M_{r2} = \left(1 - \frac{I}{I^{max}} \right)^{\gamma} \quad (5)$$

where, $C_p^{new}(l)$ is the new l^{th} gene of pth chromosome after DMP, $C_p(l)$ is l^{th} gene of pth chromosome after crossover operator, I and I^{max} are the current iteration count and the maximum number of iterations, respectively.

Example : Let us consider an offspring obtained from crossover operator [5 4 2], where the number of genes is 3 (number of elements in the offspring). Assume that each gene has same minimum and maximum limits [-5, 5] and the current iteration I is 10 and maximum iteration $I^{\max}$ is 1000. Since M_{r1} is constant, it can be assumed as 0.1, whereas γ is assumed as unity.

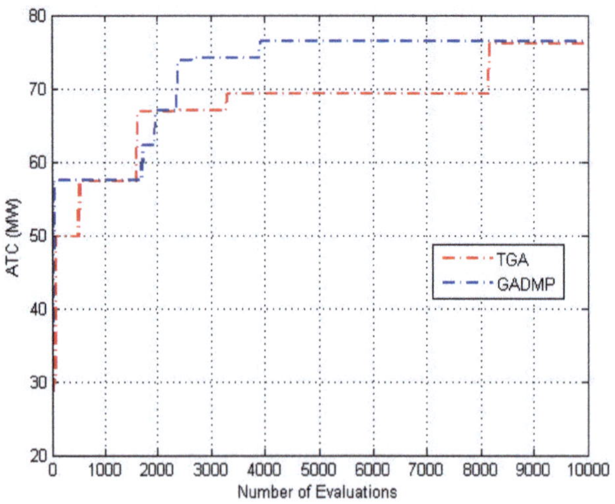

Figure. 6: Convergence behavior of TGA and GADMP

According to Eq. (5), M_{r2} is calculated as 0.999. Based on this value, eq. (4) is determined for first gene as $\hat{C}_p(l)=0.99[-5+(5-(-5))0.5]=0$, where r_2 is considered here as 0.5.Since r_2 is arbitrarily generated for every gene, the rest of the genes get different values as -2.97 and -1.98, when r_2 is set as 0.3 and 0.2 for gene 2 and 3, respectively.

Eq. (3) switches value between the new and old gene values to determine the new offspring based on r_1. Consider the generated r_1 takes the value 0.5, 0.05 and 0.1 for genes 1, 2 and 3, respectively. For the first gene, the first constraint of Eq. (3) has been satisfied (i.e. $r_1 > M_{r1}$) and so 0 replaces 5 of the first gene. In contrast, the first constraint has not been met by r_1 generated for the second gene and hence, the actual value (4) has been preserved. Similarly, the third gene also accepts second constraint as r_1 equals M_{r1}. Hence, the new offspring takes the form [0 4 2]. Hence produced new offspring is positioned in the population pool for next iteration, if the termination criterion is not met. Once the termination criterion is met, the best solution is obtained in which the line index where the TCSC has to be connected and its associated compensation level.

5 Simulation Results

The proposed GADMP is simulated in MATLAB and the experimentation is carried out in IEEE 24 RTS system, IEEE 30 bus system and IEEE 54 bus system. In each test system, the experimentation is conducted for finding optimal locations and reactance for 2 TCSCs. For the similar experimental setup, TGA is also executed and the results are tabulated in Table 1-3.

Table 1. Performance of ATC enhancement on IEEE 24 RTS bus system

Algorithms		TGA	GADMP
TCSC 1	Line no (From bus/ to bus)	13 (8/10)	23 (14/16)
	Compensation (p.u)	-0.043332	-0.0070141
TCSC 2	Line no (From bus/ to bus)	31 (17/22)	19 (11/14)
	Compensation (p.u)	0.039642	0.014148
ATC (MW) with TCSC		76.1	76.4
ATC (MW) without TCSC		19	19

Table 2. Performance of ATC enhancement on IEEE 30 bus system

Algorithms		TGA	GADMP
TCSC 1	Line no (From bus/ to bus)	10 (6/8)	10 (6/8)
	Compensation (p.u)	-.0060091	-0.0058855
TCSC 2	Line no (From bus/ to bus)	20 (14/15)	4 (3/4)
	Compensation (p.u)	-0.022319	-.0012871
ATC (MW) with TCSC		81.7	81.7
ATC (MW) without TCSC		72.7	72.7

MATPOWER 5.1 is used for performing continuous power flow analysis and hence the ATC is calculated. Apart from the test cases, the converging

Table 3. Performance of ATC enhancement on IEEE 54 bus system

	Algorithms	TGA	GADMP
TCSC 1	Line no (From bus/ to bus)	50 (37/38)	49 (36/37)
	Compensation (p.u)	-0.32783	-0.37944
TCSC 2	Line no (From bus/ to bus)	32 (21/22)	3 (3/4)
	Compensation (p.u)	0.028945	-0.017776
ATC (MW) with TCSC		31.5	37.2
ATC (MW) without TCSC		3	3

performance of GADMP and TGA is presented in Fig. 6. The GA parameters such as N_p, C_r, M_{r1} and $I^{\max}$ are set as 10, 0.5, 0.1 and 10000, respectively. The results given in Table 1 – 3 demonstrate the outperformance of GADMP over TGA. The enhanced ATC is higher than TGA result as well as when the system is without TCSC.

Conclusion

This paper presented an improved GA, termed as GADMP to enhance the ATC by optimizing locations and the capacity of TCSCs. Both static and adaptive mode of mutation operator has converged GA faster than TGA. Three test case systems such as IEEE24 RTS system, IEEE30 bus system and IEEE57 system are subjected to experiment the proposed algorithm. The experiments have also been extended for multiple TCSC connections. The experimental results have demonstrated that the TCSC connections recommended by GADMP have produced higher ATC than the connections recommended by TGA.

References

1   L. Gyugyi, "A unified power flow control concept for flexible AC transmission systems", in IEEE Proceedings of Generation, Transmission and Distribution, Pittsburgh, USA, Jul. 1992, vol. 139, no. 4, pp. 323-331.
2   S. M. Swamy, B. R. Rajakumar, and I. R. Valarmathi, "Design of hybrid wind and photovoltaic power system using opposition-based genetic algorithm with Cauchy mutation", in Proceedings of IET Chennai 4th International Conference on Sustainable Energy and Intelligent Systems (SEISCON 2013), Chennai, Dec. 2013, pp. 504-510.

3   T. Nireekshana, G. Kesava Rao, and S. Siva Naga Raju, "Enhancement of ATC with FACTS devices using Real-code Genetic Algorithm", International Journal of Electrical Power & Energy Systems, vol. 43, no. 1, pp. 1276-1284, Dec. 2012.

4   M. Rashidinejad, H. Farahmand, M. Fotuhi-Firuzabad, and A. A. Gharaveisi, "ATC enhancement using TCSC via artificial intelligent techniques", Electric Power Systems Research, vol. 78, no.1, Jan. 2008, pp. 11-20.

5   N. G. Hingorani, and L. Gyugyi, "Understanding FACTS, Concepts and Technology of Flexible AC Transmission Systems", IEEE, New York, 2000.

6   T. Jain, S. N. Singh, and S. C. Srivastava, "Dynamic ATC enhancement through optimal placement of FACTS controllers", Electric Power Systems Research, vol. 79, no. 11, Nov. 2009, pp. 1473-1482.

7   F. D. Galiana, K. Almeida, M. Toussaint, J. Griffin, D. Atanackovic, B. T. Ooi, and D. T. McGillis, "Assessment and control of the impact of FACTS devices on power system performance", IEEE Transactions on Power Systems, vol. 11, no. 4, Nov. 1996, pp. 1931-1936.

8   H. Farahmand, M. Rashidinejad, A. Mousavi, A. A. Gharaveisi, M. R. Irving, and G.A. Taylor, "Hybrid Mutation Particle Swarm Optimisation method for Available Transfer Capability enhancement", Electrical Power and Energy Systems, vol. 42, no. 1, Nov. 2012, pp. 240-249.

9   E. S. Ali, and S. M. Abd-Elazim, "Coordinated design of PSSs and TCSC via bacterial swarm optimization algorithm in a multimachine power system", International Journal of Electrical Power & Energy Systems, vol. 36, no.1, Mar. 2012, pp. 84-92.

10 M. A. Khaburi, and M. R. Haghifam, "A probabilistic modeling based approach for total transfer capability enhancement using FACTS devices", International Journal of Electrical Power & Energy Systems, vol. 32, no. 1, Jan. 2010, pp.12-16.

11 Naresh Acharya, and N. Mithulananthan, "Locating series FACTS devices for congestion management in deregulated electricity markets", Electric Power Systems Research, vol. 77, no. 3–4, Mar. 2007, pp. 352-360.

12 C. A. Canizares, and Z. T. Faur, "Analysis of SVC and TCSC controllers in voltage collapse", IEEE Transactions on Power Systems, vol. 14, no. 1, Feb. 1999, pp. 158-165.

13 Transmission transfer capability task force, Available transfer capability Definitions and determination, North American Electric Reliability Council, NJ; Jun. 1996.

14 C.-Y. Li, and C.-W. Liu, "A new algorithm for available transfer capability computation", International Journal of Electrical Power & Energy Systems, vol. 24, no. 2, Feb. 2002, pp. 159-166.

15 C. Vaithilingam, and R. P. Kumudini Devi, "Available transfer capability estimation using Support Vector Machine", International Journal of Electrical Power & Energy Systems, vol. 47, May 2013, pp. 387-393.

16 H. Sheng, and Hsiao-Dong Chiang, "CDFLOW: A Practical Tool for Tracing Stationary Behaviors of General Distribution Networks", IEEE Transactions on Power Systems, vol. 29, no. 3, May 2014, pp. 1365-1371.

17 G. C. Ejebe, J. Tong, J. G. Waight, J. G. Frame, X. Wang, and W. F. Tinney, "Available transfer capability calculations", IEEE Transactions on Power Systems, vol. 13, no. 4, Nov. 1998, pp. 1521-1527.

18 D. E. Goldberg, "Genetic Algorithms in Search, Optimization, and Machine Learning",. Addison-Wesley, 1999.

19 M. Mitchell, "An Introduction to Genetic Algorithms", London: MIT Press, 1999.

20 F. Wang, J. Li, S. Liu, X. Zhao, D. Zhang, and Y. Tian, "An Improved Adaptive Genetic Algorithm for Image Segmentation and Vision Alignment Used in Microelectronic Bonding", IEEE/ASME Transactions on Mechatronics, vol. 19, no. 3, Jun. 2014, pp. 916-923.

21 W. Li, Y. Hei, J. Yang, and X. Shi, "Optimisation of non-uniform time-modulated conformal arrays using an improved non-dominated sorting genetic-II algorithm", Microwaves, Antennas & Propagation, IET, vol. 8, no. 4, Mar. 2014, pp. 287-294.

22 X. Zuo, C. Chen, W. Tan, and M. Zhou, "Vehicle Scheduling of an Urban Bus Line via an Improved Multiobjective Genetic Algorithm", IEEE Transactions on Intelligent Transportation Systems, vol. 16, no. 2, Apr. 2015, pp. 1030-1041.

23 A. Panichella, R. Oliveto, M. Di Penta, and A. De Lucia, "Improving Multi-Objective Test Case Selection by Injecting Diversity in Genetic Algorithms", IEEE Transactions on Software Engineering, vol. 14, no. 4, Apr. 2015, pp. 358-383.

24 B. R. Rajakumar, "Static and adaptive mutation techniques for genetic algorithm: a systematic comparative analysis", International Journal of Computational Science and Engineering, vol. 8, no. 2, 2013, pp. 180-193

25 B. R. Rajakumar, "Impact of static and adaptive mutation techniques on the performance of Genetic Algorithm", International Journal of Hybrid Intelligent Systems, vol. 10, no. 1, 2013, pp. 11-22

26 Ibraheem and Naresh Kumar Yadav, "Implementation of FACTS Device for Enhancement of ATC Using PTDF", International Journal of Computer and Electrical Engineering, vol. 3, no. 3, 2011, pp. 343-348

Mohanaradhya [1] and Sumithra Devi K.A [2]

WSN Lifetime Enhancement through Residual energy balancing by Avoiding Redundant Data Transmission

Abstract: In WSN energy utilization may be optimized by avoiding redundant data transmission from sensing nodes to Base station. In dense WSN, nodes in the same sensing rage observe the same data and hence it is not required for all the nodes to transmit the data to the Cluster Head. This paper proposes a method which avoids the duplicate data transmission from different nodes in the same sensing range there by conserving energy. The proposed method checks the residual energy among the nodes having the same data and allows the node with more residual energy to transmit the data to the cluster head. This leads to balanced and distributed utilization of node energy thereby the network lifetime is enhanced.

Keywords: Energy balancing, Redundant data, Residual Energy, Threshold distance, WSN

1 Introduction

In Wireless Sensor Networks (WSN) minimizing the network operation without compromising the user requirements is necessary to prolong the network life. Wireless Sensor Networks is a network of autonomous nodes capable of sensing (getting the required information of interest), processing and transmitting the data to the targeted place. To perform these multi functionalities, each node contains Sensing unit, Processing Unit, transceiver Unit. The Sensing unit gathers the required data and converts it to digital form. Processing unit controls other units based on the algorithms deployed. Transceiver unit interconnects the nodes by providing communication between nodes. For proper functioning of node power unit supplies energy to all the units of a node,

1 Assistant Professor, Dept., of MCA, R.V College of Engineering.
mohanaradhya72@gmail.com
2 Principal, GSSS Institute of Technology, Mysore, India
sumithraka@gmail.com

but power unit is of limited capacity as each node is battery powered. Wireless
Sensor Networks are used in different categories of applications like Military,
Environmental, Healthcare, Home and Industrial applications [1]. In human
unattained applications, it is very important to conserve the energy to prolong
the network operation period. Network lifetime is dependent on the power unit
and the power will be utilized in WSN for transmitting, receiving data and processing of sensed data. [3] Nearly 80% of node energy is used for data transmission Data length   can be reduced by compression which will also consume
energy. Hence, overall energy consumption can be reduced by decreasing the
number of bits of data transmitted and processed. The energy of the network
can conserve by reducing the number of data packets transmitted between the
nodes. In data collection wireless sensor networks, as the nodes are densely
deployed the data collected are normally redundant and similar data. If every
node sends data to the sink, network energy will drain out quickly by wasting
energy in transmitting redundant data [2]. The aggregation methods decrease
the amount of data transmission between cluster head and base station by
aggregating the data received at the cluster head. This paper proposes a method which avoids the redundant data transmission between the normal nodes
and cluster head so that both the transmission and aggregation energy can be
conserved.

2 Related Work

This section gives the analysis and understanding the mechanisms for cluster
head selections, the different scenarios of redundant data and the use of residual energy criteria. With these issues, this paper proposes a method which
reduces the redundant data transmission between the sensing node and the
cluster head by considering the node residual energy as criteria. So that the
node residual energy utilization is balanced among all the nodes, redundant
data are avoided and prolongs the network lifetime.

2.1 Low-Energy Adaptive Clustering Hierarchy (LEACH)

It is a self-organizing and adaptive clustering protocol proposed by Heinzelman.
The operation of LEACH [4] is divided into rounds, where every round starts
with a setup phase for cluster formation, followed by a steady-state phase;
when data is conveyed to the sink node occur. Though LEACH uses the random

election of cluster heads to achieve load balancing between the sensor nodes, LEACH still has some problems which are listed as follows.

- In LEACH, a sensor node is elected as the cluster head, according to a distributed probabilistic approach. The sensor node gets associated with the cluster head to form a cluster, based on the signal strength i.e., by considering the distance between the cluster head and the sensor node. Nodes join the cluster head, which is nearer to them. Once the clusters are formed, cluster heads allocates the time slot to transmit the data. This approach insures lower message overhead, but cannot guarantee that cluster heads are distributed over the entire network uniformly and the entire network is partitioned into clusters of similar size, and the load imbalance over the cluster heads can result in the reduction of network lifetime.

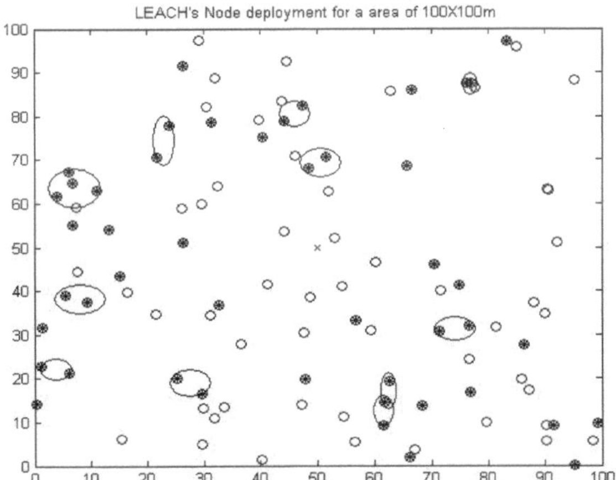

Figure. 1: Cluster head selection in LEACH

The figure 1 shows uneven distribution of the cluster heads in the network area which leads to load imbalance over the cluster heads. To achieve energy efficiency in WSN, utilization of energy in the network is to be balanced among the nodes which are selected as cluster heads[9]. To do this the role of cluster head is rotated among the nodes. But the rotation of cluster head role will not ensure the distributed consumption of energy among the cluster member nodes.

- If the probability of selecting cluster head increases in LEACH, then it can give improve its efficient than its previous probability. But there is chance of selecting the cluster head nearby and also the nodes near to the sink are selected as cluster head, which can affect the energy efficiency of the sensor networks.

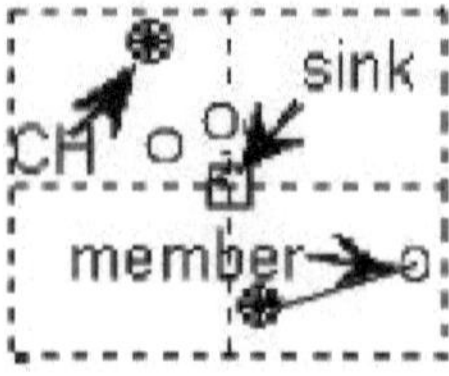

Figure. 2: Cluster heads near Sink

Figure-2 shows selection of cluster heads near to the sink where data can be transmitted directly to the sink by conserving energy without CH.

The above two issues were addressed by us in Distanced based cluster head section for WSN.

Due to dense deployment of nodes in WSN there is a chance of transmitting same data by more than one node to the cluster head which leads to waste of energy due to redundant data transmission from normal nodes and data aggregation energy at the cluster heads.

2.2 QIBEEC (Quality of Information Based Energy Efficient Clustering) algorithm

Manjuprasad [5] proposed a method to avoid redundant data transmission to the cluster heads by the normal nodes. This algorithm operates in three phases, selection of cluster heads phase, and formation of cluster and Redundant data avoidance phase.

Cluster heads are selected based on the threshold distance and threshold energy. The nodes having the minimum energy and capable of transmitting to minimum distance will be selected has cluster head. Threshold distance and energy is calculated by using the equations.

The equation gives the threshold distance and Energy.

$$M_E = (E_{tx} * E_{da}) * k * E_{fs} * k(T_d * T_d)$$ Where Threshold distance

$$T_d = \sqrt{M_{d^2} + M_{d^2}}$$

$$M_d = \frac{A}{4L}$$ Where A=Area of a square network,   L= length of Network

area, E_{tx} = transmission energy

   E_{da} = Data aggregation energy, k = Packet size.

The nodes satisfying the threshold energy and distance condition will be selected as cluster heads.

In cluster formation phase nodes within the cluster head range will be attached to the cluster head to form cluster. This cluster setup phase operation is same as in LEACH.

In redundant data avoidances phase, before transmitting the data, node will check whether it is in the same sensing range of any other node which is already active then it goes to sleep mode and transmission time slot will not be allocated to the node and the node which is already active transmits the data. The nodes decide whether to transmit or not based on the threshold range $T_{(r)}$.

$$T_{(r)} = \frac{S_r}{N}(L)$$ Where, S_r = Sensing Range of the nodes, N=Number of

Nodes, L=Length of Network area.

QIBEEC algorithm   reduces the redundant data transmission by not allowing all nodes within the threshold sensing range to transmit data to cluster head, but this algorithm is not considering balanced utilization of node residual energy in avoiding redundant data transmission. Considering residual energy as parameter only for choosing the cluster head is not sufficient to balance energy consumption across the network. Distributed, balanced utilization and proper management of node residual energy, prolongs the network life time.

3 Proposed Work

The proposed method works in the following steps.

Step1: Calculates the threshold transmitting distance based on the minimum transmission range of the nodes deployed in the network as in [6]. This threshold distance between the nodes is considered as the criteria to select the Cluster head.

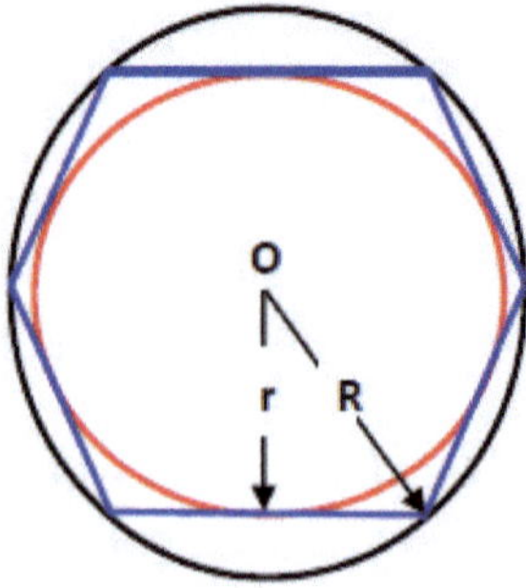

Figure. 3: Threshold transmitting distance

In figure-3 **O** represents a node and R is the minimum transmitting range of
the of the node which depends on type of node, r is the threshold distance,
here r is calculated as $r = 1/2R\sqrt{3}$. Here hexagon divides the network
without leaving any uncovered area.

Step2: Once the threshold transmission range is calculated, The next step is
to select the No Cluster head Region (NCR). No Cluster head Region is based on
the threshold distance with respect to the base station. Nodes within this region
that is nodes within the threshold distance from sink forms the NCR region.
Nodes in NCR region can directly transmit data to the sink.

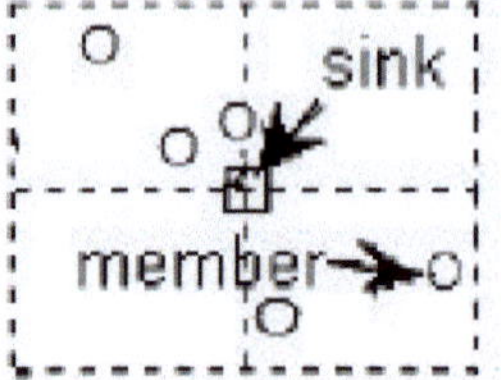

Figure. 4: No Cluster head Region

Step3: Initially one node from outside the NCR
region is randomly selected as cluster head. The next eligible node will check
either it is within the threshold distance of any other cluster head or beyond. If it
is beyond the threshold distance then node will become cluster head otherwise it
will join as the cluster member of the acquired region.

Step4: This step avoids the redundant data transmission to the cluster head from normal nodes based on the threshold sensing range and the residual energy of the cluster member node. The threshold sensing range for the normal node is calculated based on the sensing range of individual node, assuming that all the deployed nodes have the same sensing range. The $T_{(r)}$ is calculated as in [5].

After setting up the clusters when the normal nodes gets the communication slot every node will verify either it is in the threshold sensing range of other node to check for redundant data, assuming that nodes within threshold sensing range will have same data. If node is in the sensing area of other, each node will exchange their residual energy information. The node with more residual energy will become active and transmits the data. Node with less residual energy will become idle. Considering residual energy parameter in avoiding redundant data transmission from cluster   members to cluster head   leads to balanced, distributed consumption of node's remaining energy and also conserves data aggregation energy at cluster head which affects the network life time.

4 Results and Discussion

The performance of the proposed work is evaluated based on the number of dead nodes and number of rounds. Simulation is carried out with a network size of 200 nodes, 0.05J initial energy and maximum of 200 rounds in an area 100X100m. The transmitting range and the sensing range are respectively set to 25m and 5m.

The obtained results are compared with LEACH [4], DBCHS (Distance Based Cluster Head section in sensor networks for Efficient Energy Utilization) [6], QIBEEC (Quality of Information Based Energy Efficient Clustering) algorithm [5]. All the methods are simulated with same node initial energy, transmitting, receiving energy and the sink position.

As shown in figure 5, it may be observed that in LEACH, 2Level QIBEEC methods, the number of failure nodes are 200 and 124 nodes respectively on the final round. This shows that 2Level QIBEEC method is more energy efficient compared to LEACH and Single Level QIBEEC methods. However, the proposed method offers improved results even compared to 2 Level QIBEEC. It may be observed from figure 6 that there are 101 failure nodes in proposed method in the

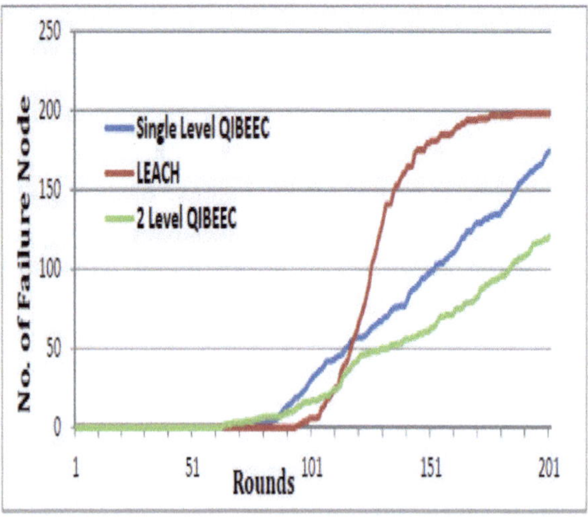

Figure. 5: Failure nodes in 2Level QIBEEC (Source [5])

200th round. It shows that, proposed method is having 11.5% less failure nodes compared to 2Level QIBEEC [5]. Hence the proposed method is more energy efficient than LEACH, DBCHS and 2Level QIBEEC.

From the graph in Figure 6, it may also be observed that during initial rounds, the number of dead nodes is very less in LEACH and once the first node dies, number of dead nodes increases drastically. Whereas in proposed method due to distributed and balanced consumption of node energy, the network lifetime is prolonged.

From the Table 1 it is observed that at the completion of 200th round LEACH have 200 failure nodes, whereas proposed algorithm have 101 failure node which is less than the LEACH. In proposed method 55.5% of deployed nodes are active and all the nodes are dead in LEACH.

Conclusion and Future work

Balanced and distributed consumption of residual energy in individual node by avoiding redundant data transmission between cluster members and cluster head, slowdowns the death rate of sensor nodes. There by extends the operating

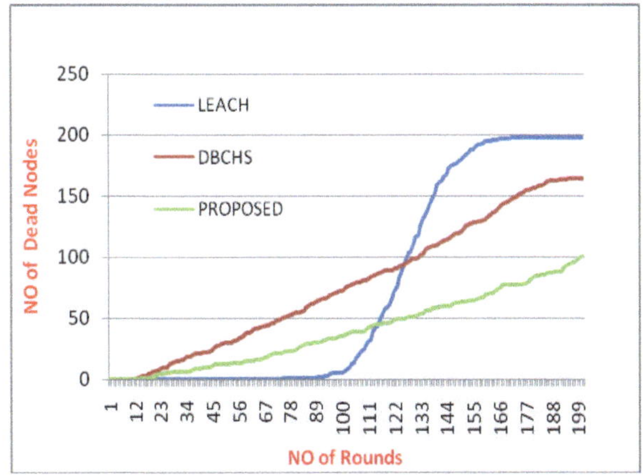

Figure. 6: Number of failure nodes in each round

Table 1. Residual Energy And Number Of Dead Nodes

Rounds	Dead nodes	
	LEACH	*PROPOSED*
1	0	0
10	0	0
20	0	2
30	0	6
40	0	9
50	0	12
60	0	15
70	0	20
80	1	24
90	2	30
100	5	35
110	28	40
120	61	46
130	111	51
140	159	59

150	180	63
160	195	69
170	197	77
180	198	83
190	199	88
200	200	101

time of the Wireless Sensor Network by efficient utilization of existing energy resource.

During cluster formation numbers of member nodes attached to each cluster head are unequal. Further work can be done to form clusters with equal number of member nodes so that nodes may be equally distributed and node energy can be consumed equally among all member nodes.

References

1   Ian F. Akyildiz and Mehmet Can Vuran "Wireless Sensor Networks" 1st Edidition Wiley Publications, 2010.
2   Rubeena Doomun, "READA: Redundancy Elimination of Accurate Data Aggregation in WSN", doi:10.4236/wsn.2010.24041 Published Online April 2011, 300-308.
3   Naoto Kimura and Shahram Latifi "A Survey on Data Compression in Wireless Sensor Networks", Proceedings of the International Conference on Information Technology: Coding and Computing (ITCC'05) 0-7695-2315-3/05 $ 20.00 IEEE.
4   W. R. Heinzelman, A. Chandrakasan, and H. Balakrishnan, "Energy efficient communication protocol for wireless micro networks," in Proc. of the 33rd Annual Hawaii International Conference on System Sciences (HICSS), Maui, HI, Jan. 2000, pp 3005 – 3014.
5   Manju Prasad and Andhe Dharani "AQOIBASEDENERGY EFFICIENT CLUSTERING FOR DENSE WIRELESS SENSOR NETWORK" International Journal Of Advanced Smart Sensor Network Systems (IJASSN), Vol 3, No.2, April 2013
6   Mohanaradhya, Sumithra Devi K A, Andhe Dharani, "DISTANCE BASED CLUSTER HEAD SECTION IN SENSOR NETWORKS FOR EFFICIENT ENERGY UTILIZATION", International Journal of Advanced Research in Engineering and Technology (IJARET), ISSN 0976 – 6480(Print), ISSN 0976 – 6499(Online) Volume 4, Issue 1, January- February (2013), © IAEME.
7   Snehal Kole, Mr.K.N.Vhatkar, Mr.V.V.Bag, "Distance Based Cluster Formation Technique for LEACH Protocol in Wireless Sensor Network", International Journal of Application or Innovation in Engineering & Management (IJAIEM), Volume 3, Issue 3, March 2014, ISSN 2319 – 4847.

8   Guihai Chen • Chengfa Li • Mao Ye • Jie Wu, "An unequal cluster-based routing protocol in wireless sensor networks", Springer Science + Business Media, LLC 2007.

9   T. V. PADMAVATHY, M. CHITRA, "Extending the Network Lifetime of Wireless Sensor Networks Using Residual Energy Extraction—Hybrid Scheduling Algorithm", Int. J. Communications, Network and System Sciences, 2010, 3, 98-106.

10 Nagendra Nath Giri, Sudha H Thimmaiah, Sheetalrani R. Kawale, Ramya and Priya Esther B, "Enhancing Energy Efficiency in WSN using Energy Potential Concepts", International Conference on Advances in Computer and Electrical Engineering (ICACEE'2012) Nov. 17-18, 2012 Manila (Philippines).

11 Giuseppe Anastasi , Marco Conti, Mario Di Francesco , Andrea Passarella, "Energy conservation in wireless sensor networks: A survey",AdHocNetworksjournale: www.elsevier.com/locate/adhoc.

Rutooja D. Gholap [1] and Saylee Gharge [2]

Neural Network based Facial Expression Recognition System

Abstract: In the field of computer science much effort is put to explore the ways of automation the process of face detection, feature extraction, face recognition etc. Over the last two decades, face recognition has been a heavily studied topic in computer vision. Paralleled to this facial expression recognition has also become an active research area with potential applications in human-computer interaction and cognition of human emotion. In this paper neural network based facial expression recognition system has been proposed. The First stage involves the pre-processing. After pre-processing 32 fiducial points on the face are chosen randomly using graphical method. In the next stage two parallel methods are used for feature extraction i.e. Gabor filter and LBP. The feature vectors obtained using both the methods are combined and applied to the ANN. Neural Network will classify the expression using 7 universal expressions. The recognition accuracy achieved using this method is 94.08%.

Keywords: Facial Expression Recognition, Graphical Method, Gabor Filter, LBP, PCA, Artificial Neural Network

1 Introduction

Face plays significant role in social communication. This is a 'window' to human personality emotions and thoughts. Facial expression recognition is the demanding task in computer vision. It helps the human beings to deliver their emotions to others. It has wide variety of applications such as virtual reality, video-conferencing, user profiling, and customer satisfaction studies for broadcast and web services, require efficient facial expression recognition in order to achieve the desired results. Despite the fact that the recognition of face and facial expression are two independent fields they facilitate the development of each other as they share many of their methodologies and even theories. For

1 M.E. Student, Mumbai University, Department of EXTC, VESIT, Chembur, Mumbai, Maharashtra,India
rutooja412@gmail.com
2 Associate Professor, Department of EXTC, VESIT, Chembur, Mumbai, Maharashtra, India

instance, facial feature extraction is commonly applied in both areas. Also, as there is a facial expression space corresponding to each face space, their classifiers are likely to be comparable. The universal categories of human emotions which are used in human-computer interaction are; Anger, Joy, Surprise, Disgust, Sad and Fear. A typical facial expression recognition system usually consists of the three stages: 1) Face Detection 2) Feature Extraction 3) Expression Recognition. The underlying approaches for all of the existing techniques of facial expression recognition are same. First, digital images of the subjects are acquired using digital camera. The image pre-processing techniques are used. Then useful features that are necessary for further analysis of these images are extracted. Finally several analytical techniques are used to classify the images according to the specific problem at hand.

2 Proposed Method

The approach is proposed using two methods for feature extraction. The First method used for feature extraction is Gabor filter and second is LBP. First the system has been developed by implementing the feature extraction methods independently and then in order to improve the recognition accuracy features extracted using both the methods is combined. Fig. 1 shows the block diagram of proposed system. The proposed approach aims to identify facial expression in HCI.

2.1 Input Image

Indian database is considered as input image. The basic emotional expressions of concern include anger, disgust, fear, happiness, sadness, and surprise. This dataset consists of 210 images of seven expressions (neutral plus six basic emotional expressions) from different male and female expressers. On this images image pre-processing is done to classify the facial expression found in each image.

2.2 Pre-processing

The input image is selected from the database. The original image is rescaled and resized to 256 × 256 so that the dimension of each observation is reduced to 65536. Since 256× 256 is the standard image size used in many applications. The resizing remains the central part of facial expression and no more complicated techniques of locating eyebrows and mouth are utilized in the pre-processing. Output of this step is used to extract the features.

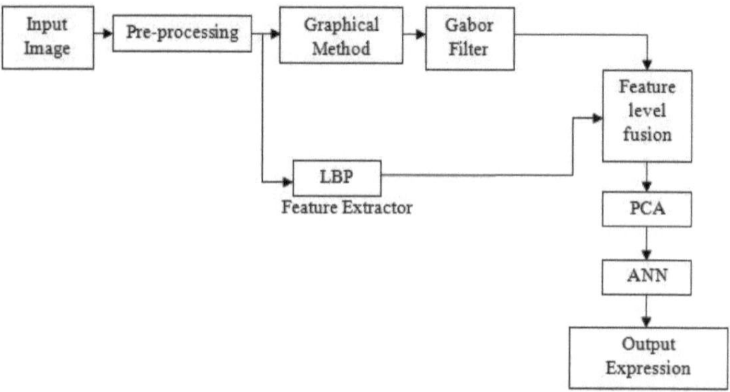

Figure. 1: Block Diagram of Proposed System

2.3 Graphical Method

On the output of previous step, Graphical method is applied. Basically graphical method is a method in which the location of any point is re-presented by using X-Y co-ordinates. Thus, to select any points on the face by clicking mouse on face image. After selection of the points this method will give the X-Y co-ordinates of the selected points. These points are called as fiducial points. Using this method 'N' number of points can be selected. For this project 32 fiducial points on the face images will be considered. These fiducial points can be then characterized in the next step in order to extract peculiar texture around them. The image coordinates of these points (geometric positions) will be used as features in this study.

2.4 Gabor Filter

The Gabor filter is applied on the output of graphical method. Gabor filter is a linear filter. Frequency and orientation representations of Gabor filters are similar to those of the human visual system and hence it is appropriate for texture representation and discrimination. Texture is often characterized by its responses to a set of orientation and spatial-frequency selective linear filters. For extracting useful features from an image a set of Gabor filters with different frequencies and orientations is used. Features extracted using Gabor filters represent the local information in the image. The Gabor filters can be defined by,

$$\Psi_{u,v}(z) = \frac{||ku,v||2}{\sigma 2} e^{(-||Ku,v||2\,||Z||2/2}\sigma 2)[e^{iku,vZ} - e^{\wedge(-}\sigma 2/2)] \tag{1}$$

Where, $z = (x, y)$ and u and v denote the orientation and scale of Gabor kernels, respectively. The number of scales and orientations is selected to represent the facial characteristics of spatial locality and orientation selectivity.

For proposed method, a discrete set of Gabor kernels which comprise 5 spatial frequencies i.e. scales and 8 distinct orientations from 0 to 180 degree is used. Each image is convolved with both even and odd Gabor kernels at the location of the fiducial points selected in previous step. Thus, total 40 complex Gabor wavelet coefficients at each fiducial point are achieved.

2.5 LBP (Local Binary Pattern)

The second method used for feature extraction is LBP. In LBP the face image is divided in to regions (blocks) and each region corresponds with each central pixel. Then it examines it's pixel neighbours based the grey scale value of central pixel to change it's neighbour to 0 or 1. The input image is transformed in to LBP representation by sliding window technique where value of each pixel in the neighbourhood is threshold with value of central pixel. The LBP code for the central pixel tc is given by

$$LBPm,r(tc) = \sum_{i=0}^{m-1} U(ti - tc)\, 2^{i} \tag{2}$$

Where, $U(ti - tc) = 1$ if $ti > tc$ and $U(ti - tc) = 0$ if $ti \leq tc$.

After obtaining the LBP code an occurrence histogram, as a nonparametric statistical estimate is computed over a local patch.

2.6 Feature Level Fusion

Since, two different feature vectors of the same input image were achieved. One final vector has to be considered while giving input to the ANN which is next step. For that purpose, the feature vectors are fused. This technique is known as feature level fusion.

2.7 PCA (Principal Component Analysis)

After feature extraction PCA is applied in order to reduce computational complexity. Principal component analysis is appropriate when you have obtained measures on a number of observed variables and wish to develop a smaller number of artificial variables (called principal components).

2.8 Artificial Neural Network

Artificial neural network is used for classification of expression. Artificial neural networks working are similar to the human neural system. A very large variety of networks have been constructed. ANN is composed of neurons and weights. Together they determine the behaviour of the network. One ANN is used for each person in the database in which face descriptors are used as inputs to train the networks. During training of the ANN's, the faces descriptors that belong to same person are used as positive examples for the person's network and negative examples for the others network. For proposed method, one input layer, one hidden layer and one output layer are used with 7 different classes which are related to the expressions.

3 Results and Discussion

Above algorithms are applied on around 210 images. The results obtained are discussed in this section.

3.1 Step I: Execution of the Proposed Method using MATLAB

The system has been implemented using MATLAB. Fig. 2 shows the GUI after execution of the program written in MATLAB.

3.2 Step II: Select Input Image (Anger)

The input image has been selected from the database. Fig. 3 shows the GUI after selecting original input image which is selected from the Indian database which has anger expression.

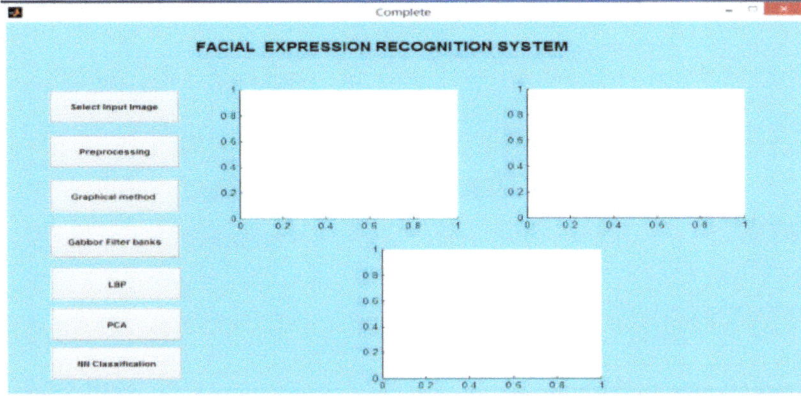

Figure. 2: GUI

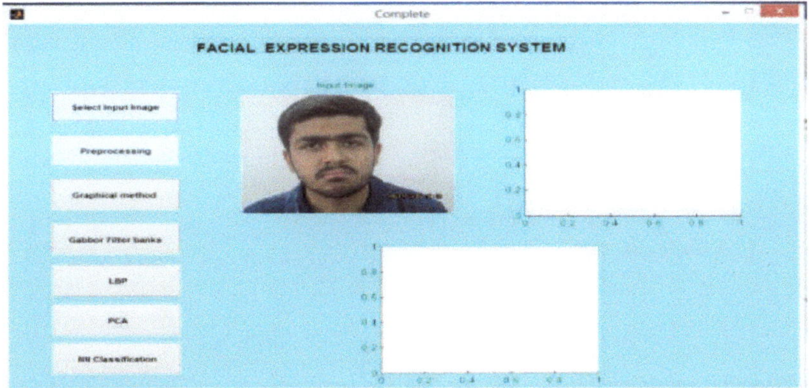

Figure. 3: GUI After Selecting Input Image

3.3 Step III: Pre-Processing

In pre-processing step the image is resized. Fig. 4 shows the GUI for the same image after pre-processing. In this step original image is resized to 256×256.

3.4 Step IV: Graphical Method

Graphical method is used to select 32 fiducial points on the face. The output of this step is as shown in Fig. 5

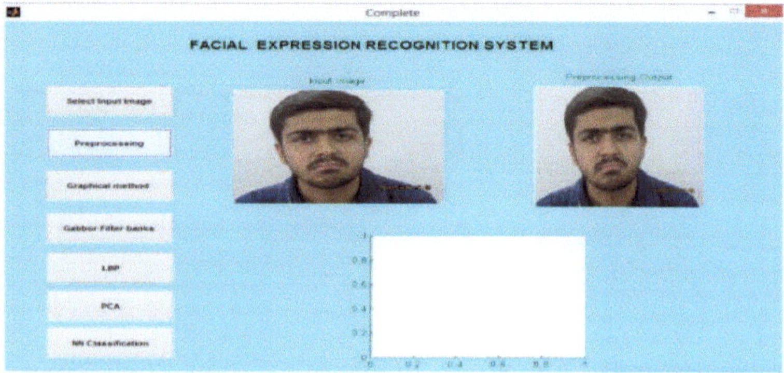

Figure. 4: GUI For The Pre-processed Output

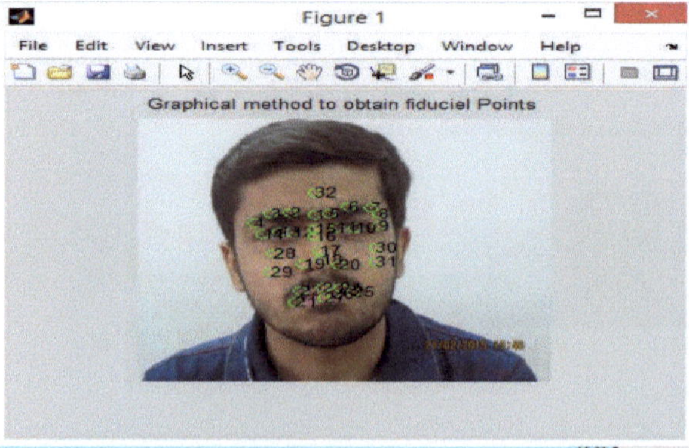

Figure. 5: Graphical Method to Obtain Fiducial Points

3.5 Step V: Gabor Filter

The output of previous step is applied to Gabor filter in order to extract the features. Fig. 6 shows the 40 distinct Gabor kernels extracted at the location of fiducial points.

3.6 Step VI: LBP

This is the second method used to extract features. For proposed method three local patches are considered in order to extract the features. The features extracted over a local patch are as shown in Fig. 7,9,11 and their relative histograms are as shown in Fig. 8,10,12 respectively.

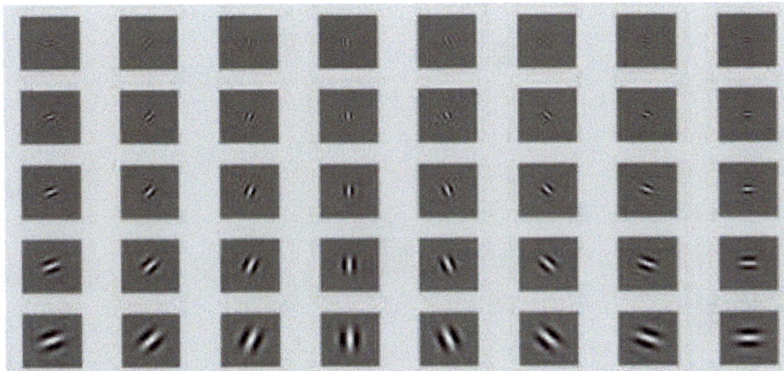

Figure. 6: Features Extracted Using Gabor Filter

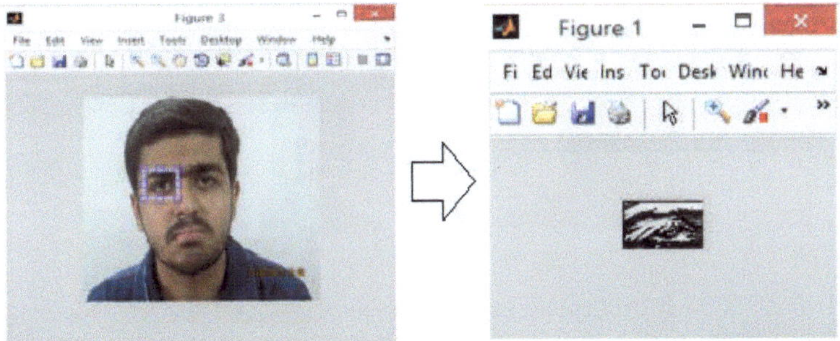

Figure. 7: Features Extracted over 1st Local Patch

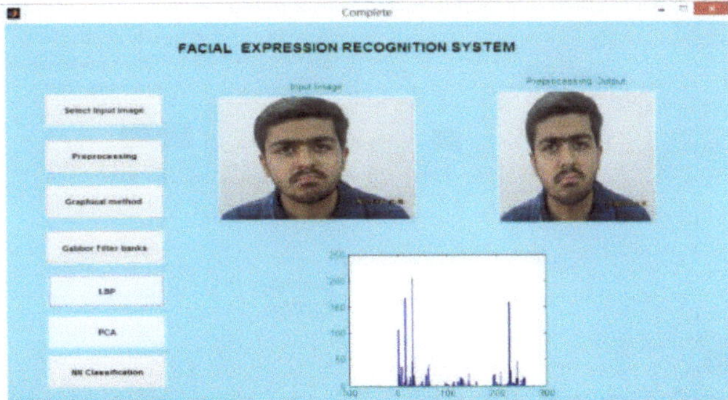

Figure. 8: Relative Histogram for Features Extracted over 1st Local Patch

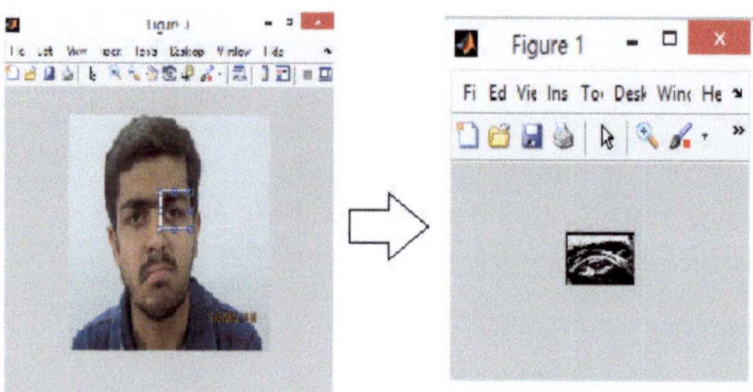

Figure. 9: Features Extracted over 2nd Local Patch

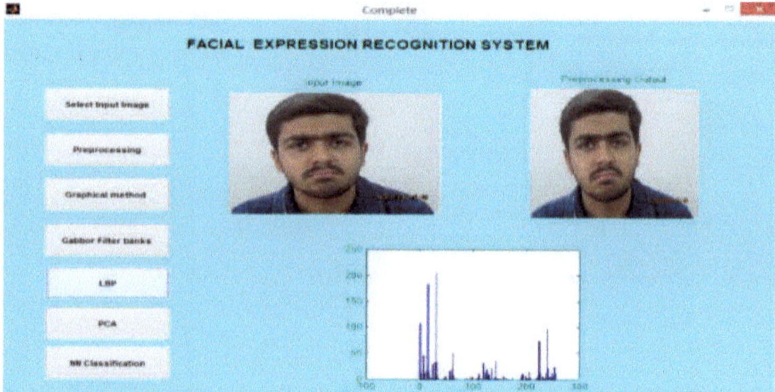

Figure. 10: Relative Histogram for Features Extracted over 2nd Local Patch

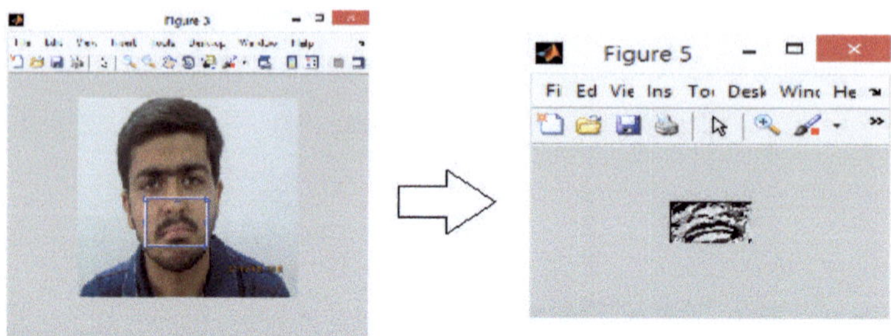

Figure. 11:. Features Extracted over 3rd Local Patch

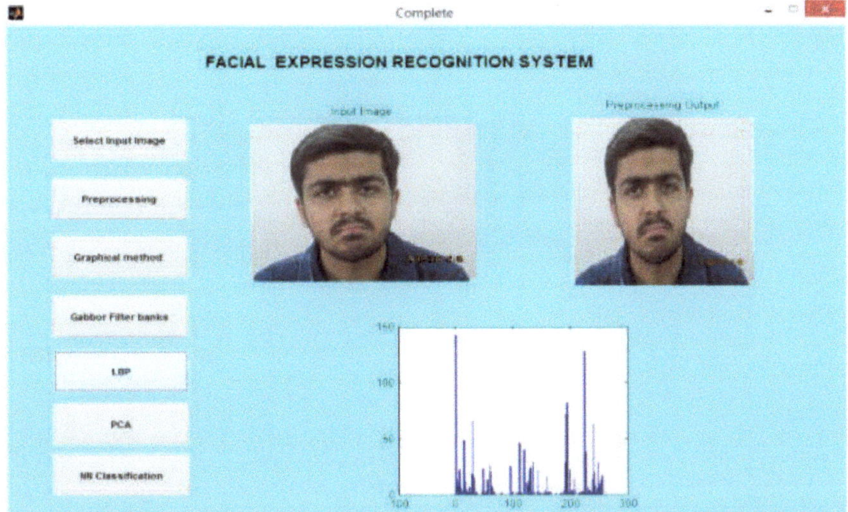

Figure. 12: Relative Histogram for Features Extracted over 3rd Local Patch

3.7 Step VII: Feature Level Fusion

The features extracted using Gabor Filter and LBP are combined in order to give a single vector as input to neural network.

3.8 Step VIII: PCA

The combined features vector of Gabor filter and LBP is used to as an input in this step. PCA is used to reduce the size of feature vector. This feature vector is given as an input to the neural network classifier.

3.9 Step IX: NN Classification

The neural network is used for the classification of the expression found in the facial image. The output of neural network is as shown in Fig. 13.

The output of neural network has shown the same expression that is given as an input. Thus, the original expression is recognised which is anger. Similarly the same procedure is applied to the images of same subject with different expressions and results obtained are as shown in Table 1.

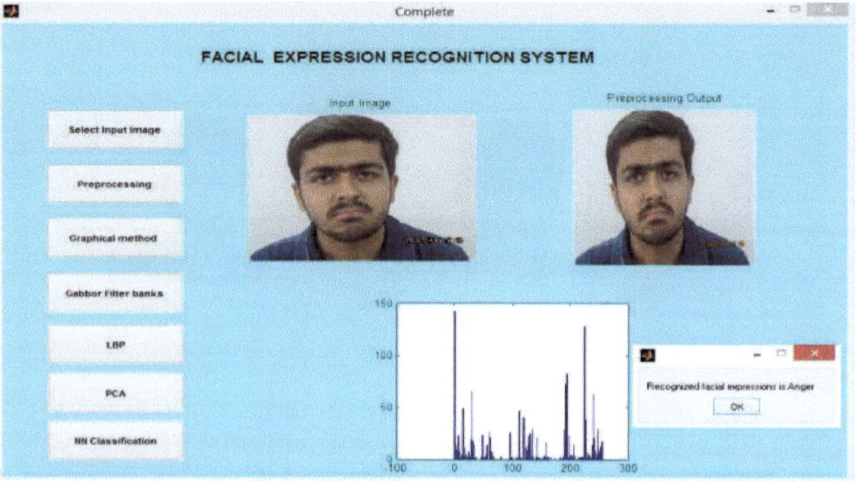

Figure. 13: Recognised Facial Expression is Anger

4 Performance Evaluation

Table 1 shows the analysis of the results based on the accuracy measurements. The rows in table shows the input image applied to the neural network while column shows the output of neural network for all seven classes. E.g. when input image was anger the 90.7% the result obtained as a anger, 1.70% it was disgust, 4.50% it was sad and 3.10% it was neutral. From results it is observed that expression mostly affects the eye, mouth and nose region. As expression changes intensity of pixel at that particular location also changes.

Table 1. Accuracy Measurements for Every Expression

	Anger(%)	Disgust(%)	Fear(%)	Happy(%)	Sad(%)	Surprise(%)	Neutral(%)
Anger	90.7	1.70	0	0	4.50	0	3.10
Disgust	0	97.5	0.80	1.70	0	0	0
Fear	0	1.0	90.9	0	3.10	1.0	4.0
Happy	0	0	0	97.5	0.4	0	2.10
Sad	3.8	0	0	0	92.7	0	3.5
Surprise	0	0	1.30	0.90	0	97.3	0.50
Neutral	1.2	0	0.80	3.60	2.40	0	92

From the results the overall accuracy of the proposed system is calculated. The features are extracted using both the methods separately and results for individual method are observed. When features are extracted using only Gabor filter the accuracy achieved is 85.50 % and when features are extracted using only LBP the accuracy achieved is 75.84 %. Using proposed method the accuracy of the system is improved to 94.08%. Fig. 14 shows the accuracy measurements for feature extraction using Gabor filter, feature extraction using LBP and Proposed method.

Conclusion

The proposed system has been developed to process the images of facial behaviour and recognize the displayed action in terms of six universal expressions i.e. Happy, Sad, Angry, Fear, Disgust and Surprise and one Neutral expression. It is observed that accuracy of the system is different for every expression. When features are extracted using only Gabor filter the accuracy achieved is 85.50 % and when features are extracted using only LBP the accuracy achieved is 75.84 %. Using proposed method the accuracy of the system is improved to 94.08%. Major strengths of this system are feature extraction methods and the method has been implemented using Indian database which was created by us.

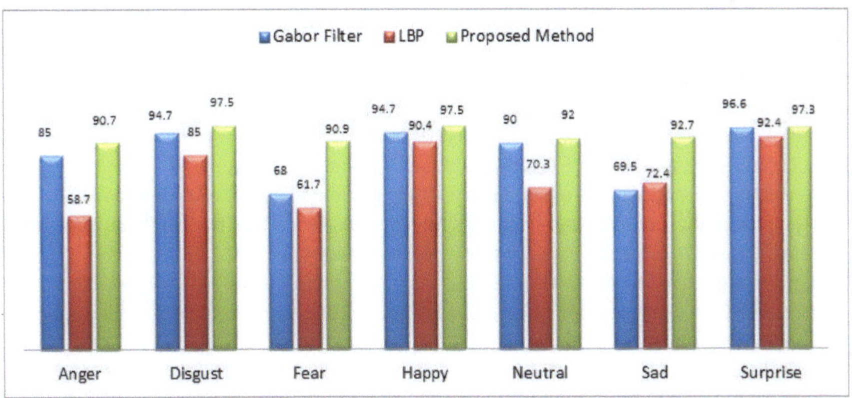

Figure. 14: Accuracy Measurements for Every Expression

Acknowledgement

We would like to thank the faculties and students of Vivekanand Education Society's Institute of Technology, Chembur, Mumbai, Maharashtra and A. C. Patil College of Engineering, Kharghar, Navi Mumbai, Maharashtra for providing pictures for the database and permitting us to use that pictures for the Proposed method for publication.

References

1   A.S. Dhavalikar, Dr. R.K. Kulkarni, "Face Detection and Facial Expression Recognition System", IEEE International Conference on Electronics and Communication Systems (ICECS-2014).

2   Andrew Ng., "Machine learning courses", Stanford University (2013).

3   Chao Li, Armando Barreto, "An Integrated 3D Face-Expression Recognition Approach",Page Number..1132-1135, 1-4244-0469-X, IEEE (2006).

4   C.L. Huang and Y.M. Huang, "Facial Expression Recognition Using Model-Based Feature Extraction and Action Parameters Classification," J. Visual Comm. and Image Representation, Volume Number. 8, No. 3, Page No.. 278-290(1997).

5   C.Shan, S. Gong, P. McOwan, "Facial expression recognition based on Local Binary Patterns: A comprehensive study", Image and Vision Computing, Volume Number 27, Page number. 803-816, 2009.

6   Cheng Zhong, Zhenan Sun and Tieniu Tan,"Robust 3D Face Recognition Using Learned Visual Codebook", 1-4244-1180-7, IEEE(2007).

7   Ewa Piatkowska, "Facial Expression Recognition System", Master's Thesis Technical Report.

8   Fei Cheng, Jiangsheng Yu, and Huilin Xiong, "Facial Expression Recognition in JAFFE Dataset Based on Gaussian Process Classification", 1686 IEEE TRANSACTIONS ON NEURAL NETWORKS, Volume Number 21, NO. 10 (OCTOBER 2010).

9   G.C. Littlewort, M.S. Bartlett, J. Chenu, I. Fasel, T. Kanda, H. Ishiguro, J.R. Movellan,"Towards social robots: Automatic evaluation of human-robot interaction by face detection and expression classification", Advances in Neural Information Processing Systems, Volume Number 16, Page Number. 1563-1570, (2004).

10 G.J.Edwards, T.F. Cootes, and C.J. Taylor, "Face Recognition Using Active Appearance Models," Proc. European Conf. Computer Vision, Volume Number 2, Page Number 581-695 (1998).

11 H. Kobayashi and F. Hara, "Facial Interaction between Animated 3D Face Robot and Human Beings," Proc. Int'l Conf. Systems, Man, Cybernetics, Page Number 3,732-3,737 (1997).

12 I. Essa and A. Pentland, "Coding, Analysis Interpretation,Recognition of Facial Expressions", IEEE Trans. Pattern Analysis and Machine Intelligence, Volume Number 19, No. 7, Page Number 757-763 (July 1997).

13 Irene Kotsia and Ioannis Pitas,"Facial Expression Recognition in Image Sequences Using Geometric Deformation Features and Support Vector Machines", 172 IEEE TRANSACTIONS ON IMAGE PROCESSING, Volume Number 16, NO. 1 (JANUARY 2007).

14 J.F. Cohn, A.J. Zlochower, J.J. Lien, and T. Kanade, "Feature-Point Tracking by Optical Flow Discriminates Subtle Differences in Facial Expression," Proc. Int'l Conf. Automatic Face and Gesture Recognition, Page Number 396-401 (1998).

15 Mayank Agrawal, Nikunj Jain, Manish Kumar and Himanshu Agrawal, " Face recognition using Eigen Faces and Artificial Neural Network", International Journal of Computer Theory and Engineering, Volume Number 2, No. 4 (2010).

16 M.J. Black and Y. Yacoob, "Recognizing Facial Expressions in Image Sequences Using Local Parameterized Models of Image Motion," lnt'l J. Computer Vision, Volume Number 25, No. 1, Page Number 23-48 (1997).

17 M. Pantic and L. Rothkrantz, "Expert System for Automatic Analysis of Facial Expression", Image and Vision Computing J., Volume Number 18, No. 11, Page Number 881-905 (2000).

18 Maja Pantic and Ioannis Patras, "Dynamics of Facial Expression: Recognition of Facial Actions and Their Temporal Segments From Face Profile Image Sequences", IEEE TRANSACTIONS ON SYSTEMS, MAN, AND CYBERNETICS—PART B: CYBERNETICS, Volume Number 36, NO. 2 (APRIL 2006).

19 Ms.Rutooja D. Gholap, Dr.Mrs. Saylee Gharge, "Facial Expression Recognition System using Gabor Filter" International Journal of Advanced Information Science And Technology, Volume Number 40, Page Number. 5-9 (August 2015).

20 Seung Ho Lee, Konstantinos N. (Kostas) Plataniotis, and Yong Man Ro, "Intra-Class Variation Reduction Using Training Expression Images for Sparse Representation Based Facial Expression Recognition", IEEE.

21 S. L. Happy and Aurobinda Routray, "Automatic Facial Expression Recognition Using Features of Salient Facial Patches", IEEE TRANSACTIONS ON AFFECTIVE COMPUTING (2014).

22 Seyed Mehdi Lajevardi and Hong Ren Wu,"Facial Expression Recognition in Perceptual Color Space", IEEE TRANSACTIONS ON IMAGE PROCESSING, Volume Number 21, NO. 8 (AUGUST 2012).

23 Thuthi D.,"Recognition of facial expression using action unit classification technique" IEEE   International Conference on Recent Trends in Information Technology (ICRTIT-2014).

24 Wei Li, Chen Chen, Hongjun Su, and Qian Du, "Local Binary Patterns and Extreme Learning Machine for Hyperspectral Imagery Classification", IEEE TRANSACTIONS ON GEOSCIENCE AND REMOTE SENSING.

25 Xue Yuan, Jianming Lu, and Takashi Yahagi, "A Method of 3D Face Recognition   Based on Principal Component Analysis Algorithm", Page Number.3211-3214, 0-7803-8834-8,IEEE(2005).

26 Yizhen Huang, Ying Li, "Robust Symbolic Dual-View Facial Expression Recognition With Skin Wrinkles: Local Versus Global Approach", 536 IEEE TRANSACTIONS ON MULTIMEDIA, Volume Number 12, NO. 6 (OCTOBER 2010).

27 Yan Tong, Wenhui Liao and Qiang Ji,"Facial Action Unit Recognition by Exploiting Their Dynamic and Semantic Relationships", IEEE.

28 Z. Zeng, Y. Fu, G. I. Roisman, Z. Wen, Y. Hu and T. S. Huang, "Spontaneous Emotional Facial Expression Detection", Journal of Multimedia, Volume Number 1, No. 5, Page Number. 1-8 (2006).

29 Zhengyou Zhang, Michael Lyons, Michael Schuster, Shigeru Akamatsu, "Comparison Between Geometry-Based and Gabor-Wavelets-Based Facial Expression Recognition Using Multi-Layer Perceptron" , INRIA   (2004)

Author's Profile

Ms. Rutooja Dilip Gholap is pursuing **M.E. in Electronics and Telecommunication** from Vivekanand Education Society's Institute of Technology, University of Mumbai, Maharashtra, India under the guidance of Dr. Mrs. Saylee Gharge. She has done **B.E. in Electronics and Telecommunication** in the year 2013 from Ramrao Adik Institute of Techology, university of Mumbai, Maharashtra, India.

Dr. Mrs. Saylee Gharge working as a **associate professor** in the department of **Electronics and Telecommunication** in Vivekanand Education Society's Institute of Technology, Chembur, Mumbai, Maharashtra, India. She has 14 years of teaching experience and has published 50 plus papers in national and international conferences.

Vinay Kumar S.B [1], Naveen Kumar S [2], Monica B.V [3] and
Ravi .U.S [4]

Indoor Surveillance Robot with Live Video Transmission

Abstract: Now a day's terrorist are targeting and attacking many indoor areas all around the world. Their main target includes office, school, and corporate head quarter. We can't eliminate the risk of terrorist attack, but we can take necessary action to reduce the risk. Presently the surveillance of indoor area is done using fixed CCTV cameras, these cameras are fixed at one place and they cover limited area and we can't change the camera view in real time.in this paper we are developing an embedded system for surveillance and safety purpose robot using Zigbee communication and wireless camera which sends the in front information of the robot to the local system for surveillance purpose.

Keywords: PIR sensor, ultrasonic sensor, gas sensor, camera, zigbee, pic

1 Introduction

A robot is a machine capable of replicating or resembling the human actions with the collection of components like sensors, power supplies and controls. The main characteristics of the robot are Sensing, Movement, Energy, and Intelligence. Robots become more advanced and sophisticated it can be used to perform the desired tasks where men cannot be in a state to carry out his tasks.

1 Assistant Professor, Department of ECE, School of Engineering and Technology, Jain University, Bangalore
vinaysb.omm@gmail.com
2 M.Tech in Embedded System Design , Department of ECE, School of Engineering and Technology, Jain University, Bangalore
naveensrinivas.cr@gmail.com
3 3rd Sem B.Tech, Department of ECE, School of Engineering and Technology, Jain University, Bangalore
monikabv719@gmail.com
4 Faculty, Institute for Academic Service, Harohalli, Kanakapura Main Road Ramanagara District, Bangalore
ravius5786@gmail.com

With automation and application of electronics in the field of robotics ensures safety for individuals and desired task can be achieved. There are many places where people cannot go due to various reasons. For example if a person wants to know what is on a ceiling of his house, he has to climb on to the ceiling. But that is not safe and there might be practical issues such the person may not fit in the available space. All these problems focus on a common solution that is a surveillance robot which can provide information of its surrounding environment to a remote user. In this paper contains a robotics based solution to gather information to solve the problems by designing a robot which can provide information about surrounding environment to its user via a computer. This robot has its own camera, to provide visual aid to the user. User can control the robot through laptop. Using this robot, people will be able to get information about the place. This is a very useful design when it comes to surveillance. Surveillance is the process of monitoring a situation, an area, or a person. Hence robots which continuously monitor the place and provides security is developed [5]. The traditional surveillance systems take a long time to detect whether there is any intruder or not. If there is no intruder, the sensing device which continuous to work and consumes much power.In this paper a robot is designed in such a way that it provides a high level surveillance as required using automation. The main objective of the paper is to provide an efficient surveillance wherever high level security is needed. An automatic patrolling vehicle acts as a security patroller in the security. System, which can monitor those dead zones of the traditional fixed surveillance system. The remote monitoring capabilities can also be enhanced by using the wireless network. And the face detection system is adapted to record and analyze the invaders. The proposed system is an embedded based robotic module, with the proposed system; humans can feel extreme comfort and can experience automation to the maximum.

2 System Architecture

The block diagram of the hardware implementation is show in the below fig. 1 and fig. 2. This robot is wirelessly operated. Wireless camera sends the real time video signals which can be seen on monitor.

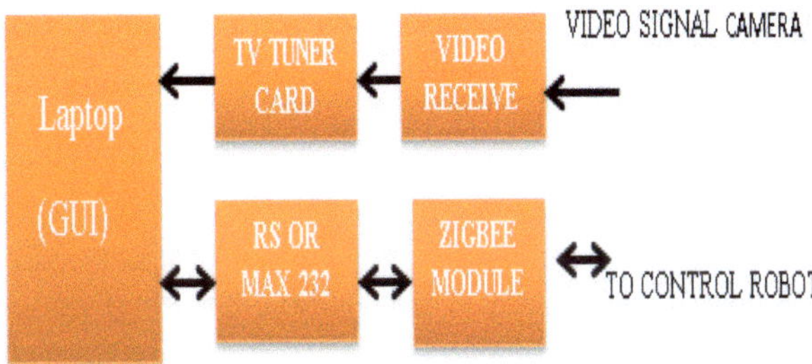

Figure. 1: The controlling part

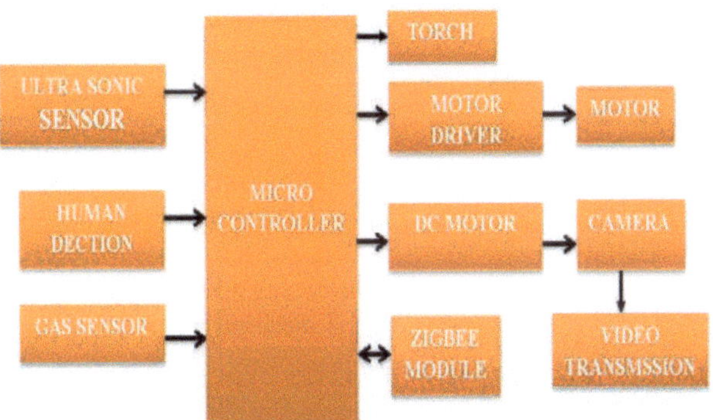

Figure. 2: The Surveillance Robot

3 Methodology

This autonomous robot will be left in desired area, robot will move randomly and have human detection sensor is connected to the microcontroller. This sensor is mainly used to sense the movement of the people. Here the sensor keep on checking for human motion ,in case if any movement of the Human is detected then it will send an alert message to control unit (pc) using Zigbee module. It will also detect the gas present in the left area and sends alert message to the control unit using Zigbee module When the human motion is detected it will activate the camera module to capture video. The captured video is sent to the control unit and the direction of the robot will be controlled according the received video to achieve the desired task.

3.1 PIC Microcontroller

PIC18 series is based on 16-bit instruction architecture that incorporates advanced RISC architecture which makes it the utmost performer among all the PIC families. The PIC18 series is integrated with new age communication protocols like USB, CAN, LIN, Ethernet (TCP/IP protocol) to communicate with local and/or internet based networks. Further it has flexible oscillator option with four Crystal modes, including High-Precision PLL for USB and two External Clock modes, providing up to 48MHz.

3.2 PIR Sensor

Passive infrared sensor is used to detect human at a certain distance as human emits heat at a wavelength of 8 to 12 micro meters the sensor detects the human.

3.3 Ultrasonic Sensor

This sensor has been used for the   obstacle detection and movement. it is been programmed in such a way thatwhen it detects the obstacle it makes movement based on which has been programmed.

3.4 Dc Motor

For the movement we are using DC motor.it requires 12 volt power supply.DC power supply. In any electric motor, operation is based on simple electromagnetism. A current carrying conductor generates a magnetic field; when this is then placed in an external magnetic field, it will experience a force proportional to the current in the conductor, and to the strength of the external magnetic field.

3.5 Motor Driver

L293D is dual H bridge Motor driver, it can drive and control the motor both in clock and counter clock wise direction.it is a 16 pin IC the maximum voltage VSS is 36 volts; it can supply a maximum current of 600mA per channel.

3.6 Gas Sensor

This is a simple-to-use liquefied petroleum gas (LPG) sensor, suitable for sensing LPG; it can detect gas concentrations anywhere from 200 to 10000ppm.Sensor has a high sensitivity and fast response time.

3.7 Wireless Camera

Wireless camera sends the video signals to the user side a camera is wired to transmitter and the signals travels between the camera and the receiver. And the real time video signals are received using laptop.

3.8 Zigbee Module

Zigbee is a low data rate communication. Zigbee protocol is mainly used to control robot wirelessly to get the sensor information to the local system. It supports data tare up to 250kbps with 10-100 meter range.

3.9 Torch

A LED light which runs on 5v DC supply in included to give night vision facility; the torch can be switched ON and OFF from local system through Zigbee.

4 Functional Description

The fig. 3 shown above explains the software implementation of the proposed system.

Condition1: if object detected take right, left or backward direction.
Condition2: if Human detected send a message to the local system.
Condition 3: if gas detected send an alert message to the local system.

5 Comparison and Experimental Results

The table1 shows the comparison of features with the referred paper. We have included PIR sensor and gas sensor for the human and gas detection, the robot wheel is provided with belt so that it can climb slope up to 50 degree therefore the system is robust, Both automatic and manual modes are included to

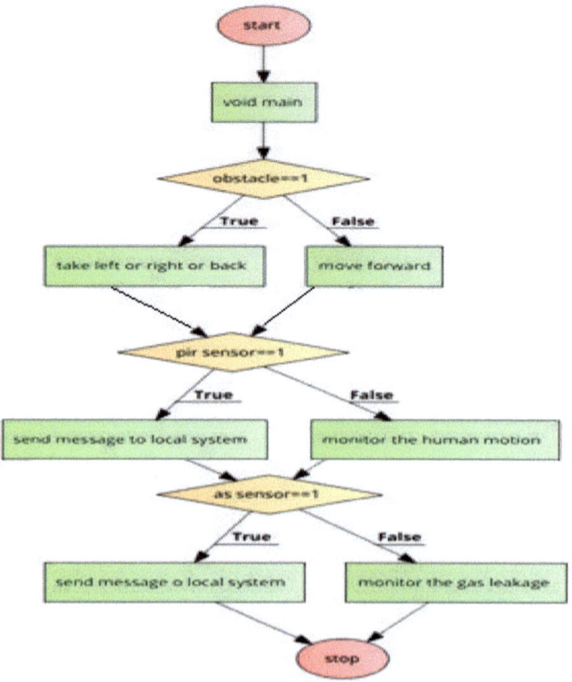

Figure. 3: The flow diagram of the system

Table 1. The comparison of the proposed system with refernce paper

Year	Tittle	PIC18F controller	PIR sensor	GAS sensor	Ultras onic sensor	Zigbee	Wheels provided with belt	Can climb slopes	DC motor for camera rotation	Torch
2014	Android based robot for border security	YES	YES	NO	NO	YES	NO	NO	NO	NO
2014	Wireless controlled surveillance robot	NO	NO	NO	NO	YES	NO	NO	NO	NO
2014	Mobile Controlled Robot For Surveillance Application	NO	NO	NO	NO	NO	NO	NO	NO	NO
2015	Indoor surveillance robot with live video transmission	YES	YES	YES	YES	YES	YES	YES	YES	YES

increase the performance. The output of PIR motion sensor in volts for different distances in meters between sensor and human is tabulated in Table2.

Table 2. The output of PIR motion sensor in volts for different distances

Meters(m)	Output Voltage (V)
0.5	4.84
1	4.86
1.5	4.85
2	4.86
2.5	4.87
3	4.86
3.5	4.85
4	4.85
4.5	4.86
5	1.2

The below fig. 4 shows the sensitivity characteristic of gas sensor, Here sensitivity of the gas sensor is decided by the value of sensor resistance at 1000ppm and resistance at various gas concentration and the below fig. 5 shows the Hardware implementation of the system.

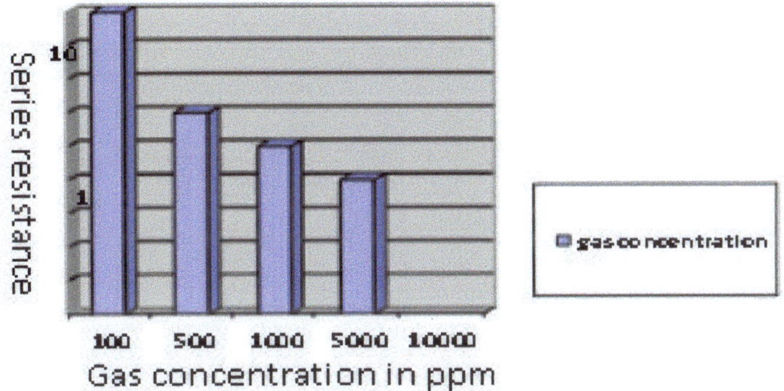

Figure. 4: The sensitivity characteristic of the gas sensor

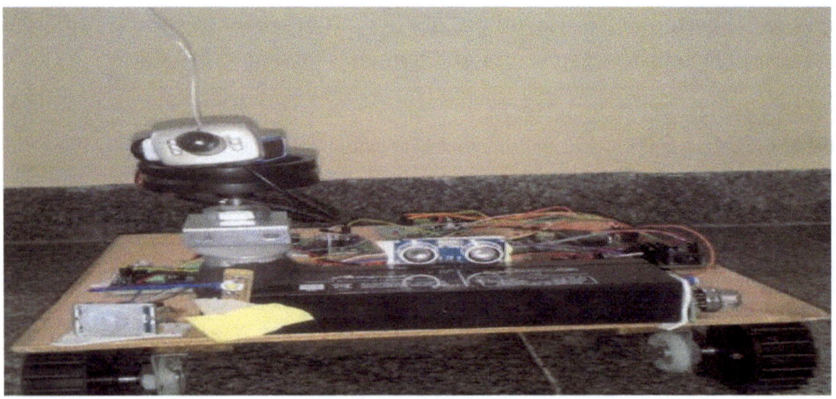

Figure. 5: The front view of the Surveillance robot

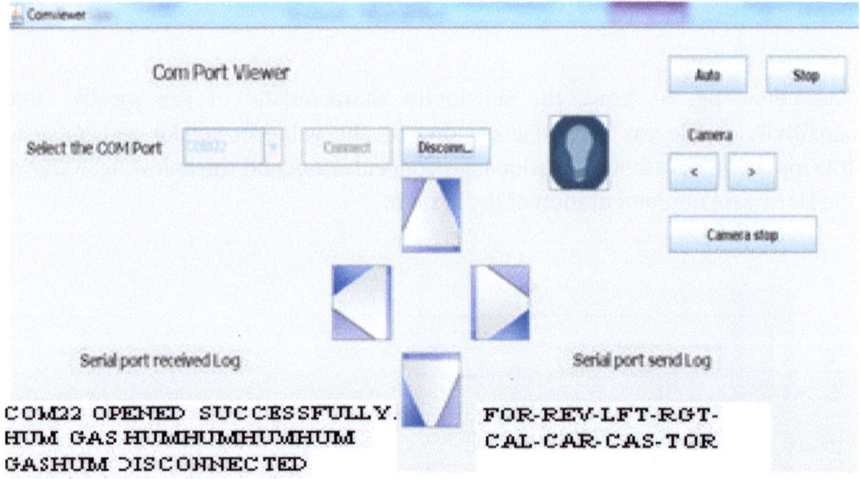

Figure. 6: The graphicle user interface to contrl robot movement

The above fig. 6 shows the graphical user interface created using java API, and it is used to control the robot movement remotely through Zigbee. The command sent to control the robot is displayed in serial port send log, and to rotate the camera two right and left buttons are provided, it receive the sensor data from robotic side and displayed in the serial port received log.

Conclusion and Future Work

The robot gets the signal from the laptop regarding itsdirections and moves accordingly. The robot sends the streaming video back to the PC and can be viewed by the user sitting in a control room. Wireless control of the robot up to 40 feet without any interference. Wireless video streaming up to 50 feet with a resolution of 1 megapixel. Battery can last up to 2 hours if used continuously. Ultrasonic sensor detects the objects within 30cm.PIR sensor detects the human motion. Gas senor detects the leakage of gas.A camera of higher resolution with night vision facility can be used for future work. Moving object recognition software using MATLAB and Video can be recorded and uploaded to server for storage purpose and 4 wheel drives could be an added advantage.

References

1   C.M.Naveen Kumar, Dr.B.Ramesh2, Prof. G. Shivakumar3, J.R.Manjunath, "Android Based Autonomous Intelligent Robot for Border Security",IJISET - International Journal of Innovative Science, Engineering & Technology, Vol. 1 Issue 5, July 2014.

2   Anjali Parade, Apurva Parad, Anurag Shinde, Jasmin Untawale, "Mobile Controlled Robot for Surveillance Application,"International Journal of Engineering Research and Applications (IJERA) ISSN: 2248-9622,International Conference on Industrial Automation and Computing (ICIAC- 12-13 April 2014).

3   Wai Mo Mo Khaing, Kyaw Thiha, "Design and Implementation of Remote Operated SpyRobot Control System," International Journal of Science, Engineering and Technology Research (IJSETR), Volume 3, Issue 7, July 2014.

4   Md Athiq UR Raza Ahamed M., Wajid Ahamed, "A Domestic Robot for Security Systems by Video Surveillance Using ZigbeeTechnology," International Journal of Scientific Engineering and Technology (ISSN : 2277-1581) Volume 2 Issue 5, pp : 448-453 1 May 2013

5   Khushwant Jain and Vemu Suluchana, "Design and Development of Smart Robot Car forBorder Security, "International Journal of Computer Applications (0975 – 8887)Volume 76– No.7, August 2013.

6   Kunal Borker, and Rohan Gaikwad, "Wireless Controlled Surveillance Robot," IJRSCSMS Volume 2, Issue 2, February 2014.

Author's Details

Mr. Vinay Kumar S.B is an Assistant Professor in the Department of Electronics and Communication Engineering, School of Engineering and Technology, Jain University, Bangalore. He obtained his Bachelor degree in Electronics and Communication Engineering from Coorg Institute of Technology, ponnampet in 2009, Visvesvaraya Technological University, Belgaum and Master degree (M.tech) in Signal processing and VLSI, Jain University, Bangalore. He is pursuing Ph.D. in Electronics and Communication Engineering, Jain University, Bangalore. My research interest includes VLSI, Reverse logic, DSP and Embedded Systems. He has altogether 5 international journals to his credit and also he presented 2 international conference and 8 technical papers in national conference. He has altogether 2 sponsored projects funded by KSCST (Karnataka State Council for Science & Technology), IISc, Bangalore.

Mr. Naveen Kumar S is a student in the Department of Electronics and Communication Engineering, School of Engineering and Technology, Jain University, Bangalore. He obtained his Bachelor degree in Telecommunication Engineering from APS college of Engineering, Bangalore in 2012, Visvesvaraya Technological University, Belgaum and He is pursuing M.tech(Embedded System Design) in Electronics and Communication Engineering, Jain University, Bangalore. My research interest includes Embedded, Micro-Controller.

Miss. Monica B V is a student in the dept. of ECE, School of Engineering and Technology, Jain University, Bangalore. She is perusing her engineering degree. Her area of interest includes VLSI, Real time embedded system, DSP.

Mrs. Ravi.U.S is a faculty of Institute for Academic Service, Bangalore. He obtained his Bachelor degree in Electronics and Communication Engineering from Jnana vikas institute of Technology in 2010,Bidadi, Visvesvaraya Technological University, Belgaum and currently pursing master degree (M.Sc) in mathematics from KSOU, Mysore.

Leena Giri G [1], Praveen Gowda I V [2], Manjula S H [3],
Venugopal K R [4] and L M Patnaik [5]

In-Page Semantic Ranking of Snippets for WebPages

Abstract: Snippets are textual information present on web pages and are the basic foundation for the index building process of most search engines. Previous works in webpage ranking use techniques for ranking webpages while considering snippets to be plain text, without exploiting the rich semantic information available from the HTML content. An HTML document not only contains snippets, but also describes semantics of that snippet and the structure of the entire document. In this paper we propose a snippet ranking algorithm which uses semantic HTML tags, that embeds the snippet, and the underlying structure of the webpage for assigning ranks. This allows a more precise ranking compared to considering snippets as plain text.

Keywords: Semantic HTML tags, snippets, ranking

1 Introduction

The primary source of information for the process of ranking a webpage by a search engine is the snippets on that page. Many techniques that power search engines rely only on the textual data that the snippet represents. But, these techniques do not exploit the rich metadata from the semantics and structure of the webpage.

A new technique called in-page semantic ranking is proposed which assigns ranks to snippets based on the semantic tags that embed that snippet and

1 Department of Computer Science and Engineering, University Visvesvaraya College of Engineering, Bangalore University, Bangalore – 560 001
Email: leenagiri@gmail.com
2 Department of Computer Science and Engineering, Dr. Ambedkar Institute of Technology, Bangalore -560 056
3 Department of Computer Science and Engineering, University Visvesvaraya College of Engineering, Bangalore University, Bangalore – 560 001
4 Department of Computer Science and Engineering, University Visvesvaraya College of Engineering, Bangalore University, Bangalore – 560 001
5 Honorary Professor, Indian Institute of Science, Bangalore 560 001, India

the position of that snippet in the hierarchy of the web page structure. With the growth of HTML5 [4], semantic tags and sectioning elements [5] have captured the attention of web developers like never before. With increasing HTML5 support by browser engines, web developers are quickly adopting HTML5 semantic tags and sectioning elements to make their sites accessible and friendly to a wider range of web users. Semantic tags not only make the site accessible by helping screen readers understand the structure of a page, but also present an added advantage for a search engine to understand the semantics of a page.

The entire process of extracting snippets from the page can be conceptually realized in three steps as detaching a sub tree from the parse tree, traversing the sub tree where the snippet forms the leaf of that sub tree, and storing the semantic tags encountered during the traversing phase as meta information along with the snippet.

1.1 Motivation

Since its inception, HTML supported basic semantic mark-up. This has become more prominent with the introduction of HTML5 semantic tags and sectioning elements. A web page is not just a collection of textual data. It has structure, hierarchy and inherent meaning that defines the data. Semantic mark-up implies the meaning and prominence of a snippet. Not only can these nuances be understood by humans and screen readers, but even search engines can also be made to respond to this subtle intelligence which gives them an advantage over other search engines that only depend on the textual information available on a web page. Semantic tags have been used widely by web developers for search engine optimization and the in-page semantic ranking algorithm provides a quantitative algorithm to rank snippets using these semantic tags. This allows the author of a web document to directly communicate the intent and the context of the document to the search engine.

1.2 Contribution

A snippet ranking algorithm which utilizes the semantic HTML tags that embeds the snippet and the underlying structure of the webpage for assigning ranks to the snippets is proposed. The ranking algorithm proposed uses the meta information arising out of semantic tags and their position in a document hierarchy to assign ranks to snippets. This ranking infers the context and the implied meaning of the snippets which can be used for ranking web pages and generating clusters. Understanding the inherent nuances of the context and importance of a snippet for a web page is significant for precision of ranking.

1.3 Paper Organization

Section II discusses a few previous work related to page ranking and HTML5 elements that are useful for in-page semantic ranking. Section III defines the problem and section IV explains the proposed ranking algorithm and the implementation details. Section V concludes the work with future enhancements.

2 Related Work

The idea of PageRank is that every web page is attached an initial score which is equally propagated through the pages it is linked with. At each iteration, in the algorithm, the score of every page is updated to the sum of all incoming scores and is re-distributed to the pages pointed to by it. The algorithm converges to a PageRank value for every page. The result of running PageRank in a set of web pages is that highly interconnected pages (good hubs and authorities) are ranked higher than others. Kritikopoulos A. et.al., [1] present WordRank, a page ranking system which is most appropriate in topic based searches, since it prioritizes strongly interconnected pages, and at the same time is more robust to the multitude of topics and to the noise produced by navigation links. They present preliminary experimental evidence from a search engine they developed for the Greek fragment of the World Wide Web, and introduce a new metric (SI score) which is based on implicit user's feedback. Explicit evaluation, where available, is considered for evaluation purposes. They conclude that popular search engines will use analogically the same time to calculate WordRank as they do with PageRank.

Po-Hsiang Wang, Jung-Ying Wang, Hahn-Ming Lee [2] propose a novel ranking method called QueryFind that is based on learning from historical query logs. They propose to reduce the seeking time from the search result list. Their method uses not only the users' feedback but also the recommendation of a source search engine. Based on these, the users' feedback is accounted to evaluate the quality of web pages implicitly. They apply the meta-search concept to give each Webpage a content-oriented ranking score because of which the time users spend for seeking out their required information from search result list can be reduced and the more relevant Web pages can be presented. They propose a novel evaluation criterion to verify the feasibility of QueryFind ranking method. The criterion is to capture the ranking order of Web pages that users have clicked from the search result list. Results show that the time users spend on seeking out their required information can be reduced significantly when compared with other ranking algorithms.

Lara Srour et. al., [3] in their work propose a novel approach to personalize the results of search engines to the opinion and interests of the specific user based on his/her behaviour in the past, and similarity and trust towards other users. They restrict similarity computation to neighborhoods of trusted peers due to the excessive computational cost of determining the similarity between any two users in real-time. Since users' incentive to provide explicit ratings tend to be low, they provide a way that allows the system to rate pages implicitly based on the length of time spent on a page. By simulation, their results show that combining trust and similarity is the best approach to get accurate personalized recommendations for web pages.

[3] presents a knowledgebase of websites using HTML5 DocType which is a required preamble for HTML5 websites. The page displays a chart which specifies the amount of websites within the top 10K, 100K and 1 million sites groups that are categorized as being in the specific verticals of Business, Education, Health, News, Shopping, Technology and Travel.

[4] discuss the new semantic elements. HTML5 offers some new elements, primarily for semantic purposes. The elements include section, article, aside, header, footer, figure, figcaption and such others. The page depicts a chart that displays the usage of popular browsers relatively.

[5] gives a detailed description of the usage of the new elements mentioned in [4] and discusses the DOM parser which is an XML parser.

3 Problem Definition

Semantic HTML supplemented by HTML5 Semantic tags not only define the presentation of a web document, but also reinforce the semantics of content on that page which is exploited by the in-page semantic ranking to rank snippets based on their inherent meaning and their importance to the given page.

A new paradigm to web document and snippet extraction that tries to combine textual snippet extraction supplemented with semantic information is considered. Every snippet that is extracted from a webpage contains meta information of the tags that encloses the snippet. Every HTML tag contains significant semantic information that reinforces the data it encloses and is of significance for the rest of that document. The meta information that the search engine requires can be visualized in the form of a parse tree that replicates the DOM tree, with respect to representing the HTML tags along with their content while maintaining the hierarchy of their occurrence. It is required, to navigate the tree just once, to extract all the snippets on the page along with their meta information about the tags that they are embedded in. The entire process of extracting snippets from the page can be conceptually realized as:

- detaching a sub tree from the parse tree.
- traversing the sub tree, where the snippet forms the leaf of that sub tree.
- storing the semantic tags encountered during the flattening phase as meta information along with the snippet which will finally be used for assigning a rank to the snippet.

4 Algorithm and Implementation

The implementation is done at three different levels, where each level contains a collection of rules. The rules either match a HTML tag or an attribute of a HTML tag. For example, when we encounter a <img> tag, the rule matches the *alt* attribute of the image tag and not the content of the tag itself which in case of self closing tags, is empty. Each level is assigned a score that is associated with the tags matched by the rule sets in that level. 47 different HTML tags were identified, and only a subset of most commonly occurring tags was used in the ranking process.

Examining the DocType is the quickest and easiest technique for determining the version of HTML that the page uses. Determining the HTML version is important in the sense to know the tags that will be encountered. If a page uses HTML4 or earlier and if the ranking algorithm tries to match HTML5 tags, then the ranking would lose the precision.

The tags are divided into three levels L1, L2, L3 and each of the levels are assigned scores R1, R2, R3 which is used by the ranking algorithm. Each level has a collection of rule sets that either match tags or an attribute of tags. The rule sets follow the same syntax as that of CSS selectors. Figure 1 illustrates the hierarchy of levels and the scores assigned to these levels. Table 1 presents the distribution of the rule sets among the three levels.

4.1 Ranking algorithm

In-page semantic ranking algorithm is a recursive ranking algorithm, which traverses the HTML parse tree until a text node is reached, wherein the snippet is extracted along with its calculated rank. The algorithm is invoked starting

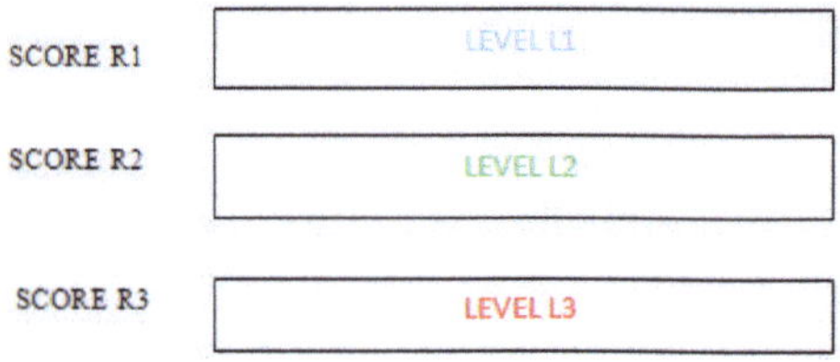

<table>
<tr><td>SCORE R1</td><td>LEVEL L1</td></tr>
<tr><td>SCORE R2</td><td>LEVEL L2</td></tr>
<tr><td>SCORE R3</td><td>LEVEL L3</td></tr>
</table>

Figure. 1: Hierarchy of levels and the scores assigned to the levels

Table 1: Distribution of the rule sets among the three different levels. The rules follow the same syntax as that of CSS selectors

Level Number	Rule sets
Level L1	<body> > <header> <section> <h1> <h2> <h3> <strong>, <b> <em>, <i>
Level L2	<section> > <header> <article> > <header> <nav> > <header> <article> <nav> <h4> <h5> <h6> <p> <a> <span> <th> <td> img[alt]

Level L3	<aside> > <header>
	<footer> > <header>
	<aside>
	<footer>
	<li>
	<blockquote>
	<pre>, <code>
	<figure> > <figcaption>
	<cite>, <abbr>, <small>, <big>

from <body> tag as the root node. *level( )* is a routine which matches the given node with the rule sets of Table 1 and returns the matching level. Since the rule sets are of the same syntax as that of CSS Selectors, *level( )* is simplified implementation of a CSS Selector matching procedure.

The ranking algorithms works as follows

1. The rank( ) begins its invocation from the <body> node with a current score of 0 and current level 0.
2. For every child node of the current node, find its level and its parent node's level.
3. If the node is a text node, then extract the snippet in the text node and return the pair (snippet, curScore( ) which is the snippet, rank pair that is appended to the snippet set.
4. If the childNode belongs to L1, then *rank( )* is called recursively with the score altered based on its parent level as illustrated in the algorithm below.
5. If the childNode belongs to L2 or L3, then *rank( )* is called recursively with the curScore divided by 2 or 3 respectively and then added with the level score R2 or R3.

If the childNode doesn't match any of the levels, then the current score remains unaltered and the *rank( )* is invoked on each of its children. This occurs with tags such as <ul> or <ol> which are not assigned to any levels since it was determined that their presence did not alter the importance of their snippets and was not significant in the ranking process.

Algorithm 1: In-page Semantic Ranking

```
rank (curNode, parentLevel, curScore)
begin
  curLevel = level(curNode)
```

```
foreach childNode of curNode
  begin
    // childNode is a text node
      if childNode is textNode
        snippet = (childNode.string, curScore)
        snippetset.append(snippet)
// childNode belongs to L1
      else if level(childNode) = L1
        if parentLevel = L2
          rank(childNode, curLevel, R1/2 + curScore)
      else if parentLevel = L3
          rank(childNode, curLevel, R1/3 + curScore)
      else if parentLevel = L1
          rank(childNode, curLevel, R1 + curScore / 2)
// childNode belongs to L2
      else if level(childNode.tag) = L2
          rank(childNode, curLevel, R2 + (curScore / 2))
          childNode belongs to L3
      else if level(childNode.tag) = L3
          rank(childNode, curLevel, R3 + (curScore / 3))

    // childNode doesn't match any level (such as the <ul> tag)
    else
    rank(childNode, parentLevel, curScore)
    endif
    endforeach
      endrank
```

Figure 2 is an example webpage that has been constructed to represent a typical HTML5 webpage that uses semantic tags for its mark-up. It is followed by an illustration of the process of running the in-page semantic algorithm on it to rank snippets on the page.

```
<!DOCTYPE html>
<html>
<head>
 <title>Webpage Title</title>
</head>
<body>
<header>
 <h1>Page Heading</h1>
</header>
```

```
<nav>
 <header>
  <h3>Navigation Heading</h3>
 </header>
 <ul>
 <li><span>Navigation Link 1</span></li>
 <li><a href="page2.html">Navigation Link 2</a></li>
 </ul>
</nav>
<section>
 <header>
  <h2>Section Heading</h2>
 </header>
 <article>
  <header>
   <h3><a href="page1.html">Article Heading 1</a></h3>
  </header>
  <p>Paragraph Snippet 1</p>
 </article>
<article>
 <header>
 <h3><a href="page2.html">Article Heading 2</a></h3>
 </header>
   <p>Paragraph Snippet 2</p>
 </article>
</section>
<aside>
<h3>Aside Heading</h3>
<ul class="tag-cloud">
 <li><a href="l1.html" rel="tag" class="w2">link 1</a></li>
 <li><a href="l2.html" rel="tag" class="w8">link 2</a></li>
 <li><a href="l3.html" rel="tag" class="w3">link 3</a></li>
</ul>
</aside>
<footer>
 <p>Footer Paragraph</p>
</footer>
</body>
</html>
```

Figure. 2: HTML5 webpage considered

Step 1: Doctype is <!DOCTYPE html> which indicates the page used HTML5, use the rule sets in Table 1 without any adjustments. If the page has used any other Doctype, then the rule sets matching HTML5 sectioning elements will be removed from Table 1 before invoking the algorithm.

Given next is the parse tree of the above HTML source at various stages of running of the in-page semantic ranking algorithm along with the ranks assigned to the snippets.

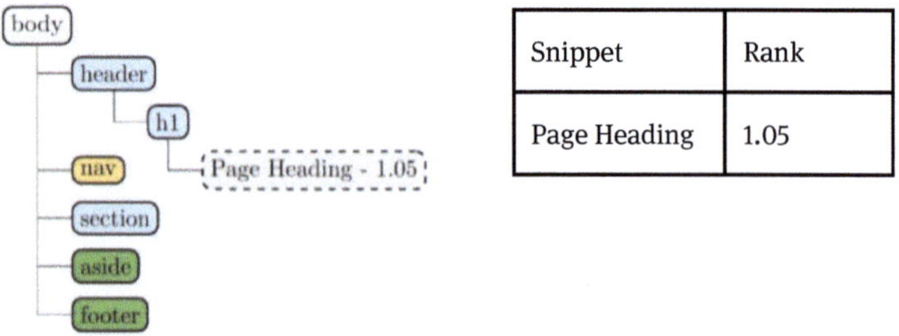

Snippet	Rank
Page Heading	1.05

Figure. 2 (a): Parse Tree with the sub tree rooted at <header> expanded

Step 2: Extract the snippet from the Title tag < title > and give it the highest preference. The content in the title tag is the most important snippet for the page and it is given the highest rank over all other snippets on the page.

Step 3: Invoke the in-page semantic ranking algorithm *rank()* beginning from the <body> node of the page. Snippet set with the snippets and their associated ranks are obtained on the completion of the execution of the algorithm.

Step 4: The final snippet set is obtained after the algorithm terminates. A snippet set is a collection of pairs - a snippet and its rank.

Conclusions

The in-page semantic ranking algorithm provides a quantitative approach to ranking snippets based on the semantic tags they are embedded in and the position of the snippet in the parse tree hierarchy. The algorithm builds a foundation for snippet ranking which can be further used by clustering or page ranking algorithms to increase the precision of their results.

It was also observed that the ranking algorithm was not effective on legacy webpages that mostly used HTML tables for layout. With the widespread adoption of HTML5 and strong usage share of HTML4, handling legacy websites is a rarely occurring case now.

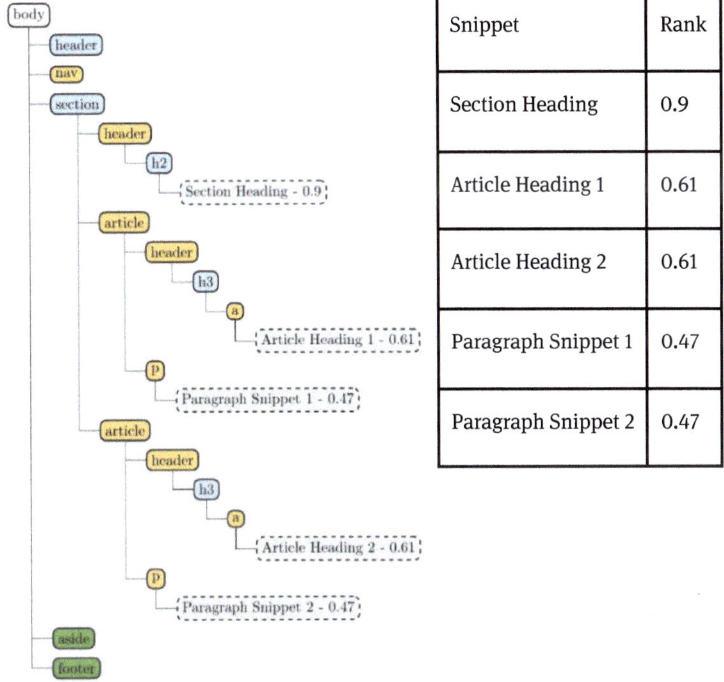

Snippet	Rank
Section Heading	0.9
Article Heading 1	0.61
Article Heading 2	0.61
Paragraph Snippet 1	0.47
Paragraph Snippet 2	0.47

Figure. 3 (b): Parse Tree with the sub tree rooted at <section> expanded

Snippet	Rank
WebPage Title	Highest Prominence (title)
Page Heading	1.05
Section Heading	0.90
Navigation Heading	0.65
Article Heading 1	0.61
Article Heading 2	0.61
Paragraph Snippet 1	0.47

Paragraph Snippet 2	0.47
Aside Heading	0.33
Navigation Link 1	0.28
Navigation Link 2	0.28
Footer Paragraph	0.25
Link Text 1	0.13
Link Text 2	0.13
Link Text 3	0.13

References

1 Apostolos Kritikopoulos, Martha Sideri, Iraklis Varlamis, "Wordrank: A Method for Ranking Web Pages Based on Content Similarity," 24th British National Conference on Databases (BNCOD'07) IEEE Computer Society, 0-7695-2912-7/07.
2 Po-Hsiang Wang, Jung-Ying Wang, Hahn-Ming Lee, "QueryFind: Search Ranking Based on Users' Feedback and Expert's Agreement," Proceedings of the 2004 IEEE International Conference on e-Technology, e-Commerce and e-Service (EEE'04), pp 299-304.
3 Lara Srour, Ayman Kayssi Ali, Chehab, "Personalized Web Page Ranking Using Trust and Similarity," IEEE/ACS International Conference on Computer Systems and Applications, 2007(AICCSA '07) pp 454-457.
4 http://trends.builtwith.com/docinfo/HTML5
5 http://caniuse.com/#feat=html5semantic
6 https://html.specwhatwg.org/multipage/sematics.html#sections.
7 O. Zamir and O. Etzioni, "Web Document Clustering: A Feasibility Demonstration," Proceedings, 21st Annual International ACM SIGIR Conference on Research and Development in Information Retrieval, SIGIR'98, New York, NY: ACM Press, pp. 46–54, 1998.
8 Steve Branson, Ari Greenberg "Clustering Web Search Results Using Suffix Tree Methods," CS276A Final Project, Stanford University.
9 Ramesh Singh, Dhruv Dhingra, Aman Arora, "SCHISM – A Web Search Engine using Semantic Taxonomy," IEEE Potentials, September/October 2010.

P. Kumaran[1] and S. Chitrakala [2]

Information Diffusion in Online Social Network: Techniques, Applications and Challenges

Abstract: Social Network plays an important role in bringing individuals together and spreads the Information in very large scale. Every Social Network like Face book, Twitter, Instagram, LinkedIn, Blogs and YouTube has millions of individuals as active members and many commercial organizations have realized that social network as a global market for their product as they perceive it as a place that connects business services and common people. A lot of efforts have been made in order to understand this event, starting from topic detection, influence spreader identification and information diffusion modeling. In this paper, we have presented an extensive survey on research work carried out in the area of topic detection, influence spreader identification and information diffusion models of Online Social Network (OSN). This survey is intended to help the researchers to understand existing works and various applications in which the information diffusion is applied to analyze the spread of rumors, computer viruses and diseases during outbreaks.

Keywords: Social Influence, Information Diffusion Models, Information Spreader.

1 Introduction

Millions of users in Online Social Networking (OSN) website use the internet to connect with their friends, family and acquaintances to produce and consume informative contents. In this manner the social network play an important role in the spread of innovative information and divergent viewpoints. Here we define information diffusion as the process by which a piece of information (knowledge) is spread and reaches individuals in the network through their

1 Research Scholar, Department of Computer Science and Engineering, CEG, Anna University
kumaran.0991@gmail.com
2 Associate Professor, Department of Computer Science and Engineering, CEG,
Anna University
au.chitras@gmail.com

interactions. They have proved that in many situations it is powerful, like In February 2013, during the third quarter of Super Bowl XLVII; a power outage stopped the game for 34 minutes. Oreo, a sandwich cookie company, tweeted during the outage: "Power out? No Problem, You can still dunk it in the dark." The tweet caught on almost immediately, reaching nearly 15,000 retweets and 20,000 likes on Face book in less than two days. A simple tweet diffused into a large population of individuals. It helped the company gain fame with minimum cost in an environment where companies spent as much as $4 million to run a 30-second advertisement. This is motivated by fact of understanding the dynamic behavior of these social networks to solve the problems (e.g. preventing the terrorist attacks, Anticipating natural disasters). It provides us with big opportunities for small businesses in Social Media Optimization (SMO) (e.g. optimizing social media campaigns), etc. Therefore, there is a lot of scope for researchers in recent years to develop various approaches, methods, models to capture and analyze information diffusion in online social network and extract the knowledge from it to predict the effective models for information diffusion. The rest of the paper is organized as follows. Section 2 gives survey on topic detection, Influence Spreader Identification and various processing models in information diffusion of OSN. Section 3 gives experimental evaluation. Section 4 gives challenges in Information Diffusion of OSN . Section 5 describes Applications. Section 6 gives Conclusion.

2 Related Works

In section 2 shows that various processes involved in information diffusion of online social network such as topic detection, Influence spreader identification and information diffusion models.

2.1 Popular Topic Detection

Popular topic detection is the primary task in information diffusion. This involves extracting "tables of content" to sum up discussions, recommending famous topic to the user, or predicting the future trending topics. The Burst topics are suggested in order to detect the topics in textual streams effectively.

Shamma et al. [32] has proposed the model called *Peaky Topics* (PT) which mimics the classical tf-idf model [30]. It relies on normalized term frequency metric. In order to measure the overall term usage, time slice of a pseudo-document composed of all the messages in the corresponding collection are considered. Using this metric, we conider "bursty topics" as individual terms and are ranked. But it is not enough to clearly identify a topic. Therefore, more scopes are there to improve these methods.

AlSumait et al. [2] has developed an online topic model, more exactly, a non-Markov on-line LDA (*Latent Dirichlet Allocation*) [3]. Gibbs sampler topic model. Basically, is a statistical generative model that relies on a hierarchical Bayesian network which matches words and messages in documents through latent topics. This method constructs an evolutionary matrix for individual topic that catches the evolution of the topic over the period of time and thus allows detecting bursty topics.

Cataldi et al. [4] has proposed a method called *Temporal and Social Terms Evaluation* (TSTE) that deals with both the temporal and social properties of stream messages along with Page Rank algorithm [24]. It allows the model to form the lifecycle for each term on the basis of biological metaphor, which is calculated based on the values of nutrition and energy that shows the advantages of user authority. Using the various learning techniques in the calculation of critical drop value based on energy, the proposed method can discover most bursty terms based on *co-occurrence* metric.

Lu et al. [27] has proposed a method called Moving Average Convergence Divergence (MACD) to predict the topics that will draw attention in near future. The aim of this method to change trend following indicator, precisely a short period and longer period, changing the average of term frequency into momentum oscillator (i) When the trend momentum value changes from negative to positive, the topic is beginning to rise; (ii) when the trend momentum value is changes from positive to negative, the level of attention given to topic is falling.

2.2 Influential Spreader Identification

Influence and homophile in online social networks have been explored by different social Scientist. In social networks, there are two broad categories identified namely influence driven and homophile driven. Therefore in online social network differentiating between influence and homophile is challenging task. Kitsak et al. [19] expressed that best spreaders are not necessarily most connected people in the network. In K-core decomposition analysis, network core is assigned to most efficient spreader [31]. In K-core decomposition, the node with the highest index value is placed in the core of the network and the node with the lowest index value is placed in periphery. Since the output of k-shell decomposition is skewed. Brown et al. [13] proposed modified methods for improving the results using logarithmic mapping in order to achieve more useful k-shell values.

Cataldi et al. [4] explored page rank algorithm [24] in accessing the distribution of influence in the network. In this algorithm, chosen node is proportional to the probability of number if times the node is visited during random walk. In this algorithm it considers only the topologies of the network and avoids the others properties such as node features and Information Processing.

Graph based approach has been developed by Romero et al. [26] for identifying influence spreader in online social network. In this approach they have designed an algorithm called Influence Passivity (IP) in which relative influence and passivity score are assigned to every user based on the ration of forward information. But in real scenario there is no possibility of universal influencer in all domains. Fig 1 shows the evidence for incorporating Influence Passivity in twitter data.

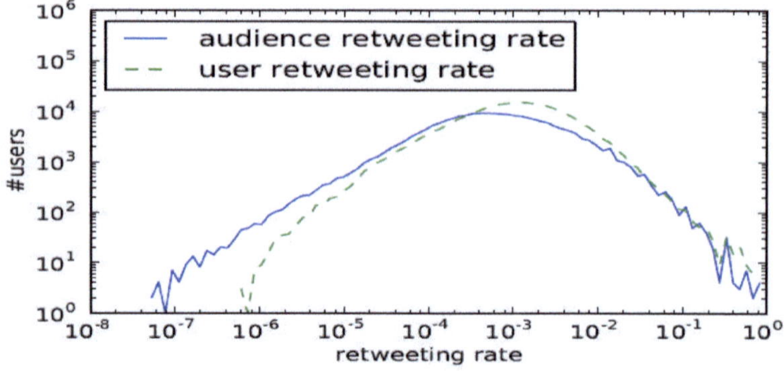

Fig. 1:
Evidence for the Twitter user passivity [26]

To resolve the problem in above graph based approach is handled by Pal et al. [25]. Here they have developed non-graph based topic sensitive method. To do so, Network members are described by nodal and topical feature. In the network most influential and authoritative nodes are ranked based on within – cluster ranking procedure for some particular topics is purely based on using probabilistic clustering over feature space. Weng et al. [34] also proposed a new version of Page Rank Algorithm for topic-sensitive modeling which is committed to twitter.

Kempe et al. [18] follows different methods and use Information Cascade (IC) and Linear Threshold (LT) models to understand influence maximization problem. Here the k value is used for finding the k-node set for the purpose of maximize the influence node in the network. The node having higher influence is initial activated node for information diffusion according to IC & LT approach. Finally sub modular functions relays on greedy hill climbing strategy, which is used for handling influence maximization problem. Therefore, summary of above method is relays on either graph based approach or non-graph based approach.

2.3 Information Diffusion Models

Information Diffusion model deals with capture or predict spreading process in OSNs. In these models it predicts the way to stopping viruses, misinformation spread and business campaign etc. The diffusion process is based on two aspects: one is its structure and another one is temporal dynamics. Usually, diffusion models developed in the circumstance of OSNs assume that influence people are identified based on their connections. Because of this assumption it can predict only when and where the information propagated in the OSN but not identify how and why it propagated. Therefore there is a need for developing diffusion models for predicting the hidden patterns in it. In general there are two categories of Information Diffusion Models: (i) Explanatory Models and (ii) Predictive Models.

Explanatory Models: The main aim of explanatory models is to retrace the paths of already propagated piece of information. Gomez et al. [12] proposed a method for predicting the structure of the diffusion patterns are inferred from the reciprocal relationships of nodes infections. Assume that Influence of nodes with its neighbor independently with some probability. They developed an iterative algorithm called NETINF, it is based on sub modular function optimization for predicting the spreading pattern that increase the likelihood of discovered data.

Gomez et al. [11] further they propose the new diffusion model which extends the NETINF algorithm as a spatially discrete network of continuous and with some conditions based temporal processes happing at different rates. The extended algorithm is called NETRATE, which infers transmission rates in pair wise and the graph of diffusion is organizing and solving a convex maximum likelihood problem [7].

Above mentioned algorithms are worked in the static network. But the OSN topology is keeps on evolving both at the edge and node wise. For that purpose, Gomez et al. [11] further extend the NETRATE algorithm in to NETPATH by, just incorporating into time – varying mechanism in to it. NETPATH uses stochastic gradients to gives online calculated structure of a network that changes over a period of time.

Choudhury et al. [5] discovered the data sampling strategy for information diffusion. Based on its experiment results it shows that, sampling methods for information diffusion are consider in its account for both network topology and user's attribute and gives lower error when compared with naïve strategies.

Assume that when there is a missing of data, Sadikov et al. [28] developed a model to predict the information flow based on k-tree model. In K-tree model only fraction of activation sequence are given, and its gives the complete properties of information spreading patterns such as depth. In Table 1 summarizes the different algorithms used in explanatory model.

Table 1: Summary of methods used in Explanatory Model

Various Models	Parameters involved in Explanatory Models				
	Static	Dynamic	Transmission Probability	Transmission Rate	cascade
NETINF	✓	×	✓	×	✓
NETRATE	✓	×	✓	✓	✓
INFOPATH	✓	✓	✓	✓	✓
K-tree	✓	×	×	×	✓

Predictive Models: Graph based Approaches- There are two basic models in this approach, namely Independent Cascades (IC) [10] and Linear Threshold (LT) [14]. In IC model edges are associated with diffusion probability but in LT model influence degree is assigned to each edge and influence threshold for each node. Hence, IC model diffuses the information to its neighbor node based probability value associated with edges and in LT model activated the inactive neighbor node based on threshold value. By this way the IC is works based on sender-centric model and LT is works based on receiver-centric model.

Galuda et al. [8] proposed a method which uses the LT model to predict the information diffusion pattern in graph structure; it relies on some parameters such as virality, influence degree, probability value of user for accepting any information. The LD model uses the gradient ascent method to describe the beginning of diffusion process. However, realistic temporal dynamics are can't be reproduced in LT model.

Saito et al. [29] proposed a method with the assumption of asynchronous extensions in IC and LT model namely AsIC (Asynchronous Independent Cascade) and AsLT (Asynchronous Linear Threshold).These models works on continuous time axis along with the parameters of synchronous model plus time – delay. However it won't provide the solution to the practical data and it works only to the synthetic data.

Guille et al. [16] developed a model called T-BaSIC (i.e. Time Based Asynchronous Independent Cascades) it propagates the diffusion process in asynchronous IC. The functions used in this model are depending on time in which it parameters are not fixed and it uses logistic regression to explore the parameter values from semantic, social and temporal nodes. For example Adrien Guille [1] used the T-Basic model to predict the next release date of iphone and the experiment results are shown in figure 2 using with the result of 32% precision value on an average.

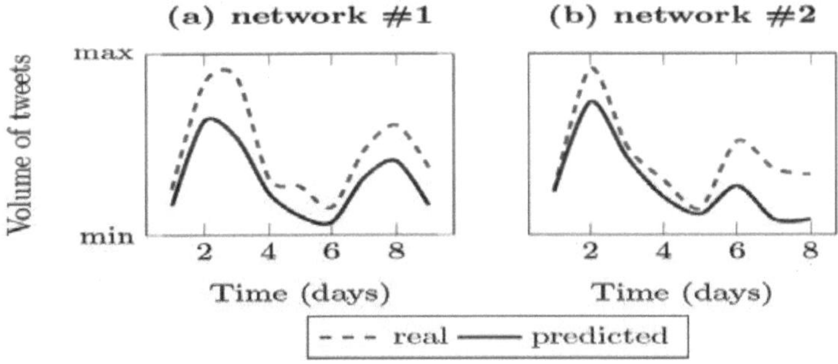

Fig. 2: Comparison result for real and predicted topic ("iphone","release") [1]

Non - Graph based Approaches: In Non-Graph based approaches there is no pre assumption of graph structure and its mainly focusing on constructing a model for epidemiological processes. SIR (Susceptible Infected Recovered) and SIS (Susceptible Infected Susceptible) are the two primary models [17] [23] used in non graph based approach. In both the models there is common assumption (i.e.) connection to another node having same probability and connections within the group made at random.

Leskovec et al. [21] proposed a basic SIS model with the assumptions of the entire node having same probability which is denoted by parameter (β) and the node who adopted the information it becomes susceptible. In Online social network the influence feature are not evenly distributed to all the nodes. Therefore there is an importance for considering these assumptions in the development of complex modeling.

Yang et al. [35] proposed a model is named as Linear Influence Model (LIM) with the primary assumptions of starting the information diffusion from the influence nodes in the network. This model predicts information diffusion in the unit of number of nodes adopted with in some particular time-period. Here representation of Influence functions in non-parametric way and which is calculated by solving non negative least squares problem using Reflective Newton Method [6].

Wang et al. [33] developed a model based on Partial Differential Equation (PDE) used to predict the information diffusion. In general topological and temporal dynamics are predicted using diffusive logistic equation. Here they consider distance as the term for topology concern and the temporal dynamic processes are assigned to logistic equation. Cubic Spline Interpolation [9] method is used for estimating the parameters of this model. Table 2 summarizes the different approaches involved in Predictive models.

Table 2: Summary of Predictive Model Approaches

Various Models	Parameters involved in Predictive Models					
	Social	Time	Graph based	Non-Graph Based	Parametric	Non-Parametric
LT-based	✓	×	✓	×	✓	×
AsIC, AsLT	×	×	✓	×	✓	×
T-BaSIC	✓	✓	✓	×	✓	×
SIS -based	×	✓	×	✓	✓	×
LIM	✓	✓	×	✓	×	✓
PDE	✓	✓	×	✓	✓	×

3 Experimental Evaluation

Implementation of Information Diffusion in Online Social Network based on some well known network measures based on centrality value, which optimize the information diffsuion process.

3.1 Network Measures

In OSN Node importance is measured through centrality. Nodes in network having high centrality have higher influence than nodes with less centrality. This centrality has further classified into three types 1.Degree Centrality 2.Closeness Centrality 3.Betweenness Centrality

Definition 1 Degree Centrality: The number of connections an actor (node) has is called as degree centrality, when ties are directed; from node it calculates total number of ties sent (out-degree) and ties received (in-degree). Out-degree typically indicates influence; in-degree indicates prestige or popularity

$$C_D = \frac{\sum_{i=1}^{N}\left(C_D\,(n^*) - C_D\,(n_i)\right)}{(N-1)(N-2)} \tag{1}$$

$$\text{Where } C_D\,(n_i).\ =\sum_j \alpha_{ij}$$

(C_D=Degree Centrality) explained by Freeman's formula for graph degree centralization.C_D (n_i) is the actor degree index for node n_i. $C_D(n^*)$ is the largest actor degree index. α_{ij} is the direct or adjacent link between actor i and actor j.

Definition 2 Closeness Centrality: The closeness centrality emphasizes the distance of an actor to all others in the network. The closer one is to others in the network, the more favored is that actor's position "Farness" is the sum of the geodesic distances from each ego to all others in the network "Closeness" is the reciprocal of farness = 1/farness.

$$\text{Closeness} = \frac{1}{\text{Farness}} \tag{2}$$

Definition 3 Betweenness Centrality: A measure of accessibility that is the number of times a node is crossed by shortest paths (geodesic path) of other pairs of actors in the network. Anomalous centrality is detected when a node has a high betweenness centrality and a low order (degree centrality), as in air transport.

$$C_B(z) = \sum_{x \neq y \neq z} \frac{\sigma_{xy}\,(z)}{\sigma_{xy}} \tag{3}$$

σ_{xy} Denotes the number of shortest paths from x to y and σ_{xy} (z) denotes the number of shortest paths from x to y passing through z. In Fig 3 shows the results with incorporated the different centrality value for computing the influence node in the network (GT Dataset).

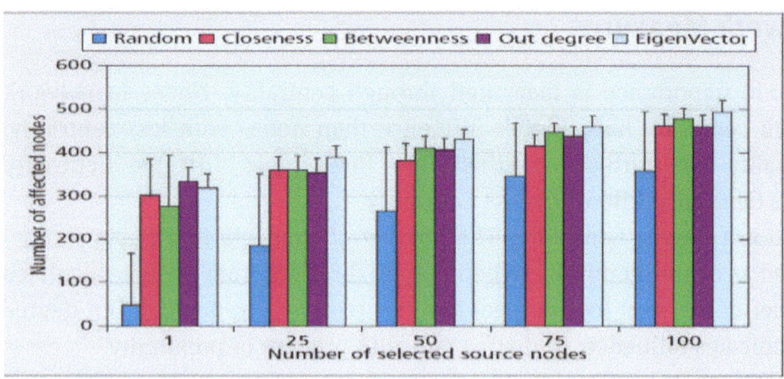

Fig. 3: The graph shown the result of centrality values [22]

3.2 Datasets

Implementation of Information Diffusion in online social network used the dataset from various social networking sites such as Twitter, Yahoo! Meme, mobile phone call network, Enron email collection, and Facebook. In general the data's available in these networks in the form of blogs, text, URL, images, audio and videos. In fig 4 shows the various social connections between the social network users.

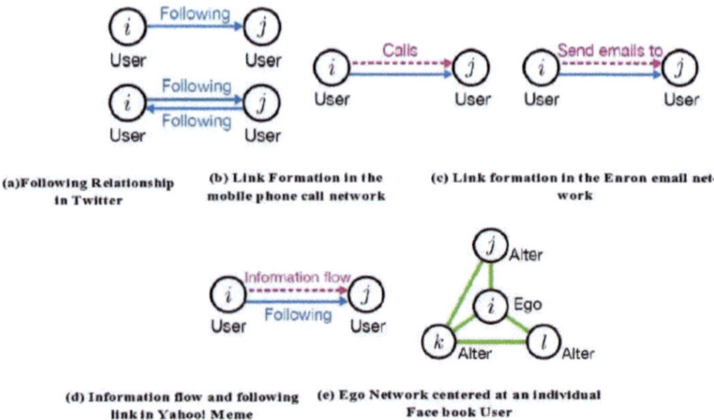

Fig. 4: Illustrations of various social connections on (a) Twitter, (b) Yahoo! Meme and (c) Face book [20]

4   Challenges of Information Diffusion In Osn

Topic Detection, Influential spreader identification and Information Diffusion modeling are the challenging areas in the field of OSN.The implementation of Information diffusion in OSN raises several reseaarch challenges.Some of these are:
- Topic Definition and scalability
- Social Dimension
- Data complexity
- Closed world assumption
- Cooperating and competing diffusion processes
- Topic –sensitive modeling
- Dynamic Networks
- Opinion Detection

5 Applications

In Information Diffusion in online social network need a lot of contribution for analysis its dynamic behaviour .As per survey result shows that very few implementaions are reused,because most of the inputs are in different formats and it uses various implementaion languages with respect to the application, which makes very difficult to evaluate or compare with existing techniques.Therefore there is an well know tool for Information Diffusion is SONDY [15] ,which gives the platform for implenting various techniques in the areas of topic detection , influence spreader Identification and Information diffusion model. Some of the applications in which the information diffusion are widely used such as ,
- Preventing the terrorist attacks
- Anticipating natural disasters
- Grassland biological risk
- Optimizing social media campaigns
- Epidemiology

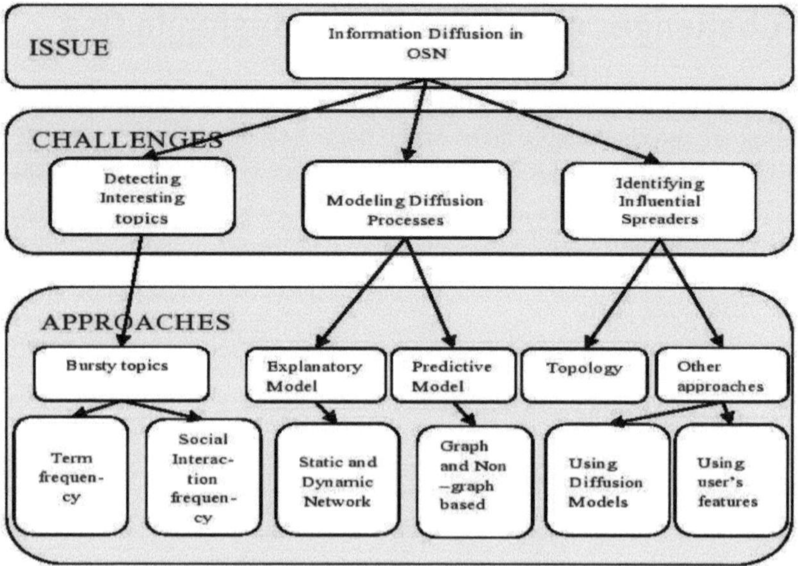

Fig. 5: The above taxonomy presents information diffusion in online social networks [1]

Conclusion

In this paper discussed different methods, models and approaches involved in topic detection, influence spreader identification and information diffusion models of OSN. And also specified many challenging issues, applications, network measures and some common social network dataset used in information diffusion on OSN. But most of the implementation results are based on some common underlying assumptions such as equal influence probability to nodes, closed world assumptions and Independent topics etc. In real scenario OSN in dynamic and time variant behavior. Therefore lot of scopes is available for researchers by adopting dynamic features of OSN into their account for more improving their performance in their result.

References

1   Adrien Guille. Information Diffusion in Online Social Networks SIGMOD'13 PhD Symposium ,ACM 2013

2 L. AlSumait, D. Barbar'a, and C. Domeniconi. On-line lda: Adaptive topic models for mining text streams with applications to topic detection and tracking. In ICDM '08, pages3–12, 2008.

3 D. Blei, A. Ng, and M. Jordan. Latent dirichlet allocation. The Journal of Machine Learning Research, 3:993–1022, 2003.

4 M. Cataldi, L. Di Caro, and C. Schifanella.Emerging topic detection on Twitter based ontemporal and social terms evaluation. In MDMKDD '10, pages 4–13, 2010.

5 M. D. Choudhury, Y.-R. Lin, H. Sundaram,K. S. Candan, L. Xie, and A. Kelliher. Howdoes the data sampling strategy impact the discovery of information diffusion in social media? In ICWSM '10, pages 34–41, 2010.

6 T. F. Coleman and Y. Li. A reflective newton method for minimizing a quadratic functionsubject to bounds on some of the variables.SIAM J. on Optimization, 6(4):1040–1058,Apr. 1996.

7 I. CVX Research. CVX: Matlab software for disciplined convex programming, version 2.0beta. http://cvxr.com/cvx, sep 2012.

8 W. Galuba, K. Aberer, D. Chakraborty, Z. Despotovic, and W. Kellerer. Outtweetingthe twitterers - predicting information cascades in microblogs. In WOSN '10, pages 3–11, 2010.

9 C. F. Gerald and P. O. Wheatley. Applied numerical analysis with MAPLE; 7th ed.Addison-Wesley, Reading, MA, 2004.

10 J. Goldenberg, B. Libai, and E. Muller. Talk of the network: A complex systems look at the underlying process of word-of-mouth. Marketing Letters, 2001.

11 M. Gomez-Rodriguez, D. Balduzzi, and B. Sch"olkopf. Uncovering the temporal dynamics of diffusion networks. In ICML '11, pages 561–568, 2011.

12 M. Gomez Rodriguez, J. Leskovec, and A. Krause. Inferring networks of diffusion and influence. In KDD '10, pages 1019–1028, 2010.

13 M. Gomez-Rodriguez, J. Leskovec, and B. Sch"okopf. Structure and dynamics of information pathways in online media. In WSDM '13, pages 23–32, 2013.

14 M. Granovetter. Threshold models of collective behavior. American journal of sociology, pages 1420–1443, 1978.

15 A. Guille, C. Favre, H. Hacid, and D. Zighed. Sondy: An open source platform for socialdynamics mining and analysis. In SIGMOD '13, (demonstration) 2013.

16 Guille and H. Hacid. A predictive model for the temporal dynamics of information diffusion in online social networks. In WWW '12 Companion, pages 1145–1152, 2012.

17 H. W. Hethcote. The mathematics of infectious diseases. SIAM REVIEW, 42(4):599–653, 2000.

18 D. Kempe. Maximizing the spread of influence through a social network. In KDD '03, pages 137–146, 2003.

19 M. Kitsak, L. Gallos, S. Havlin, F. Liljeros,L. Muchnik, H. Stanley, and H. Makse. Identification of influential spreaders in complex networks. Nature Physics, 6(11):888–893, Aug 2010.

20 Lilian Weng. Information Diffusion on Online Social network, April 2014.

21 J. Leskovec, M. Mcglohon, C. Faloutsos, N. Glance, and M. Hurst. Cascading behaviorin large blog graphs. In SDM '07, pages 551–556, (short paper) 2007.

22 Miray Kas, L. Richard Carley, and Kathleen M. Carley :Monitoring Social centrality for peer to peer social network.IEEE Communication Magazine 2013.

23 M. E. J. Newman. The structure and function of complex networks. SIAM Review, 45:167–256, 2003.

24 L. Page, S. Brin, R. Motwani, and T. Winograd. The pagerank citation ranking: Bringing order to the web. In WWW '98, pages 161–172, 1998.

25 A. Pal and S. Counts. Identifying topical authorities in microblogs. In WSDM '11, pages 45–54, 2011.

26 D. Romero, W. Galuba, S. Asur, and B. Huberman. Influence and passivity in social media. In ECML/PKDD '11, pages 18–33, 2011.

27 L. Rong and Y. Qing. Trends analysis of news topics on Twitter. International Journal ofMachine Learning and Computing, 2(3):327–332, 2012.

28 E. Sadikov, M. Medina, J. Leskovec, and H. Garcia-Molina. Correcting for missing datain information cascades. In WSDM '11, pages 55–64, 2011.

29 K. Saito, K. Ohara, Y. Yamagishi, M. Kimura, and H. Motoda. Learning diffusion probability based on node attributes in social networks. In ISMIS '11, pages 153–162, 2011.

30 G. Salton and M. J. McGill. Introduction to Modern Information Retrieval. McGraw-Hill,1986.

31 S. B. Seidman. Network structure and minimum degree. Social Networks, 5(3):269 – 287, 1983.

32 D. A. Shamma, L. Kennedy, and E. F.Churchill. Peaks and persistence: modeling the shape of microblog conversations. In CSCW '11, pages 355–358, (short paper) 2011.

33 F. Wang, H. Wang, and K. Xu. Diffusive logistic model towards predicting informationdiffusion in online social networks. In ICDCS '12 Workshops, pages 133–139, 2012.

34 J. Weng, E.-P. Lim, J. Jiang, and Q. He. TwitterRank: finding topic-sensitive influential twitterers. In WSDM '10, pages 261–270, 2010.

35 J. Yang and J. Leskovec. Modeling information diffusion in implicit networks. In ICDM '10, pages 599–608, 2010.

Shruti S. Sharma [1] and Achala Deshmukh [2]

Harmony Search Algorithm for PAPR Reduction

Abstract: Orthogonal Frequency Division Multiplexing (OFDM) is just about the finest technology used not long ago within connection CPA networks. It offers a number of rewards which include excessive spectral effectiveness, improved funnel convenience of the actual given electrical power finances and the like. Notwithstanding these kind of rewards OFDM own particular negatives likewise, such as Inter-Block Disturbance (IBI), Inter-Carrier Disturbance (ICI) along with excessive Peak-to-Average Electrical power Rate (PAPR). Nevertheless the most wrecking downside between these kind of is actually excessive PAPR because it interferes with the actual orthogonality of the process, in so doing attempts are built to reduce the PAPR. Different strategies are suitable for this particular between that a couple of the very recent along with successful strategies usually are talked about in this document.

Keywords: Orthogonal Rate of recurrence Scale Multiplexing(OFDM), Peak-to-Average Electrical power Rate (PAPR), Tranquility Research Algorithm (HSA), Threshold-based Piecewise Companding, Incomplete Transmit Routine.

1 Introduction

Orthogonal Frequency Division Multiplexing (OFDM), the name itself shows that it contains orthogonal sub-carriers, i.e. the sub-carriers are orthogonal to each other, and thereby there will be no issue of interference between the two signals. There will be overlapping of signals which helps us to provide efficient spectrum utilization. Then too, due to high PAPR the orthogonality of carriers get disturbed which dilutes the formation of the system and hence PAPR reduction is very essential requirement of OFDM systems [3].

1 PG Student (Communication Networks), Department of E&TC, Sinhgad College of Engineering, Pune (India)
dsveenashru@yahoo.com
2 Associate Professor, Department of E&TC, Sinhgad College of Engineering, Pune (India)
achala.deshmukh@gmail.com

Various techniques have been developed till now for the same. There are basically two types of PAPR reduction techniques namely: distorted and un-distorted techniques. The distorted techniques have lossless data rate but have complexity in computation while un-distorted techniques have loss in data rates but have easy computation. The distorted techniques for PAPR reduction techniques include: companding techniques while un-distorted one includes: Partial Transmit Sequence (PTS), Selective Mapping (SLM), Harmony Search Algorithm (HSA), etc. This paper will focus on un-distorted technique.

Among the un-distorted techniques PTS was mostly used, but its computational complexity was more due to continuous phase rotations required. There was a need to find a better approach than PTS, which resulted in the development of Harmony Search Algorithm technique. This technique is based on the musicians approach to select the best harmony among the generated harmonies. This technique provides better tradeoff between PAPR reduction and BER performance maintenance.

This paper thus focuses on the Harmony Search Algorithm for PAPR reduction.

2 Orthogonal Frequency Division Multiplexing

It is one of the block-based transmission systems. The technique is basically discrete Fourier based transform scheme using the multicarrier modulation scheme. The frequency selective fading channels are converted to number of flat fading channels. The advantage of this is the elimination of Inter-Symbol Interference (ISI).It also has high spectral efficiency due to overlapped carriers. OFDM converts high data rate streams into low data rate streams and uses QPSK or QAM modulation scheme. Sampling is done using the IDFT block. The signal processing algorithms are used as well so as to replace the modulators and demodulators used.

One more important parameter required is synchronization. Both frequency and phase synchronization is essential in OFDM in order to determine the OFDM signal and to adjust the required modulator-demodulator frequency. The block diagram of basic OFDM system is shown below:

The OFDM signal is given by

$$X_n(t) = \frac{1}{\sqrt{N}} \sum_{n=0}^{N-1} X_n e^{j2\Pi f n t} \qquad 0 \leq t \leq NT \qquad (1)$$

When N data blocks of OFDM signals are transmitted in parallel so as to modulate each into different subcarriers, which are orthogonal within the symbol period T.

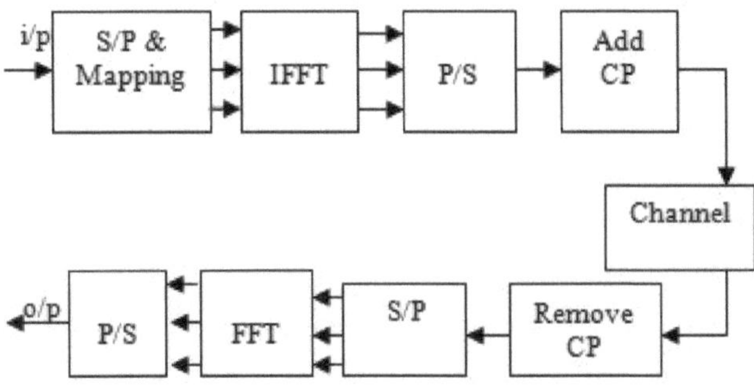

Fig. 1: Block diagram of OFDM [10]

3 Peak To Average Power Ratio(papr)

The PAPR of a OFDM signal represents the ratio of the peak power to average power of the signal. It can be seen from equation (2) that PAPR can be reduced by reducing peak power and increasing average power [1]. But, increasing the average power to reduce PAPR is not preferable that's why threshold based piece-wise companding technique is used which is further discussed in this paper. The PAPR of the OFDM signal is given as

$$PAPR = \frac{\max_{0 \le t \le NT} |Xn(t)|^2}{E[|Xn(t)|^2]} \tag{2}$$

4 Orthogonality

Two periodic signals are orthogonal when the integral of their product over one period is equal to zero. For the case of continuous time:

$$\int_0^T \cos(2\pi nft)\cos(2\pi mft)dt = 0 \tag{3}$$

For the case of discrete time:

$$\sum_{K=0}^{N-1} \cos\left(\frac{2\pi Kn}{N}\right)\cos\left(\frac{2\pi Km}{N}\right) \tag{4}$$

5 Sub-Carriers

Each sub – carrier in an OFDM system is a sinusoid with a frequency that is an integer multiple of a fundamental frequency. Each sub – carrier is like a Fourier series component of the composite signal, an OFDM symbol. The sub – carriers waveform can be expressed as

$$s(t) = \cos(2\pi f_c t + \theta_k)$$
$$= a_n \cos(2\pi n f_0 t) + b_n \sin(2\pi n f_0 t) \qquad (5)$$
$$= \sqrt{(a_n^2 + b_n^2)} \cos(2\pi n f_0 t + \varphi_n)$$

where $\varphi_n = \tan^{-1}\left(\dfrac{b_n}{a_n}\right)$

The sum of the sub – carriers is then the baseband OFDM signal

$$s_B(t) = \sum_{n=0}^{N-1}\{a_n \cos(2\pi n f_0 t) - b_n \sin(2\pi n f_0 t)\} \qquad (6)$$

6 Inter-Symbol Interference

Inter – symbol interference (ISI) is a form of distortion of a signal in which one symbol interferes with subsequent symbols. This is an unwanted phenomenon as the previous symbols have similar effect as noise, thus making the communication less reliable. In OFDM system ISI is eliminated because of the cyclic prefix, because due to cyclic prefix the symbol duration is increased which in turn reduces delay spread.

7 Inter-carrier interfernce

Presence of Doppler shifts and frequency and phase offsets in an OFDM system causes loss in orthogonality of the sub – carriers. As a result, interference is observed between sub – carriers. This phenomenon is known as inter – carrier interference (ICI). In OFDM system ICI is eliminated because OFDM leads to conversion of channel into many narrowly spread orthogonal subcarriers resulting in an immune frequency selective fading system.

8 Cyclic Prefix

The Cyclic Prefix or Guard Interval is a periodic extension of the last part of an OFDM symbol that is added to the front of the symbol in the transmitter, and is removed at the receiver before demodulation. The cyclic prefix has to two important benefits –

The cyclic prefix acts as a guard interval. It eliminates the inter – symbol interference from the previous symbol.

It acts as a repetition of the end of the symbol thus allowing the linear convolution of a frequency – selective multipath channel to be modeled as circular convolution which in turn maybe transformed to the frequency domain using a discrete Fourier transform.

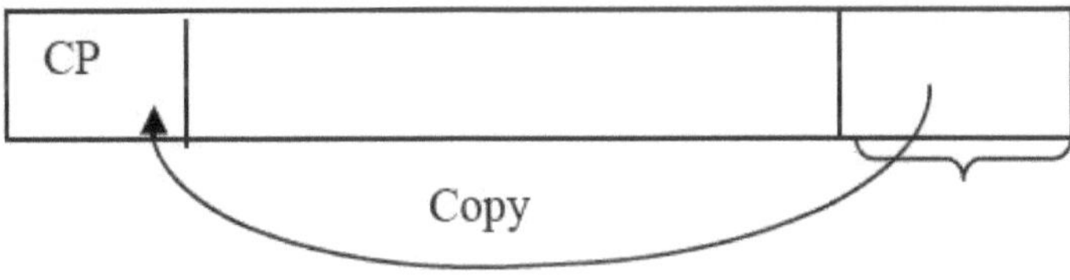

Fig. 2: Cyclic Prefix [11]

9 Harmony search algorithm (hsa)

The Harmony search algorithm works on the principle of choosing best harmony by the musician to have a pleasant harmony as the final output. The block diagram required is as shown in figure 5, which is same as that of Partial Transmit Sequence (PTS) scheme but the difference is in the process followed. The PTS technique used for PAPR reduction is complicated one due to rotation of phase factors; so as to overcome this complex method Harmony Search Algorithm is developed. An advantage of HSA is the use of stochastic random search based on harmony memory considering and pitch adjustment rates, instead of the need for an initial value as well as few mathematical requirements.[6] Following are the steps required to be followed in HSA:

Step 1 : Parameters Initialization :

Define objective function f(b) as

$$f(b) = \frac{max|x(b)|^2}{E[|x(b)|^2]} \qquad (7)$$

where, the value of unknown term is given by
$$x(b) = \sum_{i=0}^{M} b_i x_i \tag{8}$$

and b = phase factor
Step 2 : Define parameters
1. Number of phase factors
2. Pitch range
3. Harmony Memory Size
4. Harmony Memory Consideration Rate
5. Pitch Adjustment Rate
6. Distance bandwidth
7. Stopping Criterion

Step 3 : Harmony Memory Matrix is defined as :
      Where N : number of subcarriers
        L : oversampled factor

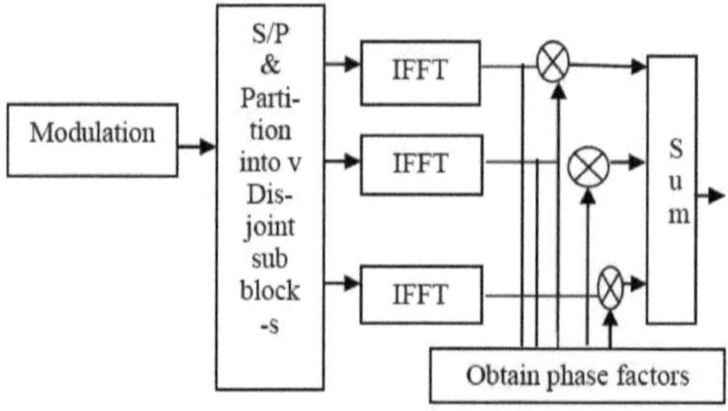

Fig. 3: Block diagram of PTS technique used for Harmony Search Algorithm.[6]

The HMS is given as

$$HM = \begin{bmatrix} b_1' & b_2' \ldots & b_{NL}' \\ \vdots & \vdots & \vdots \\ b_{NL}' & \ldots & b_{NL}^{HMS} \end{bmatrix} \tag{9}$$

Step 4: New Harmony improvisation
Here, new harmony vector is defined as

$$b' = (b'_1 b'_2 \dots b'_{NL}) \tag{10}$$

This value is compared with that in the HM and if this value is better than that in HM the value in HM is replaced by this value and then corresponding objective function is found out.

Step 5: Go to step 3 until stopping criteria is not achieved.

10 Results

Some simulations to demonstrate the performance of the HS-PTS algorithm. It is assumed that 10 random QAM modulated OFDM symbols with N=256 were generated with subcarriers, M=8 sub blocks and the phase factors.

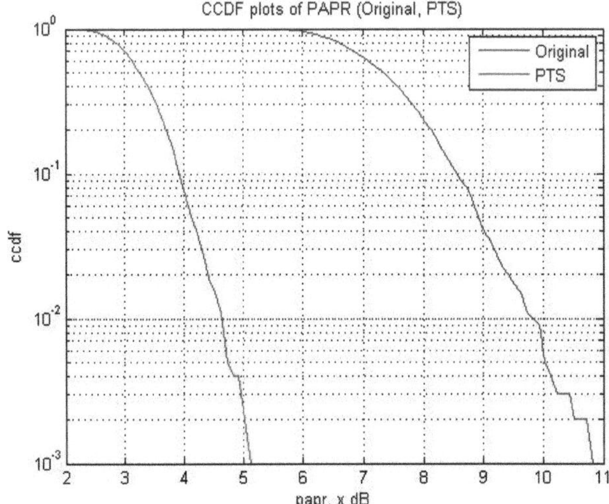

Fig. 4: Comparison between PAPR for original and PTS signal

In HS-PTS algorithm, the parameters are specified: number of phase factors M=8; pitch range of each phase factor = {+1,-1}; HMS = 10; HMCR = 0.95; PAR = 0.05;and maximum iterative number (stopping criterion) K = 10.

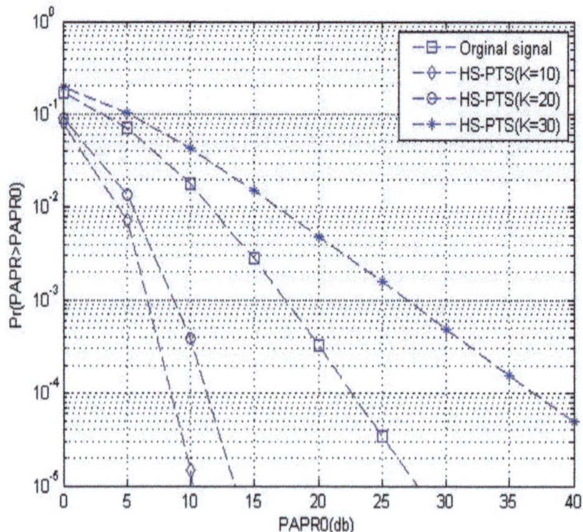

Fig. 5: Comparison between PAPR for different iterations

Thus, harmony search method is better method to reduce the PAPR value of the given OFDM signal.

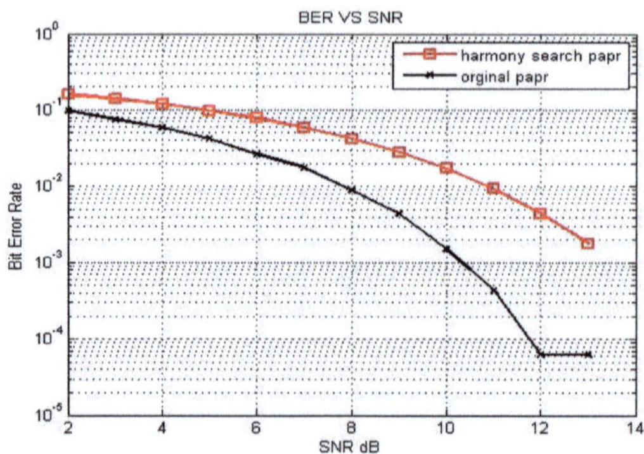

Fig. 6: BER performance of the system

The BER performance of the system is shown above, which clearly shows that harmony search algorithm provides better trade-off between PAPR reduction and BER performance of the system.

Conclusion

In this project, harmony search algorithm is used to reduce the PAPR of the signal. This technique is similar to that used by a musician to find the best possible harmony. From the result it is clear that as compare to threshold-based piecewise technique harmony search algorithm provides somewhat low PAPR reduction. But, too depending on the requirement either of the two is used.

Further work is going to improve PAPR reduction, which includes Novel Global Harmony Search Algorithm. In this technique a genetic mutation probability is considered, which modifies the harmony search improvisation step of the HS algorithm such that the new harmony mimics the global best harmony in the HM matrix. In fact it works instead of the HMCR and PAR in the pseudo code of harmony search algorithm.

References

1   Seung H.H, Jae H.L. An Overview of Peak-to-Average Power Ratio Reduction Techniques for Multicarrier Transmission,IEEE Wireless Communications, vol.12(2), pp. 56-65, 2005.

2   Hojjat Salehinejad and SiamakTalebi,Hindawi Publishing Corporation International Journal   of Digital Multimedia Broadcasting Volume 2012, Article ID 940849, 7 pages doi:10.1155/2012/940849.

3   S. H. M¨uller and J. B. Huber, "OFDM with reduced peak-to- average power ratio by   optimum combination of partial transmit sequences," Electronics Letters, vol. 33, no. 5, pp. 368–369, 1997.

4   M. G. H. Omran and M. Mahdavi, "Global-best harmony search," Applied Mathematics and Computation, vol. 198, no.2, pp. 643–656, 2008.

5   Mohsin Khan, SamimaIqbal and WaseemAsghar, International journal of Mobile Network Communications & Telematics (IJMNCT) Vol. 4, No.1, February 2014.

6   EmadMeimandKermani, SedighehAflaki, "PAPR Reduction of OFDM Signals: A Global Harmony Search Approach".

7   JingGao, JinkuanWanga , Bin Wang , Xin Song, "A PAPR Reduction Algorithm Based on Harmony Research for OFDM Systems", Procedia Engineering 15 (2011) 2665 – 2669.

8   Geem Z.W, Kim J.H, Loganathan G.V. A new heuristic optimization algorithm: harmony search. Simulation, 2001,76 (2):60–68.

9   Ramjee Prasad,"OFDM For Wireless Communication Systems"

10 T. S. Rappaport, "Wireless Communication: Principles and Practice": 2nd Edition, Prentice   Hall, 2002.

Allam Venkata Surendra Babu [1], S.Pradeepa [2] and
Siva Subba Rao Patange [3]

Closed Loop Control System based Hybrid Energy System Model

Abstract: Over the past few years there has been enormous increase in the consumption of the Electrical energy, major part of which is extracted from the conventional way of power generation using fossil fuels, due to which the resource of the fossil fuels are draining and results of the some surveys shows that fossil fuel reserves may last only more for few decades. Hence utilization of the energy generated from the renewable energy resources like Solar, Wind, Bio-Mass, etc. has to be encouraged to save the atmosphere and as an alternative fuel for the energy generation. Among the renewable energy resources Solar and Wind power generation have been increasing significantly due to the availability of the resources, ease of the generation, etc.

In this paper a closed loop control strategy has been proposed using three control circuits, among which the first control circuit is used in Solar energy system to track the operating point at which maximum power can be extracted from the Solar P-V system by tracing the power and voltage of the Array. The second control circuit is used in wind Energy System to track the maximum power point of Wind turbine generator using the wind speed and wind power under variable environmental conditions. The third control circuit is used in charge controller system to control the constant voltage charging of the battery. The detailed description of the entire closed loop control system for hybrid system has been given, along with the comprehensive simulation results. A software model is developed in Matlab/ Simulink.

Keywords: Closed loop control, P-V Array, wind turbine, Hybrid Energy System, Control Strategy.

1 Power Electronics (EEE), BMS College of Engineering, Bangalore, India
surendrallam@hotmail.com
2 Associate Professor, Power Electronics (EEE), BMS College of Engineering, Bangalore, India
Spradeepa.prasad@gmail.com
3 Principal scientist, Structural Technologies & Division, NAL, Bangalore-17
siva@nal.res.in

1 Introduction

In the recent years after the evolution of the energy extraction from the renewable energy resources like Solar, Wind, Bio-mass etc. the world started motivating towards the utilization of the renewable energy, as the energy obtained is clean, harmless to the environment etc. Solar energy is obtained due to the irradiation of the sunlight, solar cell is a p-n junction diode in which when sunlight falls on the cell the photons in it flows from the n-junction to the p-junction due to which the electric current flows, hence the solar energy is created.

Wind is also a form of solar energy which is caused due to the uneven heating caused in the atmosphere, which flows from the higher altitude to the lower altitudes. And the wind flow pattern depends on many parameters like season, water bodies, earth terrain, vegetation and geography. And wind turbine converts the kinetic energy of the wind to mechanical energy which is connected to separately excited DC Generator which converts the mechanical energy to electrical energy.

As transportation is also a major part in the utilization of the fossil fuel hence to reduce the consumption the fossil fuel usage, as a part of contribution here a hybrid energy system had been modelled. Were the project is all about a closed loop control system based hybrid energy system modelling integrating both solar power and the wind power for hybrid vehicles operated using battery.

The paper is organized as follows In chapter 1 the introduction of the project had been discussed, along brief explanation of hybrid energy system, motivation and the objectives of project is also deliberated, chapter 2 is about the literature survey which explains about the work done in the earlier stages, chapter 3 deliberates the design of solar P-V module and the design of wind turbine, chapter 4 explains how the hybrid energy system is modelled in the Matlab/Simulink, chapter 5 is about the testing and the simulation of the hybrid energy model, and the references are deliberated at last.

2 Design of the Energy System

2.1 Solar P-V Module [a]

P-V array is formed by connecting modules in series and parallel, the diode equivalent of the P-V module is shown below in Figure 1. And the solar P-V module is modeled below.

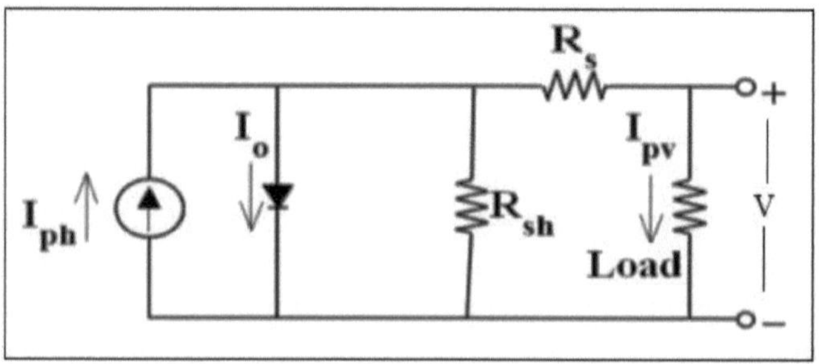

Figure. 1: Diode equivalent of P-V Module

Saturation current of the module:

$$I_o = I_{rs} \left[\frac{T}{T_r} \right]^3 \exp\left[\frac{q^* E_{go}}{Bk} \left\{ \frac{1}{T_r} - \frac{1}{T} \right\} \right]$$

(1)

The current output of P-V module:

$$I_{pv} = N_p {}^* I_{ph} - N_p {}^* I_o \left[\exp\left\{ \frac{q^*(V_{pv} + I_{pv} R_s)}{N_s kAT} \right\} - 1 \right]$$

(2)

The Voltage Output of the P-V module:

$$V_{pv} = \frac{AkT_r}{q} \ln\left[\frac{I_{ph} + I_o - I_{pv}}{I_o} \right] - R_s I_{pv}$$

(3)

This equations are used to simulate in Matlab/Simulink and the result shows the nonlinear characteristics of photovoltaic array at different irradiations and temperature [a].

Table 1. Solar Module Specification

Rating	305.5W
Current at Peak	5.58A
Voltage at Peak	54.7V
Short circuit current	5.96 A
Open circuit voltage	64.2 V

2.2 Modelling of Wind turbine [a]

Wind turbine converts the Kinetic energy of the wind to mechanical energy, this conversion depends on many parameters like velocity of the wind, pitch angle of the blade, tip speed ratio, performance coefficient of the turbine swept area of the turbine, and air density etc. Output of the wind turbine is the wind power

$$P_0 = \frac{1}{2}(\rho A V_w{}^3)$$

(4)

The above equation clearly states the power available in the wind, but the mechanical power obtained from the turbine is different from each turbine as it depends on the performance of the turbine. Hence the aerodynamic power generated from the wind is given by

$$P_m = \frac{1}{2}(\rho A C_p V_w{}^3)$$

(5)

ρ = air density (kg/m^2); A= swept area of turbine; C_p = performance coefficient; V_w = wind velocity (m/s).

$$C_p(\lambda,\beta) = C_1\left(\frac{C_2}{\lambda_i} - C_3\beta - C_4\right)exp^{\frac{C_5}{\lambda_i}} + C_6\lambda$$

(6)

$$\frac{1}{\lambda_i} = \left(\frac{1}{\lambda + 0.08\beta} - \frac{0.035}{\beta^3 + 1} \right)$$

(7)

Where $C_1=0.5176$, $C_2=116$, $C_3=0.4$, $C_4=10$, $C_5=21$, $C_6=0.0068$.

Power coefficient (C_p) is the function of tip speed ratio (λ) and pitch angle (β). The characteristics curve shows the detailed behavior of mechanical power extracted from the wind at different wind speed [b]. And the power coefficient reaches maximum of 0.48 for maximum tip speed ratio of 0.81. The mechanical power obtained from the turbine has to be converted to torque, the mechanical torque is the ratio of the power obtained from the turbine to [a]

$$T_m = \frac{1}{2}(\rho A C_M R V_w^2)$$

(8)

$$C_M = \frac{C_p}{\lambda}$$

(9)

The Eqn. 9 defines the amount of torque delivered by the turbine to the connected generator shaft, it clearly states that torque is directly proportional to the square of the wind speed.

The wind turbine specifications are tabulated in table 2.

Table 2. Wind Turbine Specifications

Rating	1.8KW
Cut-in Speed of wind	3m/s
Cut-out Speed of wind	25m/s
Rated wind Speed	12m/s
Air-Density	1.255kg/m^3

3 Design of Closed Loop Control System using Matlab/Simulink

The closed loop model is developed using Matlab/Simulink; the main block diagram of the Hybrid Energy System is shown in Figure 2. It contains solar power, Wind power, Energy monitoring and interface block and Battery storage device. The brief explanation of the each block is given below.

Hybrid energy systems, namely Solar and Wind power blocks generates power depending on the conditions on the input side. The details about these power blocks are explained in the Figure 3 and Figure 4.The output from these blocks are controlled power and is fed to the battery charge controller for monitoring the input energy [c]. In turn, depending on the requirement of voltage condition of the battery, the third controller, controls the battery voltage within the limits. The storage device shown in the Figure 2 is the battery.

The Figure 3 shows the internal circuitry of solar energy system. In which solar energy is extracted from the solar panel and is given to boost converter for DC power regulation. For the extraction of the maximum power from the solar module an Incremental Conductance Maximum Power Point Tracking (MPPT) Algorithm had been designed by tracing the maximum point of power in the solar module P-V characteristics [a]. The output power depends on two parameters of the solar module, [c] Irradiation and temperature of the module. The changes occurred in the solar module creates the change in output power from the solar module, which is given to the MPPT algorithm technique which monitors the changes and generates the pulses for triggering the gate of IGBT for regulating the output power to be in defined limit. This changes are made as per the duty cycle of the gate drive of IGBT [c]. Triggering pulses are generated using the pulse width modulation according to the changes in the duty cycle of gate drive of IGBT.

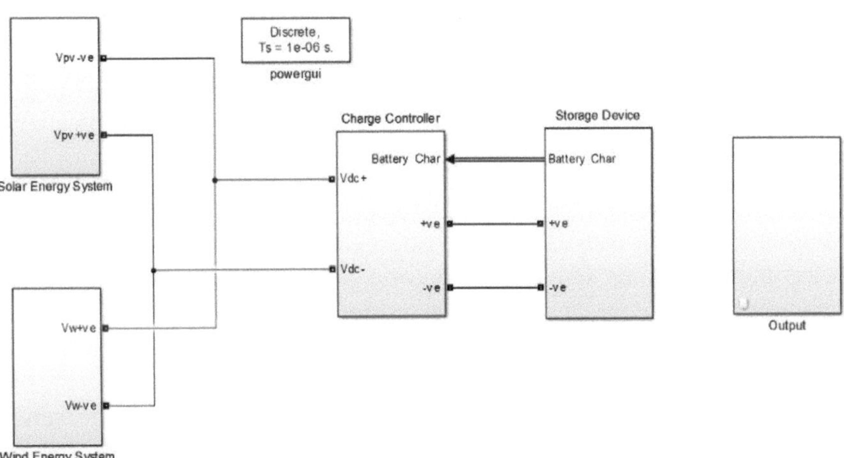

Figure. 2: Complete simulation of the Hybrid energy system [a]

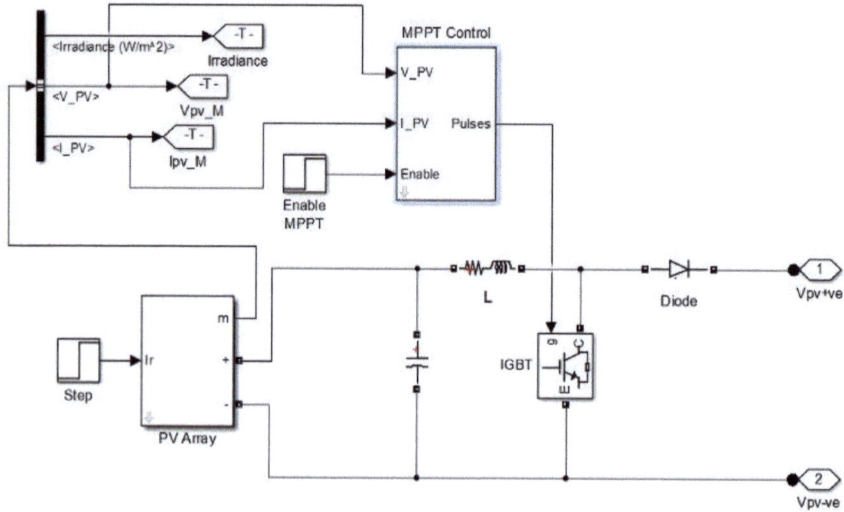

Figure. 3: Design of Solar energy system with Boost controller [a]

Figure.3 shows the design of wind energy system in which is been designed using wind turbine connected to the separately excited DC machine, and the output of the DC machine to be given to the Buck-Boost converter for the regulation. When their occurs the changes in the wind speed or in tip angle the torque obtained from the wind turbine changes which is connected to the shaft of the separately excited DC machine which generates the electricity from the input load torque.

Figure 4. is shows battery charge controller in which the power obtained from the two sources is been monitered and will take energy from any one source or from the both as per the reuirement of battery charging. Here the controller is designed such the battery starts charging when the battery volatge is less than the certain limit and it will avoid the battery usage when the battery voltage is below certain limit. Hence the controller is been designed [e].

Figure.5 shows the battery storage device in which the energy from the source chosen by battery charge controller is been connected to the battery to get it charged. The battery starts getting charged when the voltage of the battery is less than the 95% of its full charge and it will not allow the battery to get drained when the battery voltage goes below 15% of full charge.

Modeling of the Hybrid energy system is done using Matlab/Simulink and compiled, the results are obtained which is explained in the section VI.

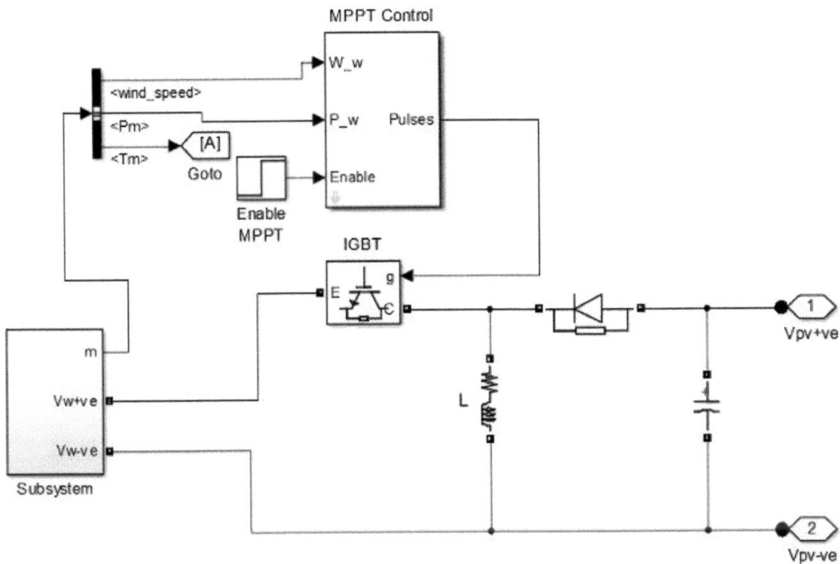

Figure. 4: Design of wind energy system with Buck-Boost controller [a]

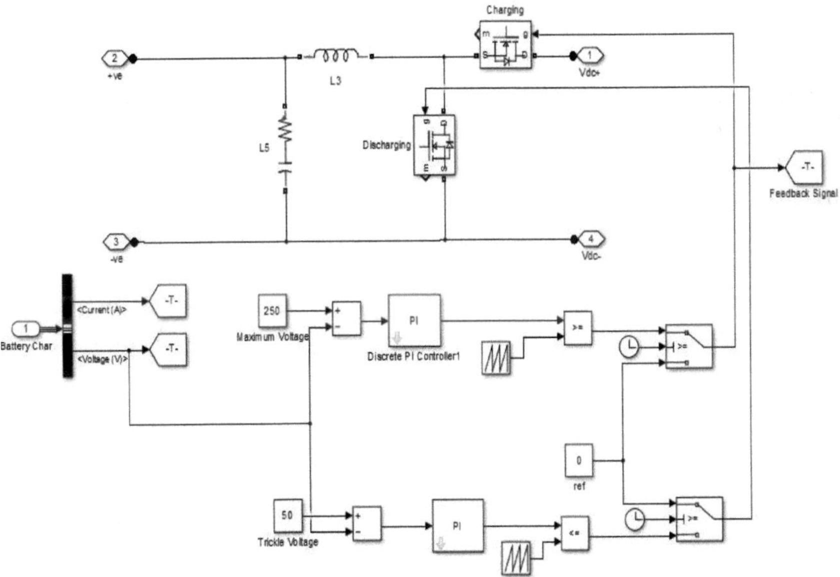

Figure. 5: Design of closed loop charge controller system

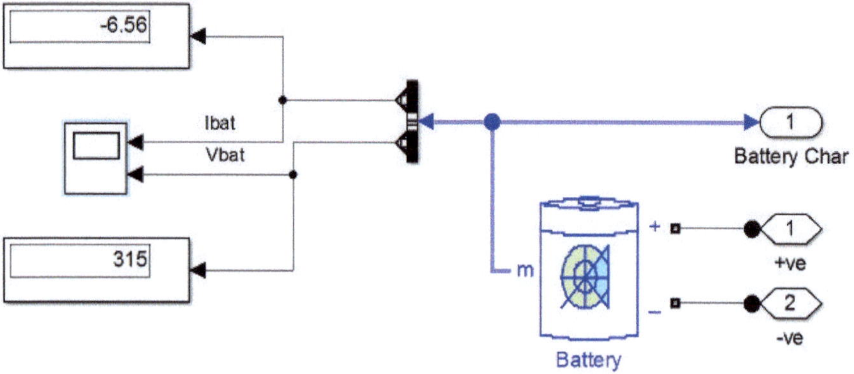

Figure. 6: Battery storage device [a]

4 Simulation Results

This chapter gives the brief explanation about the results of Hybrid energy system which is simulated using Matlab/Simulink with considering following parameters.

- Step time:- 0.025 sec,
- Sampling time:- $1*e^\wedge(-6)$,

The output waveforms obtained from the solar energy system is as shown in the Figure 7. The disturbance is given in the form of change in irradiance i.e., initially irradiance is 1000 W/m^2 and it's been increased to 1500 KW/m^2 adding 500 KW/m^2 after a delay of 0.01 Sec. With the given input the output is obtained as shown in Figure 7. From the Figure it can be observed that voltage obtained is 315+/-5% V tolerance reaches to its final value of at 0.005 Sec, current obtained reaches to final value of 5.87 A, and the power (P=V*I) output is 1846 W.

To demonstrate the wind energy system is triggered after a delay of 0.01 Sec. The output waveforms obtained from the wind energy system is as shown in the Figure 8. In Wind energy system the input disturbance is created with varying the wind speed i.e., initially Hybnd speed is 10 m/s and it's been increased to 15 m/s adding 5 m/s after a delay of 0.015 Sec. With the given input the output is obtained as shown in Figure 8. in which voltage obtained reaches to its final value of 315+/-5% V tolerance at 0.015 Sec, current obtained reaches to final value of 6.2 A, and the power (P=V*I) output is 1865 W.

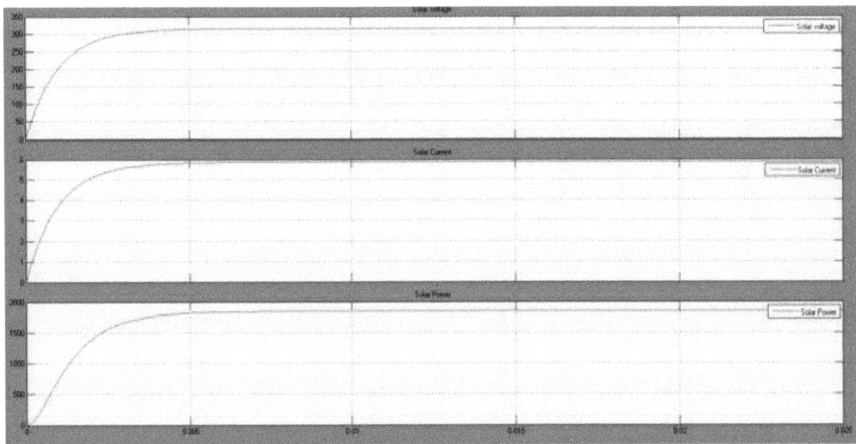

Figure. 7: Voltage, Current and Power output waveforms of Solar Energy System

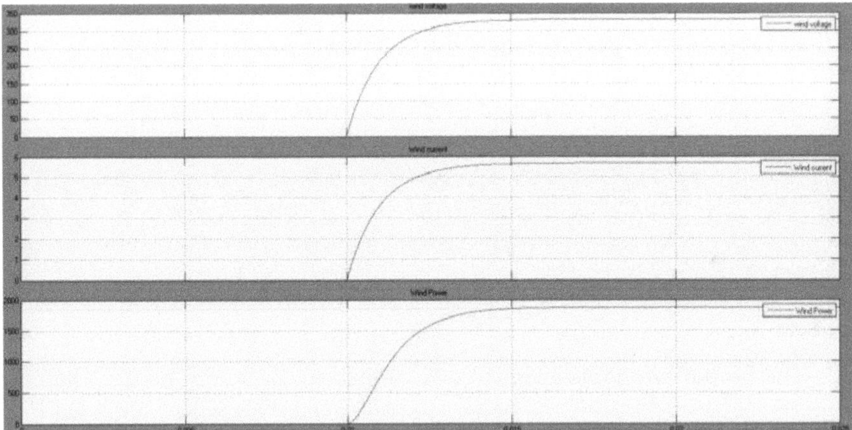

Figure. 8: Voltage, Current and Power output waveforms of Wind energy system

The output obtained from both energy systems are connected to charge controller whose function is to controller the input to the battery. The waveforms obtained from the charge controller is shown in the Figure 9. From the Figure it can be observed that for the time interval 0 to 0.01 Sec only the solar energy system is producing energy and from the time interval 0.015 to 0.02 Sec the Wind energy system is added to existing energy and then from the interval 0.02 to 0.025 Sec only Wind energy system output is taken to charge the battery.

When the input exceeds 315 V charge controller controls the input of battery by regulating the input voltage level to 315 V, hence when the input increased to 600V in time interval 0.01 to 0.02 Sec the charge controlled restricted the battery charging to 315 V only. The waveforms obtained from the input and output is as shown in Figure.5.3.

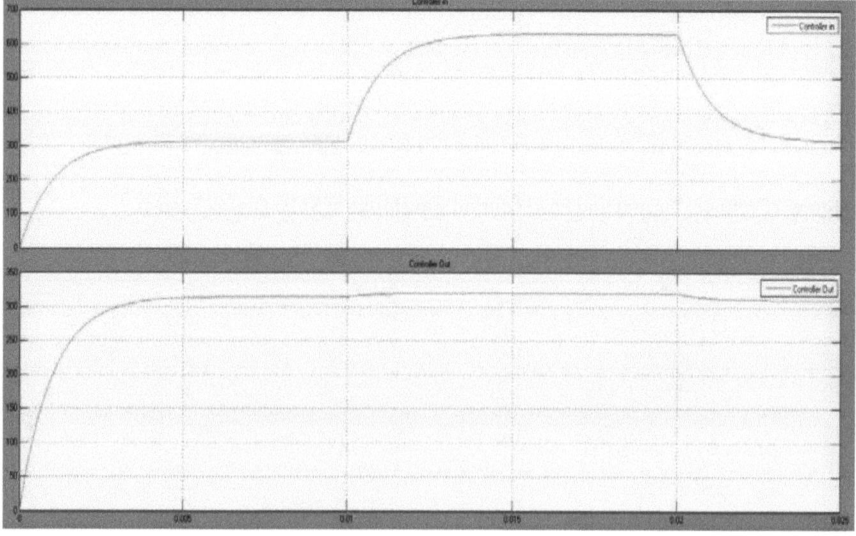

Figure. 9: Input and Output waveforms of Charge Controller

The output of charge controller is connected to battery hence when the input voltage is 315+/-5% the battery gets charged. Waveforms obtained from the battery storage is shown in Figure 10. From this figure it can observe that charging current of the battery is -6.56 A, and voltage is 315 +/- 5%. That is when the Wind energy is added the battery charging raised to 320 from 315 V, which can be regulate to 315 V as future scope of the project.

A.V.Surendra, S.Pradeepa, Siva Subba Rao gratefully acknowledge TEQIP-II part of AICTE, New Delhi, for its support & financial assistance in successful completion of the project, also would like to thank Dr. Mallikarjuna Babu, Principal, BMSCE, Bangalore and NAL/CSIR, Bangalore for its support towards the completion of the project.

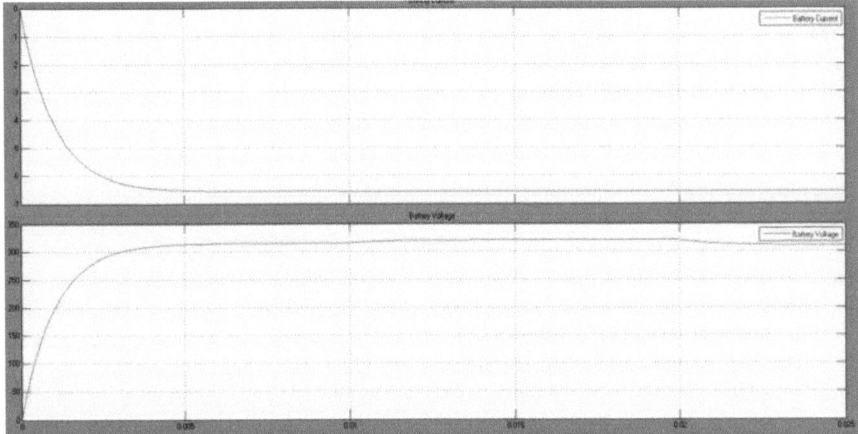

Figure. 10: Voltage and Current waveforms of battery

Conclusion

Modelling of the hybrid energy system is been done using matlab/simulink, and the results of solar energy system, wind energy system and the battery characteristics are obtained which are shown in earlier section. as a scope for improvement of the project in future we can design the hybrid energy system adding peizo electric energy and controlling the charge of the battery to 2% of tolerance

Major attributes of the project

- Solar Energy System and Wind Energy System is implemented in Simulink,
- Hybrid Energy System is designed using solar energy and Wind Energy System for battery operated vehicles,

Acknowledgement

A.V.Surendra, S.Pradeepa, Siva Subba Rao gratefully acknowledge TEQIP-II part of AICTE, New Delhi, for its support & financial assistance in successful completion of the project, also would like to thank Dr. Mallikarjuna Babu, Principal, BMSCE, Bangalore and NAL/CSIR, Bangalore for its support towards the completion of the project.

Reerences

1   A.V.Surendra, S.Pradeepa, Siva Subba Rao, "Hybrid Energy System Modelling for Battery Operated Vehicles", published in International Journal of Advance Research and Innovation, Volume 3 Issue 2 (2015) 446-449.
2   N. Pandaniarjan, Ranaganth Muth, wrote a paper titled "Mathematical Modeling of Photovoltaic Module with Simulink," which had published in an International Conference on Electrical Energy Systems, on January 3-5, (2011),
3   Daniel W. Hart, "Power Electronics", McGraw Hill Education (India) Private Limited, «2011 Edition», (2013),
4   Ned Mohan, Tore M. Undeland, William P. Robbins, "Power Electronics, Converters, Applications, and Design", John Eiley and Sons INC, «3rd Edition», (2013),
5   S.Chowdary, S.P.Chowdary, G.A.Taylor, Y.H.Song, wrote a paper titled, "Mathematical Modeling of a Stand-Alone Polycrystalline P-V Plant with MPPT Facility", published in IEEE conference on Advancements in solar energy (2014).

A. Sri Krishna[1], K. Gangadhar[2], N. Neelima[3] and
K. Ratna Sahithi[4]

Topology Preserving Skeletonization Techniques for Grayscale Images

Abstract: Thinning is an interesting and challenging problem, and plays a central role in reducing the amount of information to be processed during pattern recognition, image analysis and visualization, computer-aided diagnosis. A topology preserving thinning algorithm is used which removes the pixels from a gray scale images. First it checks for the condition whether a pixel is acyclic or not. If a pixel is acyclic, then it is removed from the image else retains in the image. A topology preserving skeleton is a synthetic representation of an object that retains its shape, topology, geometry, connectivity and many of its significant morphological properties. It has been observed in our experiments that thinning algorithm is stable and robust and yield promising performance for wide range of images.

Keywords: Skeletonization, Topology preservation, Thinning.

1 Introduction

Thinning is a widely used pre-processing step in digital image processing and pattern recognition. It is iterative layer by layer peeling, until only the "skeletons" of the objects are left. Thinning algorithms are generally constructed in the following way: first the thinning strategy and the deletion rules are figured out, then the topological correctness is proved[3]. In the case of the proposed algorithms, first consider some sufficient conditions for parallel reduction operators to preserve topology[26], and then the deletion rules were accommodated to them.

---

1 Dept. of IT, R.V.R& J.C College of Eng. Guntur, India
2 Acharya Nagarjuna University
3 Dept. of IT, R.V.R& J.C College of Eng. Guntur, India
4 Dept. of IT, R.V.R& J.C College of Eng. Guntur, India

Thinning is the process of reducing an object in a digital image to the minimum size necessary for machine recognition of that object[28]. After thinning, analysis on the reduced size image can be performed as shown in Fig.1. Thinning is essentially a "pre-processing" step used in many image analysis techniques. The thinning process reduces the width of pattern to just a single pixel. Thinning when applied to a binary image, produces another binary image[1].

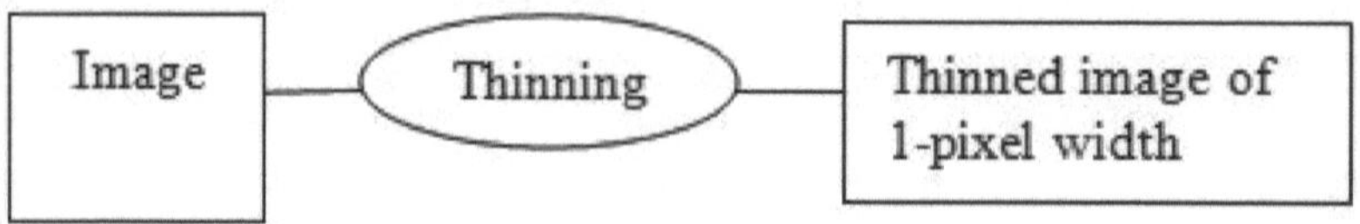

Fig. 1: Thinning block diagram

Thinning has been used in a wide variety of applications including: pattern recognition, medical imaging analysis, bubble-chamber image analysis (a device for viewing microscopic particles), text and handwriting recognition and analysis[6]. A thinning algorithm is said to be good if it posses the following set of desirable features: maintaining connectivity of resulting skeletons; producing skeletons of unit width; insensitive to noise; time efficient and preserving topology[26].

The paper is organized as follows. In Section II, prior work is discussed. Topology Preserving Skeletonization algorithm is presented in Section III. Section IV consists of results and discussions. Conclusion is given in Section V.

2 Prior Work

Skeletonization is a process for reducing foreground regions in a binary image to a skeletal remnant that largely preserves the extent and connectivity of the original region while throwing away most of the original foreground pixels. It (i.e., skeleton extraction from a digital binary picture) provides region-based shape features[21]. It is a common preprocessing operation in pattern recognition or in raster-to-vector conversion.

The three skeletonization techniques are :

- Ridges are detected in distance map of the boundary points.

- Calculating the Voronoi diagram generated by the boundary points, and
- Thinning is nothing but layer by layer peeling.

In digital spaces, only an approximation to the "true skeleton" can be extracted. The two requirements that are to be considered is shown below:

- Topological (to maintain the topology of the original object),
- Geometrical (forcing the "skeleton" being in the middle of the object and invariance under the most important geometrical transformation including translation, rotation, and scaling).

2.1 Distance transformation

Skeletonization using distance transformation consists of three steps:

- The original (binary) image is converted into feature and non-feature elements. The boundary of the object contains feature elements.
- Generating the distance map where each element gives the distance to the nearest feature element.
- The ridges are detected as skeletal points.

The idea behind our ridge detection algorithm is based on the well-known fact that the gradient at any point on a distance map generally points towards the ridge and reverses its direction as it crosses the ridge. In other words, for a point to be on a ridge, it must be a local maximum on some direction, i.e., on a line passing through the point bearing that direction. Consider a line with arbitrary orientation passing through a point of a distance map. If the point is a local maximum in the direction of the line, the distance values of the point's two opposite neighboring points must be less than that of the point, and the directions of the two opposite neighbors' gradient vectors projected onto the line must be opposite, pointing to the given point. In short, the given point generates a sign barrier between the two opposite neighbors on the line, if it is on a ridge of a distance map. To determine the minimum number of orientations, we need to understand when a particular orientation of the line succeeds or fails to detect certain ridges.

There are two ways that a ridge interacts with a projection line: a ridge either intersects or does not intersect the line. When a ridge intersects the projection line, it generates a sign barrier between the two points on the line enclosing the ridge. In other words, a ridge is guaranteed to be detected by the sign barrier on the projection line if it intersects the line. In contrast, if a ridge is nearly parallel to the projection line and does not cross it, the ridge does not produce a sign barrier on the line. Another projection line with an orientation substantially different from the orientation of the ridge (or, equivalently, from the that of the first projection line) will detect such a ridge, since the ridge parallel to the first projection line will appear perpendicular to the second line and cross it at some point, generating a sign barrier on it. For the projection lines to have sufficiently different orientations, two orthogonal lines would be the best choice.

After performing distance transformation, the distance map is generated as shown in Fig.2.

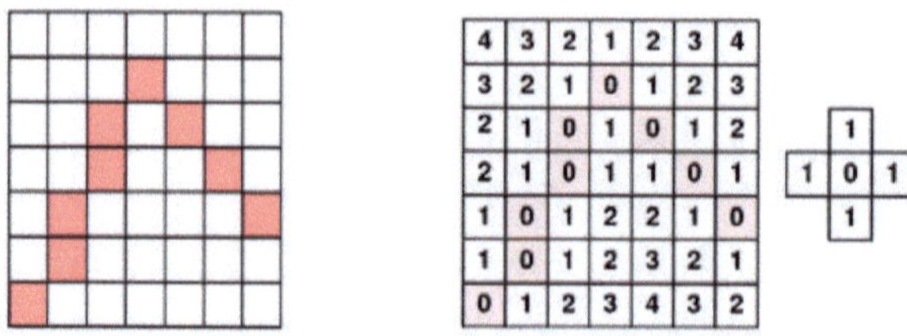

Fig. 2: Extracted feature points are marked by pink squares (left) and distance map using city block (or 4-neighbour) distance (right)

2.2 Voronoi diagram

The Voronoi diagram of a discrete set of points (called generating points) is the partition of the given space into cells so that each cell contains exactly one generating point and the locus of all points which are nearer to this generating point than to other generating points as shown in Fig. 3.

Using the Voronoi diagram to compute the skeleton of a polygonal shape is attractive because it results in skeletons which are connected while retaining Euclidean metrics Furthermore we obtain an exact medial axis compared to an approximation provided by other methods. Thus we may reconstruct exactly an original polygon from its skeleton invertibility one to one mapping. Finally

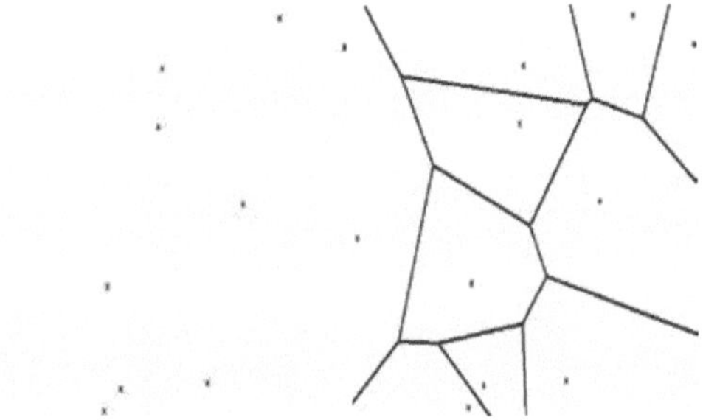

Fig.3: The 10 generating points (left) and their Voronoi diagram (right)

algorithms to compute the Voronoi diagram and hence the skeleton are much faster than approaches that compute a distance transform.

Any method that utilizes Voronoi diagrams of polygons to compute skeletons must overcome the disadvantages listed below before it can be of practical value

- Natural shapes are non-polygonal. Thus accurate polygonal approximations of such shapes are required in order to compute skeletons without loss of accuracy.
- The skeleton of a many sided polygon of very short sides will have a large number of redundant edges because of the Voronoi edges at these vertices. This results in an increase in the complexity of the skeleton without significant addition of any shape information.
- Finally robust and practical algorithms for Voronoi diagram construction of polygons are not very common Most existing algorithms make assumptions about cocircularity of no more than three points and colinearity of no more than two. These constraints are difficult to satisfy in most practical applications.

2.3 Basic Thinning Algorithms

Thinning is a frequently used method for making an approximation to the skeleton in a topology- preserving way[9]. All thinning algorithms can be classified as one of two broad categories:

1. Iterative thinning algorithms and 2. Non-iterative thinning algorithms.

In general, iterative thinning algorithms perform pixel by- pixel operations until a suitable skeleton is obtained. Iterative algorithms may be further classified as[27]:

1) Sequential Thinning Algorithms

2) Parallel Thinning Algorithms

The above algorithms are shown in Fig. 4. Non-iterative thinning methods use methods other than sequential pixel scan of the image[6].

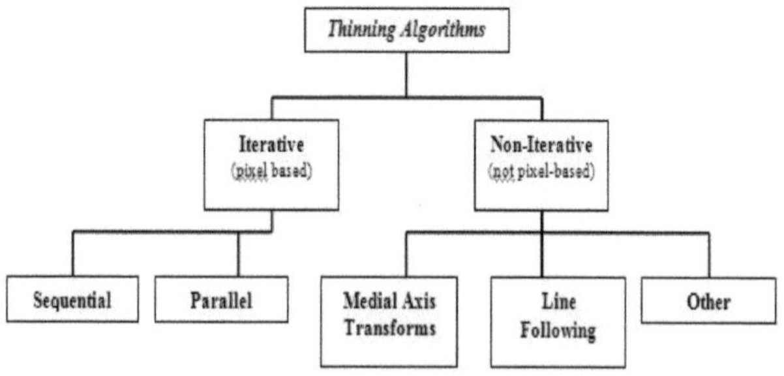

Fig. 4: Classification of Thinning Algorithms

The main difference between Sequential and Parallel thinning is that
Sequential thinning operates on one pixel at a time and the operation depends on preceding processed results where as parallel thinning operates on all the pixels simultaneously. Sequential thinning produces better skeletons and requires less memories but parallel thinning is substantially faster. Most of the thinning algorithms developed were parallel algorithms.

2.4 Sequential Thinning

Using sequential algorithms, contour points are examined for deletion in a predetermined order, and this can be accomplished either by raster scan or contour following. Contour following algorithms can visit every border pixel of a simply connected object[1], and of multiply connected picture, if all the borders of picture and holes are followed[2]. These algorithms have an advantage over raster scans because they require examination of only the contour pixels instead of all pixels in P and in every iteration[12].

When a contour pixel p is examined, it is usually deleted or retained according to the configuration of N(p). To prevent sequentially eliminating an entire branch in one iteration, a sequential algorithm usually marks (or flags) the pixels to be deleted, and all marked pixels are then removed at the end of iteration. This generally ensures that only one layer of pixels would be removed in each cycle[1].

To avoid repetition, it is assumed that a pixel p considered for deletion satisfies all the following properties unless otherwise stated:
1. p is a black pixel.
2. p is not an isolated or end point, i.e., $b(p) \geq 2$.
3. p is a contour pixel, i.e., p has at least one white 4- neighbor.

2.5 Parallel Thinning

In parallel thinning, pixels are examined for deletion based on the results of only the previous iteration. For this reason, these algorithms are suitable for implementation on parallel processors where the pixels satisfying a set of conditions can be removed simultaneously. Divide each iteration into sub-iterations or sub-cycles in which only a subset of contour pixels are considered for removal[5][8].

At the end of each sub-iteration, the remaining image is updated for the next sub-iteration. Four sub-cycles have been used in which each type of contour point (north, east, south, and west) is removed in each sub-cycle. These have also been combined into two sub-iterations with one sub-iteration deleting the north and east contour points and the other deleting the rest[24].

In fast Parallel Thinning Algorithm, the method for extracting the skeleton of a picture consists of removing all the contour points of the picture except those points that belong to the skeleton[16][18]. In order to preserve the connectivity of skeleton, each iteration is divided into two sub-iterations [17].

Table 1. Fast Parallel Algorithm for Thinning Digital Patterns

p_9 (i-1,j-1)	p_2 (i-1,j)	p_3 (i-1,j+1)
p_8 (i,j-1)	p_1 (i, j)	p_4 (i,j+1)
p_7 (i+1,j-1)	p_6 (i+1,j	p_5 (i+1,j+1)

According to Table 1, in the first sub iteration, the contour point p_1 is deleted from the digital pattern. If it satisfies following conditions[6]:

(a) $2 \leq B(p_1) \leq 6$

(b) $A(p_1) = 1$

(c) $p_2 \times p_4 \times p_6 = 0$

(d) $p_4 \times p_6 \times p_8 = 0$

In the second sub iteration, only condition (c) and (d)are changed and rest two conditions remain the same is shown below[6]:

(c') $p_2 \times p_4 \times p_8 = 0$

(d') $p_2 \times p_6 \times p_8 = 0$

where $A(p_1)$ is the number of 01 patterns in the ordered of $p_2,p_3,p_4,...p_8,p_9$ that are the eight neighbors of p_1 and $B(p_1)$ is the non-zero neighbors of p_1, which is shown below:

$$B(p_1) = p_2 + p_3 + \ldots\ldots\ldots\ldots\ldots + p_9.$$

If any condition is not satisfied then p_1 is not deleted from the picture. By condition (c) and (d) of the first sub-iteration it will be shown that the first sub iteration removes only the south-east boundary points and the north-west corner points which do not belong to an ideal skeleton[18].

	P2	
P8	P1	P4
	P6	

Fig. 5: Points under consideration and their locations

2.6 Non-Iterative Thinning

These algorithm are considered to be non pixel based; they produce a certain median or center line of the pattern directly in one pass without examining all the individual pixels.

Some algorithms obtain approximations of skeletons by connecting strokes having certain orientations. For example, four pairs of window operations are used in four sub cycles to test for and determine the presence of right, or left diagonal, vertical, horizontal, limbs in the pattern. At the same time, the operators also locate turning points and end points by a set of final point conditions, and these extracted points are connected to form a line segment approximation of the skeleton. In [1], the boundary pixels are first labeled according to the above four local orientations. For each boundary pixel, a search is made for the same kind of label on the opposite side of the boundary (within a maximum stroke width) in the direction perpendicular to that given by the label.

2.7 Extraction of Medial Axis Transform

The medial axis transform (MAT) of an image is computed by calculating the Euclidean distance transform of the given input image pattern. The MAT is described as being the locus of the local maxima on the distance transform. After the computation of Euclidean distance transform (EDT) of the input image, the EDT is represented in image representing the Euclidean distances as gray levels.

The maximum Euclidean distance is represented as maximum gray level intensity in the EDT image. The pixel coordinates of the maximum gray level intensity are extracted from the EDT image by converting the EDT image into row x column matrix. The row and column of the matrix gives the coordinates of the MAT line of the image pattern.

3 Topology Preserving Skeletonization

In this section, a simple topology preserving skeletonization technique is proposed that iteratively removes pixels of a grayscale image. A grayscale image is taken as input. The output is a set of points that belongs to its skeleton. For every pixel in the grayscale image, check the acyclicity, i.e., if a candidate pixel is 1 and its 8-neighbourhood pixels are also 1, then it is cyclic else it is acyclic. If it is acyclic, then the value of that pixel is set to 1 otherwise 0. Check for connectivity to retrieve thinned component of the original image.

3.1 Algorithm: Topology Preserving Skeletonization

Require : Binary image b_i
Ensure : $S(b_i)$ that belong to the Skeleton of b_i;
      1: Initially $S(b_i) = b_i$;
      2: For every element $T \in S(b_i)$ do

3: if(bd(T)) is acyclic , then $S(b_i)$ =1;
4: while $S(b_i)$ $\neq$ ø , do
5: For every element T $\in$ $S(b_i)$ do
6: Compute x = b_i – $S(b_i)$;
7: [l, num] = bwlabel (x, 8);
 8: initialize m=0;
 9: if b_i $\neq$ x , then
10: m=1;
11: if num $\neq$ 1 , then
12: b_i = x;
13: return $S(b_i)$);

4   Results and Discussions

The present method is applied on English alphabet set both on lower case and upper case as shown in Fig.3 and Fig.4. The English alphabets are chosen for experiment analysis because they contain different shapes. The thinned alphabet set contains no restoration and they are not affected by any border noises which are usually present in most of the skeleton approaches. One of the main problems with thinning method was loss of information due to binarisation because it could not always be possible to correctly binarise the whole character image using one threshold value.

Topology preserving skeletonization algorithm is proposed to evaluate its performance and examine its effects. The algorithms are chosen for representing different modes of operation in thinning [10]. The performance of the thinning algorithm can be compared and evaluated on the basis of following parameters:

a) Connectivity of Pattern.

b) Data Reduction Rate (DRR)

Connectivity of Pattern: By comparing the original image with its skeleton obtained by topology preserving skeletonization algorithm, we can see that the connectivity is maintained in the skeleton of the image.

Data Reduction Rate: The algorithm will guarantee the highest data reduction value producing perfect skeleton. It reveals information about how good is the algorithm in data reduction when comparing the skeleton S and the original pattern P. Formally: this can be measured as *Mdr*= |S|/|P|. Data reduction rate for different images is shown in Table2.

Table 2. Data Reduction Rate for different images

SNO.	IMAGE	% OF REDUCTION
1	A	19.35
2	C	19.02
3	G	16.08
4	H	16.95
5	U	18.27
6	X	18.14
7	Y	17.86

The present algorithm has been tested over all uppercase and lowercase alphabets. But for convenience, only some of the alphabets are shown in Fig.6 and Fig.7.

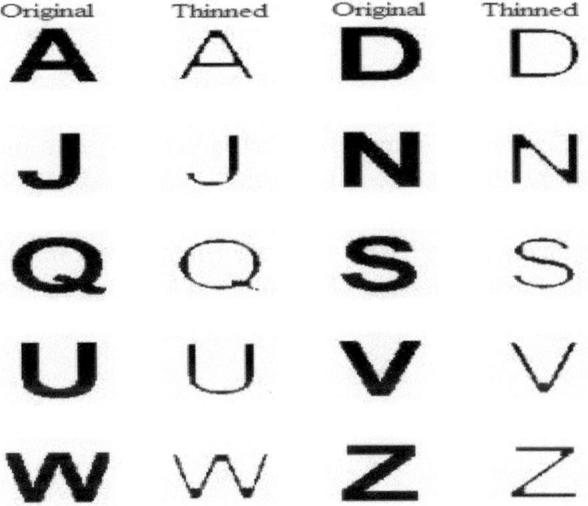

Fig. 6: Original and Thinned Image of English Upper-case alphabets

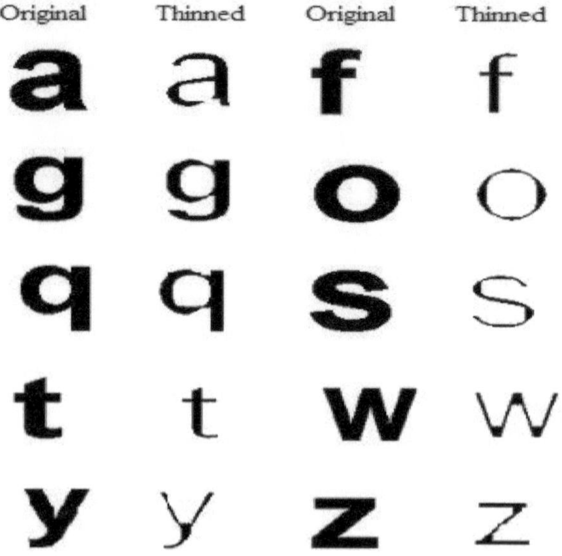

Fig. 7: Original and Thinned Image of English Lower-case alphabets

Conclusion

A topology preserving thinning algorithm based on iteratively culling pixels is introduced. The topology preserving thinning algorithm is used to retain the morphological properties of the images. It checks for the acyclicity condition and removes the pixels from a grayscale image. The shape preserving algorithm tends to retain the original shape of the image. This algorithm is very simple and easy to obtain skeletons. The performance of the skeleton is evaluated and is observed that for all the images connectivity is obtained and reduction rate is less than 20%.

References

1 L. Lam, S.-W. Lee, and C. Y. Suen, "Thinning methodologies—A comprehensive survey," IEEE Trans. Pattern Anal. Mach. Intell., vol. 14, no. 9, pp. 869–885, Sep. 1992.
2 P. Wiederhold and S. Morales, "Thinning on cell complexes from polygonal tilings, Discrete Appl. Math.," vol. 157, no. 16, pp. 3424–3434, 2009.

3   L. Liu, E. W. Chambers, D. Letscher, and T. Ju, "A simple and robust thinning algorithm on Cell complexes," 177, no. 5,pp. 975–988, 2001.

4   T. Y. Zhang and C. Y. Suen, "A fast parallel algorithm for thinning digital patterns", Comm. of the ACM 27, 1984, pp. 236-239 [5]. Zichang Guo and Richard W. Hall.

5   "Parallel Thinning with Two- Subiteration Algorithms," Communications of ACM, vol. 32(3), March 1989, pp. 359-373.

6   Harish Kumar and Paramjeet Kaur Deptt. Of CSA, Chaudhary Devi Lal University, Sirsa, Haryana, India1"A Comparative Study of Iterative Thinning Algorithms for BMP Images".

7   W.H.Abdulla,A.O.M. Saleh and A.H. Morad, "A preprocessing algorithm for hand-written character recognition", Pattern Recognition Letters 7, Jan 1988, pp. 13-18.

8   C. M. Holt, A. Stewart, M. Clint, and R. H. Perrott, "An improved parallel thinning algorithm," Commun. ACM, vol. 30, no. 2, pp. 156–160, 198.

9   . K. Jang and R. T. Chin, "Analysis of thinning algorithms using mathematical morphology," IEEE Trans. Pattern Anal. Mach. Intell.,vol. 12, no. 6, pp. 541–551, Jun. 1990.

10   Peter Tarabek, "Performance measurements of thinning algorithms", Journal of Information, Control and Management Systems, Vol. 6, (2008), No.2.

11   A Datta and S. K. Parui, "A Robust Parallel Thinning Algorithm for Binary Images." Pattern Recognition, vol. 27, no. 9, March 1994, pp.1181-1192.

12   M. Ashwin and G. D. Babu, "A new sequential thinning algorithm to preserve topology and geometry of the image," Int. J. Math. Trends Technol., vol. 2, no. 2, pp. 1–5, 2011.

13   T.A. Aberra, "Topology Preserving Skeletonization of 2D and 3D binary images," M.S.thesis, Dept. Mat., Tech. Univ. Kaiserslautern, Germany, 2004.

14   Jun-Sik Kwon, Jun-Woong Gi and Eung-Kwan Kang, 2001, "An Enhanced Thinning Algorithm Using Parallel Processing", IEEE, pp.no.752-755.

15   Gabor Nemeth and Kalman Palagyi, 2009, "Parallel Thinning Algorithm Based on Ronse"s Sufficient Conditions for Topology Preservation", Research Publishing Service, pp.no.1-12, Aug.16.

16   P. S. P. Wang and Y. Y. Zhang, 1986, "A fast serial and parallel thinning algorithm," in Proc. Eighth Euro. Meeting Cybern. Syst. Res. (Vienna, Austria), pp. 909-915.

17   C. J. Ammann and A. G. Sartori-Angus, 1985), "Fast thinning algorithm for binary images," Image Ksion Comput., vol. 3, no. 2, pp. 71-79.

18   "A comment on a fast parallel algorithm for thinning digital patterns," Comm. ACM, vol. 29, no. 3, pp. 239-242.1986.

19   S. Suzuki and K. Abe, **1987)**, "Binary picture thinning by an iterative parallel two- subcycle operation", Patt. Recogn., vol. 10, no. 3, pp. 297-307.

20   M. P. Martinez-Perez, J. Jimenez, and J. L. Navalon, 1987, "A thinning algorithm based on contours", Compuf. Vision Graphics Image Processing, vol. 38, pp. 186-201.

21   R. C. Gonzalez and P. Wintz, 1987, "The skeleton of a region," in Digital Image Processing. Reading, MA Addison-Wesley, pp. 398-402.

22   Y. Y. Zhang and P. S. P. Wang, 1988), "A modified parallel thinning algorithm," in Proc. 9th Int. Con$ Putt. Recogn. (Rome, Italy), pp. 1023-1025.

23   Richard W.Hall, 1989, "Fast Parallel Thinning Algorithms: Parallel Speed and Connectivity Preservation", Comm. CM. Jan, vol.32.no.1.pp.no.124-131.

24   Y. Y. Zhang and P. S. P. Wang, 1996, "A Parallel thinning algorithm with two subiteration that generates One-Pixel-Wide Skeletons", Proc. IEEE Int"l Conf. Pattern Recognition, vol. 4, pp. 457-461.

25 N.H.Han, C.W.La, and P.K.Rhee, 1997," An Efficient Fully Parallel Thinning Algorithm", IEEE -08186-7898-4, pp.no.137141.
26 S. Yokoi, J. I. Toriwaki, and T. Fukumura, 1973, "Topological properties in digitized binary pictures," Syst. Comput. Contr., vol. 4, no. 6, pp. 32-39.
27 Xingwei Yang, Xiang Bai, Xiojun and Wenyu Liu, 2008, "An Efficient Thinning Algorithm", IEEE, Computer Socity-2008, pp.no.475-478.
28 P.S.P.Wang and Y.Y.Zhang, 1989, "A Fast and Flexible Thinning Algorithm", IEEE, Trans. May, Vol.38. No. 5. pp. 741-745.

Authors

Dr. A. Sri Krishna received the PhD degree from JNTUK, Kakinada in 2010, M.Tech degree in Computer Science from Jawaharlal Nehru Technological University (JNTU) in 2003, M.S degree in software systems from Birla Institute of Technology and Science, Pilani in 1994, AMIE degree in Electronics & communication Engineering from Institution of Engineers, Kolkata in 1990. She has 23 years of teaching experience as Assistant Professor, Associate Professor, Professor and presently she is working as a Professor and Head, Dept of Information Technology at RVR&JC College of Engineering, Guntur. She has published 15 papers in International/ National Journals and Conferences. Her research interest includes Image Processing and Pattern Recognition. She is member of IE(I) and member of CSI.

Dr. K. Gangadhara Rao, Associate Professor in Computer Science and Engineering at Acharya Nagarjuna University, Guntur. His research interests include Computer Networks, Software Engineering, Object Oriented Anaysis and Design, Operating Systems, Cloud Computing.

N.Neelima received masters degree M.Tech(CSE) from Acharya Nagarjuna University in the year 2007, Guntur. She has 8 years of teaching experience. Presently she is working as an Assistant Professor in Department of IT, RVR & JC College of Engineering and she is pursuing Ph.D from JNTUH, Hyderabad. Her research interest includes Image and Signal Processing and computer vision. She is a member of IAENG.

K. Ratna Sahithi pursued her B.Tech Graduation in Information Technology in RVR & JC College of Engineering. She has been actively participating and presenting papers in student technical Symposium seminars at National Level. Her area of interest includes Image Processing, Computer Vision, Networks and Web Technology.

M. Apurvakeerthi[1] and J. Ranga Rao[2]

Control Office Application on Tracking Movement in Indian Railways

Abstract: Signaling system is one of the largest networks which play a vital role in Indian Railways (IR) over 150 years. Rail operations are performing efficiently with partial human intervention with utilization of signal data. Every minute rail operation can to be recorded and maintained for the motto of providing safer, accident-free operation, as well as ensuring punctual and reliable management of train movement. Data logger, an RDSO certified product, which is an advancement from Efftronics Pvt Ltd which meets the requirement of logging and reporting event information ,signal controlling, interlocking relays to verify their operation, diagnose faults and in maintenance with the help of NMDL software and other supporting applications. COA REPORTS are developing from tracking software tool with COA application, its aim of development is involved in presentation of logged data though direct reports in accordance with user interface requirements on data. At most, COA REPORTS development have been a major requirement from 77 divisional/area control offices, for the assurance of convincing evidence for accident analysis and equipment maintenance through reports. COA REPORTS, which can track the train movement information and details of that train at a selected station and at selected time and present in a user friendly, understandable way to authorized staff being involved. It is advancement to manual entries system from EXTRACTDBWITHQRY tool raw data. Front End Processor (FEP) communicates logged signal data to server system (DB) through Central Monitoring NMDL software. This database acts as input to COA REPORTS .In accordance to address the requirement and for retrieval of accurate acknowledgment of train information at a selected time and location, this developing module generates Train Information Detailed and Summary Reports, Network Performance Detailed Report between as output .The implementing module is a two tier architecture and revolves around spiral model which enhances cost and requirement constraints with automation to evolve customer benefits.

1 M. Tech, Computer Science & Engineering, VR Siddhartha Engineering College, Kanuru, Vijayawada, Andhra Pradesh, India
2 Asst. Professor, Computer Science & Engineering, VR Siddhartha Engineering College, Kanuru, Vijayawada, Andhra Pradesh, India

Keywords: FEP, CMU, NMDL, COA, COA REPORTS

1 Introduction

In computerized world, every invention is being modernized and automated for reduction of manual works and time. This evaluation of automatic evolved in Indian Railways, signal data. Signaling system is one of the largest networks which play a vital role in Indian Railways (IR) over 150 years. With the help of signaling system IR trains running larger distances and binding dispersed areas and promoting national integration. Rail operations are performing efficiently with partial human intervention and become the ninth largest commercial employer. Every minute rail operation have to be recorded and maintained for the motto of providing safer, accident-free operation, as well as ensuring punctual and reliable management of train movement. The above mentioned constraints are satisfied with the enhancement of Data Logger.

Data Logger, an RDSO certified product from Efftronics Systems Pvt. Ltd. Its invention overcome the failures and upgrades the reliability in signaling system with predictive maintenance. Data Logger, a microprocessor based acquisition device which logs the every single minute change in the status of signal system relay contacts. It acts like a "Black Box" which can scan, store and process the data for generating various user friendly reports. It is also recognized as "Event Logger".

In accordance with the technological world, the Data Logger data can be acquired through online and offline reports with NMDL software. Across India, 16 zones and 64 divisions having various stations can be interconnected with the help of OFC (optical fiber cable) and data is brought to centralized system called Front End Processor (FEP).FEP data stores in database server through Central monitoring unit (CMU)as shown in figure 1.

2 Literature Survey

In "COA Application: A Survey of the Literature" Mr. Achal Jain in the year March 2008, Currently deployed at two locations in south India, CRIS plans to roll out the application nationwide across 76 control offices. The COA application constitutes three distinct layers - the presentation layer, the business logic layer and the data access layer. A large network combined with

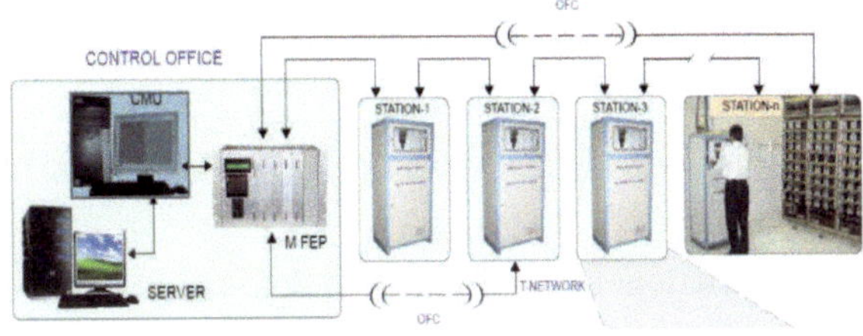

Figure. 1: Data Logger Function

heavy traffic, both passenger and freight with wide variations in running patterns, makes the task of train controlling and operations a very difficult one. Section Controllers who are assigned to control a portion of the network within each divisional control office, used to manually record the movement of trains on a train chart, while constantly keeping track of various parameters needed for efficient planning. The control chart is a time distance graph which is used to control and plan the movement of trains. It is used to ensure that scheduled passenger trains run on their paths, ensure efficient paths for unscheduled freight trains, forecast the movement of trains and provide information on arrivals and departures as per commercial requirements.

A good controller constantly optimist on resources and ensures trains run smoothly through the system, reducing detention of both scheduled and unscheduled trains. With increasing traffic levels, the Section Controllers were often constrained by the need to document every event (ie., train arrival/ departure, track failures, crew change, signal status, fuel status) and yet focus on their primary duty of planning and reporting the movement of trains. The existing system also involved a lot of paper work by support staff who were required to replicate the operational data given in the control chart to various registers As a result, the core activity of advance planning and timely reporting of train movements was getting diluted. Consequently, management also had to make special efforts to obtain the information required for good decision making. Indian Railways decided to automate the process of controlling train operations to ensure greater efficiency in train operations.

*Presentation Layer:*The presentation layer is further divided into following components: Control View, Data-Entry and Cache. The presentation layer has been developed as smart client solution with Click Once capabilities for zero-touch deployment.

Business Layer: The controllers enter the information in the presentation layer. The UI validates the information and transfers it to the business components for business validation and logic checks. The validated information is sent to the presentation layer for the controller to view the information.

2.1 Data Access Layer

The master data (e.g. board information, trains running on board etc.) is used as reference information for rendering the control chart and for making decisions. Since this information seldom changes, it is cached in memory, but when they do change, the cache is refreshed. In this approach we improve safety, and on-time operations through Control Office Application.

2.2 How It Works

At each of the 73 control offices, a cluster of database and application servers provide an intuitive GUI-based system for train controllers to record the movement of each train on a virtual coa reports. The application requires the controllers to enter only the data related to the train operations with ease. The application then draws the graph of the sectional trains together with advance forecasting, taking into consideration all operational parameters like blocks, caution orders, crossing, precedence, priority, load details, etc. It also communicates with the adjacent divisions on handover/ takeover details of trains.

2.3 Functionality

- There shall not be islands of information. The required information shall be shared electronically between various users across the geographical spread of the system
- By having a single source of information, conflicting data shall be avoided.
- The processes shall be standardized and structured reporting shall be adopted.
- The development for the functionality shall be modular, emphasize on reusable components to be built and avoid duplicity of efforts.
- An integrated approach for IR shall remove the difference in business practices and it will streamline data collection, consolidation and analysis.
- Since the required information shall be readily available, it shall take less time for other functionaries to carry out their duties. This shall help

them in better planning for train movement and resource utilization.
- The extent of dependency on control room personnel to get the required data shall be reduced.

In "COA Application: A Survey of the Literature" Dweep Bhoomi in the year 15th January 2010 describes that the classification can be done manually through offline. Controlling of Train movement is the prime function of Railway Operation. Preparation of the control chart is an integral part of train operation as it provides a visual tool for making operational decision. Computerized Control Charting system is required to handle high traffic volume in the Indian Railways efficiently. The COA is a mission critical application for the Indian railways, as it is used for train operations on 24x7 bases. CRIS has developed the software for COA and implemented.

2.4 Key Benefits

Updated train position is a feature developed by the coa Reports in which the train position is updated on a near real-time basis. This would also help avoid manual enquiry at information counters in railway stations. The Control Office Application has unbundled an immense potential in improving forecasting of train arrivals and departures at various stations, improving punctuality of trains and implementing a number of innovative passenger information systems using latest information technology aids.

2.5 Basic Methods

To monitor the train moments the COA application follows spiral model which consists of the following components. The spiral model is a risk-driven process model generator for software projects. Based on the unique risk patterns of this project, the spiral model guides a team to adopt elements of one or more process models, such as incremental, waterfall, or evolutionary prototyping. The spiral model has four phases. A software project repeatedly passes through these phases in iterations called Spirals.

Identification: This phase starts with gathering the business requirements in the baseline spiral. In the subsequent spirals as the product matures, identification of system requirements, subsystem requirements and unit requirements are all done in this phase.

Design: Design phase starts with the conceptual design in the baseline spiral and involves architectural design, logical design of modules, physical product design and final design in the subsequent spirals.

Construct: Constructing phase refers to production of the actual software product at every spiral. In the baseline spiral when the product is just thought of and the design is being developed a POC (Proof of Concept) is developed in this phase to get customer feedback.

2.6 Evaluation and Risk Analysis

Risk Analysis includes identifying, estimating, and monitoring technical feasibility and management risks, such as schedule slippage and cost overrun. After testing the construction, at the end of first iteration, the customer evaluates the software and provides feedback.

Risk management is one of the in-built features of the model, which makes it extra attractive compared to other models. High amount of risk analysis hence, avoidance of Risk is enhanced. Good for large and mission-critical projects. Strong approval and documentation control .Risk analysis Can be a costly model to use. Risk analysis requires highly specific expertise. Project's success is highly dependent on the risk analysis phase. Spiral may continue indefinitely. Pavan Kumar Murali, Maged M. Dessouky, Lu et al[10] To route and schedule trains over a large complex network can be computationally intensive. One way is to reduce complexity could be to "aggregate" suitable sections of a network. To present a simulation-based technique to generate delay estimates over track segments as a function of traffic conditions and signaling, to test our technique by comparing the delay estimates obtained for a network in Los Angeles with the delays which has been shown to be representative of the real-world delay values. Railway dispatchers could route and schedule freight trains over large network by using our technique to estimate delay across aggregated network sections.

2.7 Software

The software is divided into two parts.

2.8 System software

The system software supports all the functions in the data logger. This software reads and interprets the application software to enable the CMU to perform the appropriate functions. The software is use friendly and menu driven.

2.9 Application software

Application software generates following reports

Log of Entries Report: This report will give the list of all the log of entries either digital or Analog inputs that have been taken place in users selected time. User can take report for selected duration of time.

Fault Report: This report will give the faults that have occurred like analog voltage failures, signal failures, route failures etc.

Graph Report: The graph report of analog voltage with user selected time is available. Train simulation Report: The train movement can be simulated ON line or OFF line with set of data available.

Software support for the user to perform the following operation is also provided:

- ✓ Setting of time
- ✓ Status of Digital Inputs.
- ✓ Status of Analog Inputs.
- ✓ Printer
- ✓ Change of password

3 Proposed System

Control Office Application from CRIS is an IT application to led the train operations information to passengers and control officers .To run this project as successful application, efficient and accurate data without delays must be available to the user. Discrepancies arise in train departure and arrival timings recording by two adjacent division section controllers. To avoid this situation, automated tracking and generation of train arrival/departure information must evolve.COA is designed to give the Reports to the End user in order to display Train Tracking Information, CRIS Server posted timings. The main intension of project is designing the new module called control office application in reports Exe to generate COA reports like COA Information report, trains information summary reports and assistant station master summary reports in textual format and graphical formats.

3.1 Objective

- At present train departure and arrival timings at interchange stations and terminal stations are taken by COA server of CRIS from manual charting data of section controllers.
- On observing serious discrepancies in train departure and arrival ti-

mings recording by two adjacent division section controllers, Railway Board(CRB&ML) decided to take the train arrival and departure events from data loggers of interchange and terminal stations then post them to the COA server of CRIS(Centre for Railway Information System).

3.2 Process of data posting to server

After getting data it is imported in to excel and manual analysis is going .But Railway authority do not want manual reports. so we are going to Automize the system by generating software based reports. So we proposed COA reports to generate software based

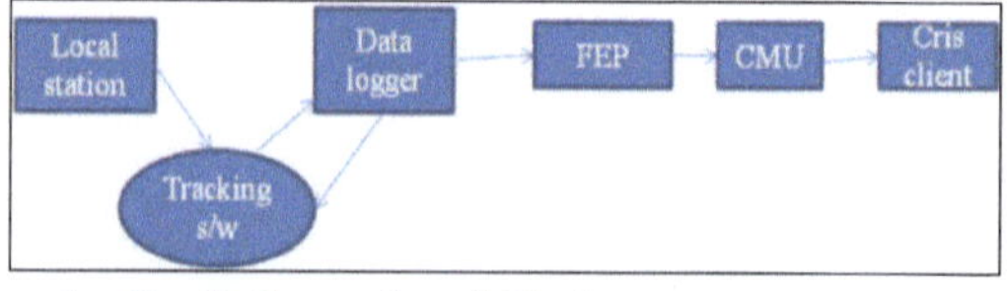

Figure. 2: Train Tracking Information Feeding To Server from A Station

The COA tracking is done the Panel is connected to the relay room and the relay supply is connected to the site where the signals are shown to train to proceed. From relay room the relays connected to the DATA LOGGERS where the packets are generated and are sent to the system where the COA tracking is done. Then it will post the data into NMDL and again it is sent into the Data Logger. This is process done in the Local Station. Now packets are sent to the next Data Logger i.e. DL2 shown in Figure 2,3 the packets are sent in the Bi-directional way. The data is sent to the FEP (Front End Processor) then it is sent to the CMU and to DATABASE server where the analysis department has a Backup to about trains running status. From CMU it is sent to the CRIS where the Railway Department will verify the trains running status.

For this we have arrangement like

- At the terminal and interchange stations a PC with train tracking software is to be installed by S&T and connecting it to data logger of the station.

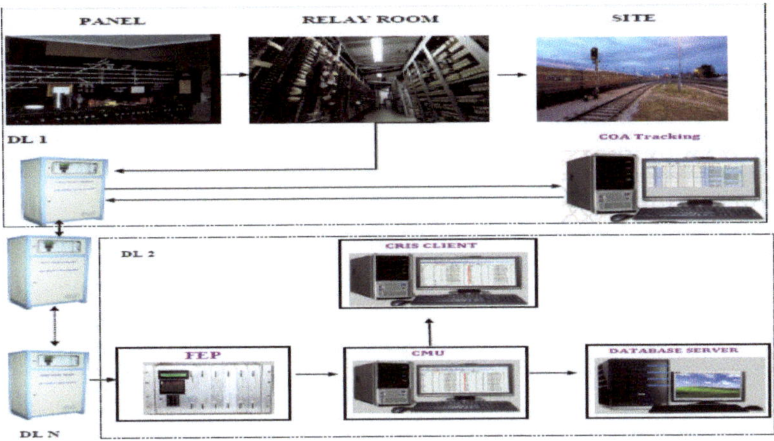

Figure. 3: Architecture Of Data Loggers

The train arrival/departure events are automatically generated by the PC with the help of train tracking software-except the train number.
The input to the tracking software is data logger Logged information and database.

- After the record generation station master enter the train number and send record by using send button.
- This data again sent to data logger network and reaches Central Monitoring Unit through Front end processor. This is posted in to server.
- Now the CMU sent the data to Cris client and Cris client post into Cris server via LAN connectivity from COA server to the PC in the control office.
- All the data stored in cris server but for representation of this data there is no format.
- For this We retrieve the data from server through Extract DB software.

After getting data it is imported in to excel and manual analysis is going .But Railway authority do not want manual reports. so we are going to Automize the system by generating software based reports. So we proposed COA reports to generate software based reports.

4 Results

4.1 Trains Information Report

Sno	Train No	Loco No	Line No	Direction	Train Type	Arr Time	Dep Time
1	12345		1	U	Passenger	19-Jan-13 10:30	19-Jan-13 10:40
2	Grt235	4567	1,2	U,D	Goods	19-Jan-13 11:30	19-Jan-13 12:40

Figure. 4: Report Output Display Screen Of Trains Information Report

If Train No & Loco No both exists then displaying train type as "Goods" If only Train No exists then displaying train type as "Passenger "Retrieving data from NewTrnArrDepinfo table and combining Arrival & Departure train movements records based on train no & displaying as a single record in the report.

4.2 Coa Information Report

The aim is for displaying the overall flow from ET - CMU- Cris Client. The internal purpose is for Analysis I can easily identify whether the Record Generated at Local Station is Posted in to Cris Server or Not with or without much delay. In COA Information Report, all the train details will be displayed at CRIS server, over all duration (in mins).

Sl No	Train No	Loco No	Train Type	Train of Date	Event Type	Event Time	Tracked Time	Entry Time by ASM	ET-CMU		CMU-CRIS CLIENT		CRIS SERVER		Over All Duration (In Mins)
									ASM Sent Time	CMU Rcvd Time	CMU Sent Tme	Cris Clnt Rcvd Time	DB Posted Tried Time	DB Posted Time	
1	1248		Passenger	19-Jan-13	A	10:30	10:31	10:32	10:33	10:34	10:35	10:36	10:37	10:38	8

Figure. 5: Report Output Display Screen Of COA Information Report

4.3 Network Performance Report

Network Performance Report For T-CMU: The internal purpose is for Analysis I can easily identify whether Delays are exists in the Network or not. In Network Performance Report For ET-CMU, the train details will be displayed. Required data for report will be retrieved from database and N/W delay values will be calculated based on the sent time and received times and also average delay value will be calculated for N/W delay columns.

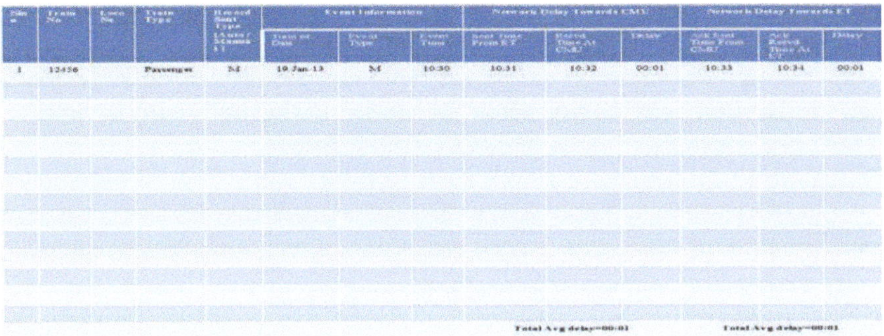

Sno	Train No	Loco No	Train Type	Record Sent Type (Admn / Manual)	Event Information			Network Delay Towards CMU			Network Delay Towards ET		
					Train or Date	Event Type	Event Time	Sent Time From ET	Recvd Date At CMU	Delay	Ack Sent Time From CMU	Ack Recvd Time At ET	Delay
1	12456		Passenger	M	19 Jan 13	M	10:30	10:31	10:32	00:01	10:33	10:34	00:01
								Total Avg delay=00:01			Total Avg delay=00:01		

Figure. 6: Report Output Display Screen Of Network Performance Report For Et-CMU

4.4 Network Performance Report for CMU-Cris Client

The internal purpose is for Analysis I can easily identify whether Delays are exists in connected network between CMU - Cris Client or Not. In Network Performance Report For CMU-Cris Client, the train information details will be displayed required data for report will be retrieved from database and N/W delay values will be calculated based on the sent time and received times and also **average delay** value will be calculated for N/W delay columns.

4.5 Trains Information Summary Report

Internal purpose is to view the Details of Tracked Trains by Our Software. In Trains Information Summary Report, data will be displayed in the form of Textual Report & Graphical Report in a single page .In Textual report, the following details will be displayed TRAIN TYPE, Count of Trains with corresponding train type as count format, Percentage of counts among total trains count as percentage.

Slno #	Train No	Loco No	Train Type	Record Sheet Type (Auto/ Manual)	Event Information			Network Delay Towards CMU/SC Root			Network Delay Towards CMU		
					Train of Date	Event Type	Event Time	Sent Time From CMU	Recvd Time At CMU/Clnt	Delay	Ack Sent Time From Clnt/Clnt	Ack Recvd Time At CMU	Delay
1	1234		Passenger	A	19-Jan-13	A	10.30	10.33	10.36	00:01	10.37	10.38	00:01
										Total Avg delay=00:01			Total Avg delay=00:01

Figure. 7: Report Output Display Screen Of Network Performance Report For ET-CMU

	Counts	Percent
Goods Trains	10	33
Passenger Trains	20	67
Total Trains	30	

Figure. 8: Report Output Display Screen Of Trains Information Summary Report

4.6 Graphical Format

These options will be implemented for both summary & detailed reports. Filter also used in grid to sort all records based on particular sorted column .Also additional option provided to view detailed report from summary report, with right click on Textual or Graphical Report popup menu will be displayed with Detailed Report as option.

4.7 ASM Summary Report

The internal purpose is to help for Railway Officials to know whether ASM is responsible to his duty or Not. In ASM Summary Report, data will be displayed in the form of Textual Report & Graphical Report in a single page. In Textual report, the following details will be displayed trains sent, Count of Trains with corresponding train sent type as count, Percentage of counts among total trains count as percentage. In Graphical Report, graph will be generated with data displayed in textual format as shown.

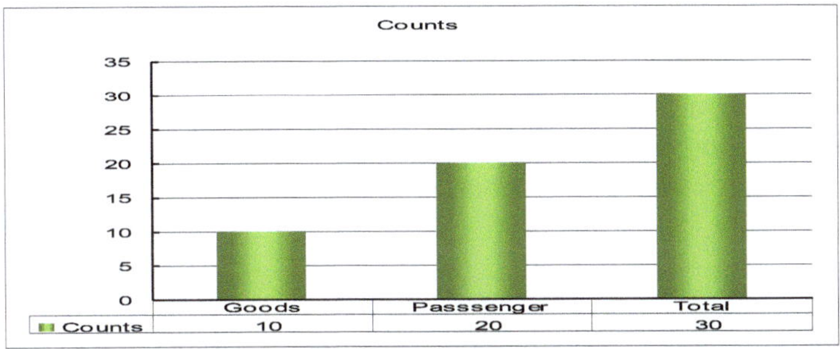

Figure. 9: Graphical format

	Counts	Percent
In Time Sent Trains	75	75
Late Sent Trains	13	13
Un Sent Trains	12	12
Total Trains		100

Figure. 10: Report Display Screen Of ASM Summary Report in textual format

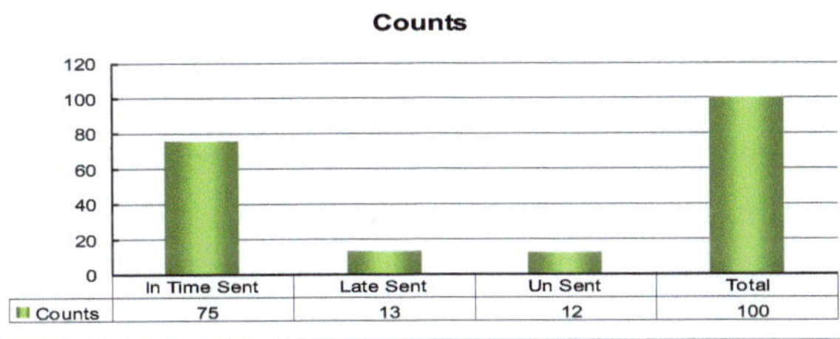

Figure. 11: Report Output Display Screen Of ASM Summary Report In Graphical Format

4.8 Coa Over All Duration Summary Report

The internal purpose is for Analysis and can easily identify to track the Train Movements with in Time or Not by Observing the Tracking Delay values. In COA

Over All Duration Summary Report, data will be displayed in the form of Textual Report & Graphical Report .In Textual report, the train details will be displayed. Count of Trains based on Overall delay from Event time to DB Posted time. Also displaying Over All Duration with different time counts as <10 Mins, 10-15 Mins, 15-30 Mins, 30-45 Mins, 45-60 Mins, >1 Hour.

Total No.of Trains	Delay Type	Over All Duration/Delay					
		<10 Mins	10-15 Mins	15-30 Mins	30-45 Mins	45-60 Mins	>1 Hour
100	Train Movement Counts	84	10	5	1		

Figure. 12: Output Display Screen Of COA Over All Duration Summary Report

Conclusion

Thus COA application provides information and tools which will enhance the decision making of rail traffic controllers. The tracking helps improve operations by increasing accuracy of arrival and departure predictions. This has empowered the division controllers and other management to make more informed strategic decisions by providing accurate, detailed, up-to-date information about rail traffic and demand. "Customer satisfaction was a key drive and Training is a critical factor for the successful roll out of COA. Integration with the National Train Enquiry System for providing near real time data generated by COA. COA has introduced to facilitate data exchange and for internal reporting purposes CRIS plans to integrate the COA across divisions and zones for better operational efficiency. CRIS plans to automate the data entry using data loggers/GPS.

References

1 JagmohanRam,RDSOSpecificationNo.IRS:S:99/2006,"DataLoggerSystemHardwareanual," EfftronicsSystemsPv.Ltd,March,(2011).

2 "DataLoggerInstallationManual,"EfftronicsSystemsPvt.ltd,http://wwwefftronics.com/datal ogger.php,April,)(2013).

3   Deep Bhoomi,"Control office application:A literature survey",15th january, (2010).

4   MrRavidramIaugurated,"computerisedtraincontrolchartingsystem"May13,(2006)

5   Dileepa Jayakody, Mananu Gunawardana, "GPS/GSM based train tracking system—
    utilizing mobile networks to supportpublicTransportation",Vol.2,Issue.10,October, (2013).

6   Kunal Maurya,Mandeep Singh, Neelu Jai,"Real Time Vechicle Tracking System using GSM
    and GPSTechnologyAnAntitheftrackinSystem",ISSN2277-1956/V1N3-1103-1107pp103-
    1107,September,(2013).

7   Mathur,"OperatingmanualforIndianrailways"governmentofIndianministryofrailways,http:
    //www.indianrailways.gov.in/railwayboard/uploads/codesmanual/operating/20manual-
    traffic.pdf,Sept,2008.

Ramesh K[1] and Sanjeevkumar K.M[2]

New Change Management Implementation Methods for IT Operational Process

Abstract: Change management is the process of planning and coordinating the implementation of all changes (any planned alteration or addition to the operating environment including application system) into the production environment in a logical and orderly fashion. it does not develop changes, nor does it implement them. In this paper we have proposed new change management meeting Agenda, change planner/Developer responsibilities and coordinator responsibilities To allow the coordination and planning of changes in order to provide a stable production environment.

Keywords: Change Management, Process, Agenda, Priority, Risk Level, Lead Time.

1 Introduction

Change Management is the process of planning, coordinating, implementing and monitoring changes affecting any production platform within Information Technology's control. Philosophically thinking, change is the only constant in the world. Same as for any thing else, this is true for business organizations as well. Every now and then, business organizations change the way they operate and the services/products they offer. There are new initiatives in organizations and the old ineffective practices are forced to leave. In addition to that, technology is constantly changing and the business organizations need to par with that as well. There are many approaches about how to change. Of course, we may all agree that the change is required for an organization, but can we all be in agreement of how the change should take place? Usually not! Therefore, deriving a change management process should be a collective effort and should result from intensive brainstorming and refining of the ideas. In this tutorial, we

1 Department of Computer Science, Department of MBA Karntaka State Women's University,Jnanashakti Campus Toravi,Vijayapura, Karnataka,India
rvkkud@gmail.com, rameshk@kswu.ac.in
2 Department of Computer Science, Department of MBA Karntaka State Women's University,Jnanashakti Campus Toravi,Vijayapura, Karnataka,India

will have a look at the change management process suggested by John Kotter. Since this process has shown results for many Fortune 500 companies, Kotter's approach should be considered with respect. The objectives of the Change Management process are to:

- Ensure that changes are made with minimum disruption to the services IT has committed to its users.
- Support the efficient and prompt handling of all changes.
- Provide accurate and timely information about all changes.
- Ensure all changes are consistent with business and technical plans and strategies.
- Ensure that a consistent approach is used.
- Provide additional functionality and performance enhancements to systems while maintaining an acceptable level of user services.
- Reduce the ratio of changes that need to be backed out of the system due to inadequate preparation.
- Ensure that the required level of technical and management accountability is maintained for every change.
- Monitor the number, reason, type, and associated risk of the changes.
- The Change Management procedure for the Information Technology Division defines how the change process is implemented in all of the IT platform environments. The objectives of the operating procedures, in addition to those detailed above are to provide documentation that allows IT management to understand, at any point in time, the configuration of the IT environment.
- Minimize the bureaucratic impact on the development community while maintaining control of the environment.
- Activities of the Change Management Process at include
- Receiving change requests from the   Request for Service process
- Determining whether or not the change is in the best interests of
- Assigning the change to resources within IT for solution identification, sizing and risk analysis
- Accepting or rejecting the requested change
- Assigning the change to solution development resources
- Reviewing the solution prior to implementation
- Scheduling the change
- Communicating change status as required to all interested parties
- Closing the change

2 Change management objectives

To minimize the adverse impact of necessary changes on system integrity, security and the service level agreements. To allow the coordination and planning of changes in order to provide a stable production environment. To maximize the productivity of person involved in the planning, coordinating and implementation of "quality" changes.

2.1 Change Management Mission

1) To ensure the documentation of all proposed changes prior to installation in the production environment.
2) To ensure verification of technical completeness:
   ➤ Accuracy of technical completeness
   ➤ Plans for final test, install, backout and recovery
   ➤ Identification and review of all technical dependencies including effects on concurrent changes
3) To ensure that the timing of changes executions does not conflict with the business cycles or   priorities.
4) To ensure appropriate management involvement and approval:
   ➤ Required sign-offs(i,e justification, user acceptance ,affected areas)
   ➤ Approval ,deferral, or disapproval of change installations
5) To ensure the verification of successful testing to degree required by organization standards   before change introduction to the production environment.
6) To ensure the documentation of actual change installations and/or change backouts:

   ➤ To enable communication of change results To provide a history of changes
   ➤ To support the maintenance of systems documentation.

2.2 Change Management Life Cycle Functions

Change request initiation: The person or group requesting of change will complete the submitter section of request for the change template (displayed later).if a user is involved, IT can help complete the form or screen.

Change priority(Impact and Risku Assessment): The impact the change will have on the production environment and services provided to the user and the business reasons for the change are determined and approved (see sample change priority Guidelines).

Change Development: As part of the change development, based on the installation's standards, a plan for installation

Is required suggested checklist of items to include in the plan is:

- Test procedures followed, with results
- Back out procedures (must be available)
- Resource plan
- Description of any training required
- Required and supplied documentation
- proposed installation schedule
- Business reason for the change
- management application

Change coordination: This function controls the day-to-day activities of the change management system with the goal of meeting the change management objectives weekly change management meeting is chaired by the change coordinator.

2.3 Change coordination functions

- ➢ Change request forms are filed and change log kept current (or this is done through a problem/change management system, if applicable).
- ➢ All items listed in the change management mission are reviewed for all major (high and medium risk) changes. Based on installation standards, all items must be acceptable for the change to be scheduled for installation. changes. Based on installation standards, all items must be acceptable for the change to be scheduled for installation.
- ➢ All approved changes are scheduled for installation.
- ➢ A ten day detailed change installation schedule and three-month major change schedule are updated and produced weekly.
- ➢ All changes are tracked and when required, follow-up is done.

other management reports such as highlights of major changes planned and results of major changes installed(history change) are produced.

3 Proposed change management methods

3.1 Change management meeting Agenda

1) To review the effects of significant change (high and medium risk) and confirm schedules for implementing these changes. The meeting in focal point for all planned change information and management controls.

2) To review planned changes for next ten days. Based on the day of the change meeting, this then-day projection should cover to (20weekends)discuss any emergency change since the last change meeting.

3) To review any schedule or procedural violation s within the past week along with the results of the changes.

4) Accurate minutes should be taken, distributed to all attendees and saved for history purpose.

Purpose: The meeting is conducted to review the suitability of changes scheduled for implementation and verify the coordination and cooperation of participating areas. The results of this meeting will be the basis for summary report for management.

It is the function of this meeting to review the business and technical assessment and to review and approve the schedule of each change, to verify that the change meets the business objectives, and will not have a negative effect on the business community. In addition, all changes are reviewed to see if and how they affect the disaster recovery plan or recovery strategy. if it is determined that the change is not needed or is untimely, it will be cancelled or rescheduled.

This meeting marks the change request deadline. All change requests that are issued after the meeting begins will be deferred until the next meeting. If the request cannot be postponed (i.e. emergency), it may be discussed at the current meeting.

Meeting participants

> Change coordinator (meeting chairperson)
> Technical support
> Operations management
> Processing services
> Application Group
> Help Desk

> Network management
> Change Requestors/Implementers
> Hardware Group-Vendor (if applicable)

3.2 Result Sample for change priority Guidelines

Change Risk level

Risk Level	Lead Time	Description
H	3 wks	Very difficult or impossible to backout. major impact if it fails. Involves Several or critical system. Large staff involved in implementation. Will Disrupt hardware, software, data if unsuccessful.
M	2 wks	back out is difficult or lengthy. Involves several systems. May disrupt Hardware, software, data if unsuccessful.
L	1wks	small impact on failure. Involves non critical systems. backout is simple or Not needed.
E	None	Immediate business requirement with no advance warning. high severity. These MUST result from a severity 1 problem and have been entered in the system with a problem number assigned.

Sample: Change priority Grid

Change problem impact

--Risk-	CRITICAL	HIGH	MEDIUM	LOW
HIGH	÷1	1	2	3 changes receive Priority for resource And Scheduling based On priority assigned.
MEDIUM	1	2	3	4
LOW	2	3	4	5
EMERGENCY	÷1	-	-	-

> Develop back out and recovery procedures so operations personnel can use them if Necessary.
> Develop test procedures.
> Schedule and run test procedures.
> Use test results as part of change documentation.
> Notify all individuals that will be impacted by the change (i.e. help Desk staff, Operations, technical support, the client community, etc.).
> update the change system (or notify change coordination) throughout the development process.
> notify management affected by the change of any adjustment to the planned Schedule.

3.3 Change planner/Developer Responsibilities

Note: The change planner/Developer's responsibilities differ for those of the change coordinator so they are listed here for comparison.

Change planner/Developer: He is the one who requests/creates a changes to be moved into production.

Objective: To supply the change management system (or change coordination function) with a Change description, dates requested and purpose of the change. This includes testing and supplying all supporting documentation. To communicate the difficulty and/or success of the development process back to the change management system or change coordination function.

Responsibilities

> Obtain approvals from their management before the request for change is scheduled.
> Review the testing results from change development.
> Develop and /or prepare the change for production.
> Complete and supply all required documentation to appropriate personnel.
> Update all affected procedures (i.e. operational, user and offsite if applicable).

3.4 Change coordinator Responsibilities

Change management coordinator: He is the person responsible for maintaining the change management process.

Objective: To ensure that policies and procedures are adhered to and that senior management Has the information (i.e. reports, feedback) needed to ensure that the change

Management approach is meeting all objectives.

Responsibilities

- Chair the change management meetings.
- Mediate all conflicts regarding scheduling, lack of approval or lack of documentation Prior to implementation of the change.
- Revise the change procedures if needed and advise management.
- Act as focal point for suggestions regarding change procedures.
- Audit the change Log or mainframe system reports to ensure that the changes are being updated properly.
- producer statistical and trend reports for management use to analyze change management system effectiveness.
- Advise management of procedural violations or omissions through management reports.
- Maintain an awareness of all projects that will introduce change to the production environment.

- Distribute the change schedule prior to the weekly meeting or instruct attendance to Print off their own copy for personnel use.
- Review documentation that accompanies each change and verify that it is complete, including management approvals.
- Verify the success of the change and its correlation to the original request of the user.

Conclusion

All substantive changes to the IT Environment must adhere to the change Management process. A substantive change has the potential to affect the ability of users and systems to interact with each other. All changes must conform to the published guidelines that dictate the "look and feel" OF THE IT environment. All changes to the production systems within IT will have a corresponding set of documentation that describes the change, the business reason for the change and the disposition of the change. This includes emergency and exception changes. The risk and/or impact ratings of the requested change will determine which of the four phases of the change workflow (Analysis, Design, Testing and Implementation) will be required to promote the change into production. The proposed methods will allow the coordination and planning of changes in order to provide a stable production environment. In future we have plan to release the second version of methods.

References

1   Carol Kinsey Goman, "The Biggest Mistakes in Managing Change," Innovative Leader, December 2000,pp. 500-506.
2   study "Making Change Work"imai, Masaaki (1986). Kaizen: The Key to Japan's Competitive Success. McGraw-Hill/Irwin
3   Prosci – People Focus of Change Management Software Engineering Institute of Carnegie Mellon University.
4   Richard Lynch, John Diezemann and James Dowling, The Capable Company: Building the capabilities that make strategy work (Wiley-Blackwell, 2003
5   Kotter, J. (July 12, 2011). "Change Management vs. Change Leadership -- What's the Difference?". Forbes online. Retrieved December 21, 2011.
6   IBM Global Services (2003). "IBM and the IT Infrastructure Library" (PDF). Retrieved 2007-12-10

7   IBM (1980). A Management System for the Information Business. White Plains, New York: IBM.
8   Schiesser, Rick, 2002. IT Systems Management. New Jersey, Prentice Hall. ISBN 0-13-087678-X
9   http://www.itsm.info/ITSM%20Change%20Management%20Best%20Practices.pdf
10 http://www.itsmcommunity.org/downloads/Sample_Process_Guide_-_Change_Management

J. Jayachandiran [1] and D. Nedumaran [2]

Enhanced Multiple-Microcantilevers based Microsensor for Ozone Sensing

Abstract: An enhanced, multiple-microcantilevers based microsensor for ozone sensing was modeled and simulated on a COMSOL Platform. In this model, silicon has been chosen for its very good mechanical properties and gold nanolayers on the top of the cantilever act as the sensing element. Resonance characteristics and stress analysis have been carried out. The eigenfrequency analysis covers the resonance modes, while the stationary studies deal with deformation and displacement of the microcantilever beams. The performance of the microsensor depends on the geometrical dimensions as well as the material properties of the microcantilevers. The simulation results, stress analysis, displacement, and eigenfrequency analysis show that the designed microsensor is the best model for ozone sensing applications.

Keywords: Multiple cantilevers, microsensor, ozone sensing.

1 Introduction

Nowadays most of the electronic devices are miniaturized in the form of a single chip due to advancement in microtechnology and nanotechnology. These technologies play an important role in miniaturizing the sensing devices. A gas sensor is an essential device for medical, environmental, and industrial applications. Many methods have been used for sensing the particular gas molecules from the atmosphere. The cantilever beam method is an effective method for real-time detection of the gas molecules. The dimensions and design of the cantilever play a vital role in modeling a cantilever based sensor. Chen et al. (1995) developed and analyzed the resonance frequency change of the V-shaped bimetallic cantilever. Apart from the dimensions of the cantilever, a suitable sensing element is needed for increasing the sensitivity of the sensor. The ozone

1 Central Instrumentation and Service Laboratory, University of Madras, Guindy Campus, Chennai 600 025

2 Central Instrumentation and Service Laboratory, University of Madras, Guindy Campus, Chennai 600 025
dnmaran@gmail.com

molecule has good interaction with gold nanoparticles (AuNPs) and exhibited very good sensing characteristics [Puckett et al. (2005), Plante et al. (2006), Pisarenko et al. (2009)]. Banerjee & Banerjee (2013) fabricated and analyzed the nanomechanical response of the resonance stage of both single and coupled nanocantilevers. Banerjee et al. (2012) also studied the unusual dimensional dependence of resonance frequencies of Au nanocantilevers and these cantilever beams respond evenly to nanomechanical disturbances. Hayden et al. (2013) performed and studied the effect of surface roughness on the deflection of a gold coated silicon micro-cantilever. When the gas molecules hit the cantilever beam, it will bend because of adsorption/absorption of molecules and hence the cantilever resonance frequency changes with respect to the amount of gas molecules adsorbed/absorbed on the sensing material, leading to the application of cantilever deflection or resonance frequency sensing method as an effective tool for making many novel gas sensors [Chen et al. (1995), Hayden et al. (2013), Wasisto et al. (2013)]. Several cantilever based sensors have been modeled and simulated using the Finite Element Method (FEA) prior to fabrication [Pakdasta et al. (2009), Gerlich et al. (2012), Yoon et al. (2012), Korsunsky et al. (2007), Firdaus et al. (2010), Bashir et al. (2000)]. In this paper, we present a two-sided, coupled cantilevers based microsensor for the detection of ozone molecules. For the simulation of the proposed model, we have used the COMSOL MULTIPHYSICS.

2 Device Modeling

2.1 Dimensions of the Proposed Microsensor

Our microsensor contains four microcantilevers, with two cantilevers fixed with one side of the Z-type frame and the remaining two cantilevers fixed on the other side of the frame. Each cantilever has the dimensions of 1 μm ×7 μm × 0.08 μm. A thin 0.001 μm gold nanolayer was coated on the top surfaces of the cantilevers. The Z-type frame was fixed in a 12 μm radius silicon disc. Both the Z-type frame and the disc have a thickness of 1.5 μm as shown in Figure 1.

Before fixing the microcantilevers to the Z-type frame, the frame and a rectangle-shaped frame were taken and analyzed under different force conditions for knowing their deformation and bending nature. The Z-type frame gives better results.

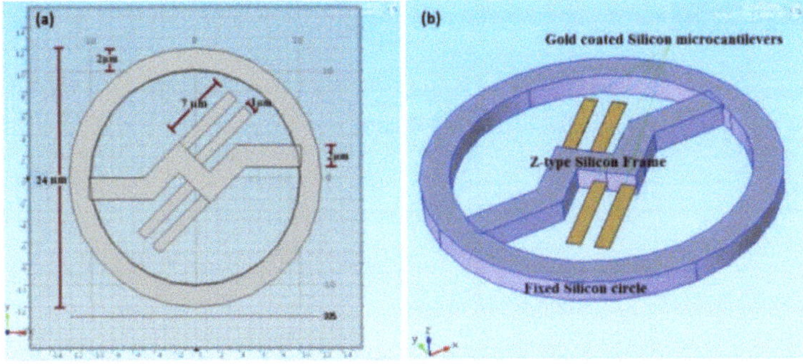

Figure. 1: (a) Dimensions and (b) material specification of the microsensor

2.2 Material properties and sensing method of the microsensor

Silicon is selected for its very good mechanical properties and most of the micro devices are made up of silicon only. Even though gold is a noble metal, gold nanoparticles (AuNPs) interact with the ozone molecules. Material properties and dimensions of the microsensor were modeled and characterized. Several dimensions have been tested in this study, and based on the performance of the model, the optimum dimensions were chosen and they are given in Table 1.

An adsorbing nanolayer was coated (placed) on the top surface of the microcantilevers. Then the targeted ozone molecules were fed on the microcantilever. The sensing material deposited on the microcantilevers interacts/reacts with the receptive nanolayer coating, resulting in an increase in the cantilever mass. Moreover, the cantilever suffers a surface stress, gets deflected, and undergoes a change in the vibrational frequency. The frequency shift is calculated and calibrated as a function of ozone concentration. The sensing mechanism of the microcantilever sensor is diagrammatically shown in Figure 2.

Chen et al. (1995) proved that there is a resonance frequency change due to the adsorption of gas molecules. The gas molecules produce a surface stress on the top of the cantilever. The resonance frequency of the cantilever can be expressed as

$$v = \frac{1}{2\pi}\sqrt{\frac{K}{m}}, \tag{1}$$

where K is the spring constant and m is the effective mass of the cantilever.

Table 1. Material properties and dimensions of the microsensor

Proposed microsensor model	Material	Width (μm)	Depth (μm)	Height (μm)
		120	120	1.5
4 Cantilevers	Silicon	1	7	0.08
Nanolayers	Gold	1	7	0.01
Z-type frame	Silicon	2	20	1.5
Disc	Silicon	Radius = 12 um		1.5

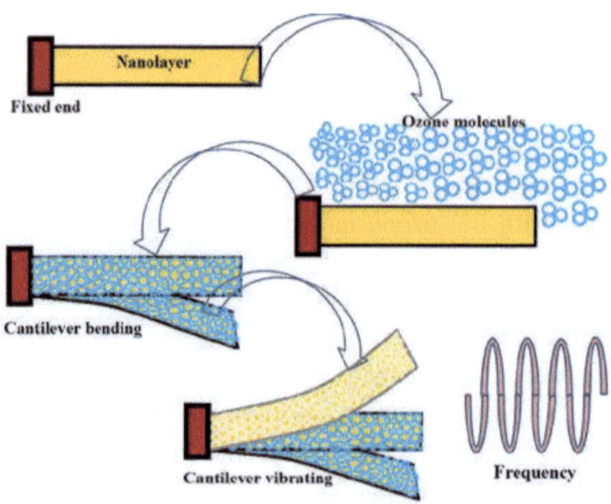

Figure. 2: Sensing mechanism of the microsensor

The small amount of mass change can be expressed as

$$\delta m = \frac{K}{4\pi^2 (v_2^{-2} - v_1^{-2})} \qquad (2)$$

where v_2 and v_1 are the change in resonance frequency before and after adding the mass with the cantilever.

The change in the resonance frequency of the cantilever is given by,

$$\frac{v_2 - v_1}{v_1} = \frac{(\delta K / K - \delta m / m)}{2} \qquad (3)$$

3 Results And Discussion

3.1 Von Mises Stress

von Mises stress is mainly used for finding whether a design can withstand a given load/stress. It is a geometrical combination of all the stresses (normal stress in the three directions, and all three shear stresses) acting at a particular area of our model.

The proposed microsensor was allowed to vibrate by applying a uniform load on the surface of the cantilevers. The given stress was equally distributed on the surface, inducing strain (deformation/bending). Then the applied load was varied up to 15 N, with a maximum displacement of 1.7×10^7 nm for both the cantilevers. Figure 3 shows the von Mises stress distribution on the surface of the cantilevers. The von Mises stress is at a maximum towards the fixed end of the cantilevers, and the value is 6.8678×10^5 Pa. As the yield point value of the gold is 100 GPa, the proposed microsensor design is safe.

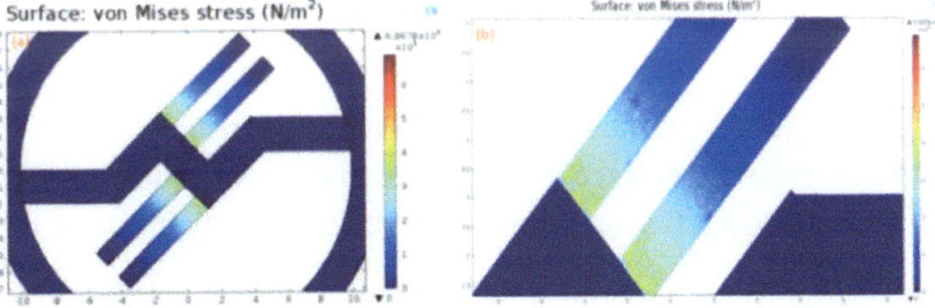

Figure. 3: von Mises stress (a) 2D view of the microsensor, (b) focused view of the microsensor

3.2 Performance of the microsensor

The stationary study was performed to find the displacement characteristics of the microcantilever. Each cantilever was vibrated under a uniform stress. The ozone gas was applied vertically to the surface of the gold coated microcantilevers. The vibrating displacement of the cantilevers was found to be small when ozone was passed, whereas it was found to be a maximum without ozone. This clearly shows that the adsorption of ozone on the gold nanolayers changes the vibrating displacement of the cantilevers, which in turn changes their resonating frequency. This exhibited that the proposed model is best suited for sensing the ozone concentration. Figure 4 (a) shows that the cantilever displacement is a maximum at 10000 nm before the passage of the ozone gas, whereas it reduces to 9000 nm, after applying the ozone gas as shown in figure 4 (b).

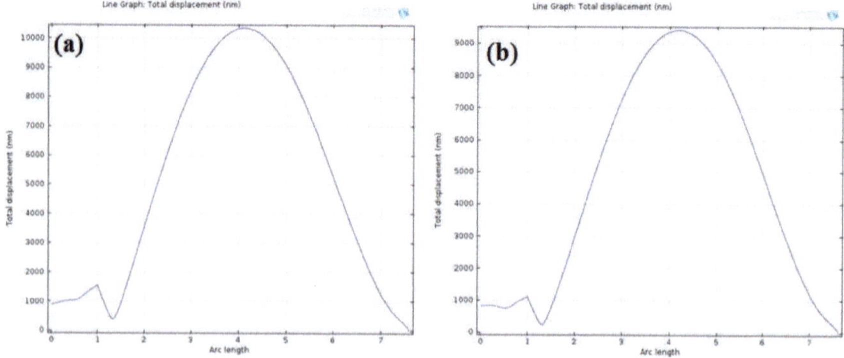

Figure. 4: Total displacement of the cantilever

The dynamic mode of operation of the proposed simulated sensor is shown in Figure 5. The ozone molecules interact with the nanolayer, increase the cantilever mass, and decrease the amplitude of vibration, showing that the vibrational frequency of the microcantilevers was changed after adsorption of ozone molecules.

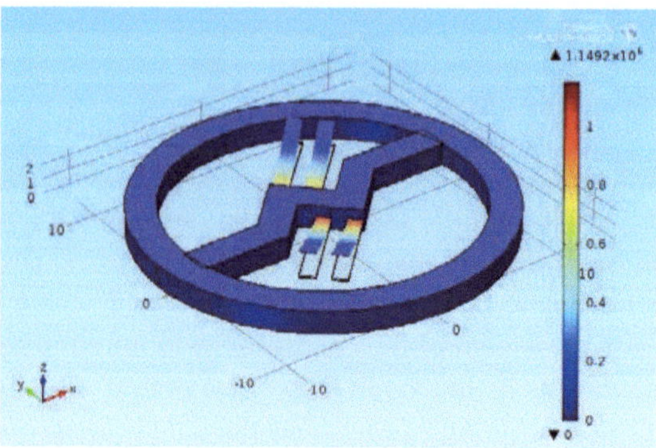

Figure. 5: Simulated microsensor

An eigenfrequency study was conducted for finding the resonating modes of the sensor and the graph of solution number vs frequency of the resonating cantilevers is given in Figure 6. The graph looks like a staircase function and has nearly flat plateaus. The eigenfrequency response for the proposed sensor model was found to give the resonance characteristics up to 4 MHz. But, for sensor fabrication, an optimum value of 2 to 2.5 MHz can be used as a detection resonance frequency range, since this range is almost half of the sensor resonance bandwidth frequency.

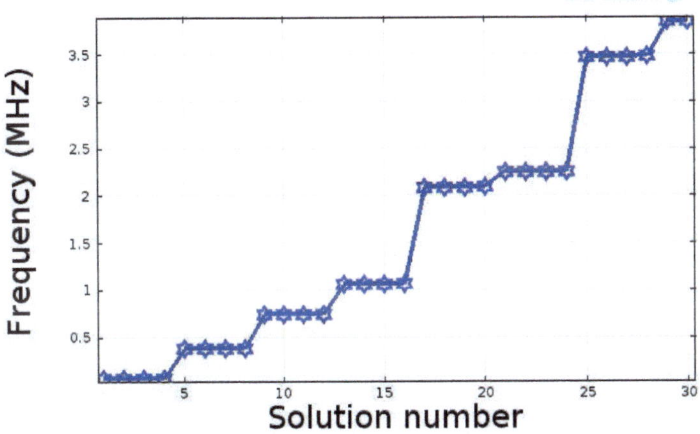

Figure. 6: Eigenfrequency resonance modes of the simulated microsensor

Conclusion

In the present simulation study, we have modeled a two-sided, coupled micro-cantilevers based microsensor for ozone sensing. The proposed sensor was simulated and analyzed in COMSOL MULTIPHYSICS. The proposed model was characterized for material as well as dimensional properties in addition to the analytical properties. From our analysis, the proposed model is found to be highly sensitive for small mass variations, as evident from the von Mises stress analysis and eigenfrequency study. Further, the cantilever displacements after ozonation show that the proposed model can be used to fabricate a small ozone sensor.

Acknowledgments

Financially this work was supported by University Grant commission (UGC), Rajiv Gandhi National Fellowship for Students with Disabilities (RGNFD).

REFERENCES

1   Chen, G. Y., Thundat, T. E., Wachter, A., & Warmack, R. J. (1995). "Adsorptioninduced surface stress and its effects on resonance frequency of microcantilevers", *Journal of Applied Physics*, 77(8), 3618–3622.
2   Puckett, S.D., Heuser, J.A., Keith, J.D., Spendel, W.U., & Pacey, G.E. (2005). "Interaction of ozone with gold nanoparticles", *Talanta*, 66(5), 1242–1246.
3   Plante, M.C., Garrett, J., Ghosh, S.C., Kruse, P., Schriemer, H., Hall, T., & LaPierre, R.R. (2006). "The formation of supported monodisperse Au nanoparticles by UV/ozone oxidation process", *Applied Surface Science*, 253(4), 2348–2354.
4   Pisarenko, A. N., Spendel, W. U., Taylor, R. T., Brown, J. D., Cox, J. A., & Pacey, G. E. (2009). "Detection of ozone gas using gold nanoislands and surface plasmon resonance", *Talanta*, 80(2), 777–780.
5   Banerjee, A., & Banerjee, S. S. (2013). "Fabrication of single and coupled metallic nanocantilevers and their nanomechanical response at resonance", *Nanotechnology*, 24(10), 1–12.
6   Banerjee, A., Rajput, N. S. & Banerjee, S. S. (2012). "Unusual dimensional dependence of resonance frequencies of Au nanocantilevers fabricated with self-organized microstructure", *American Institute of Physics Advances*, 2(3), 032105-1 – 032105-9.
7   Hayden, V. C., Saydur Rahman, S. M., & Beaulieu, L. Y. (2013). "Computational analysis of the effect of surface roughness on the deflection of a gold coated silicon micro-cantilever", *Journal of Applied Physics*, 113(5), 054501–054507.
8   Wasisto, H.S., Merzsch. S., Waag, A., Uhde, E., Salthammer, T., & Peiner, E. (2013). "Airborne engineered nanoparticle mass sensor based on a silicon resonant cantilever", *Sensors and Actuators B*, 180(1), 77– 89.
9   Pakdast, H., & Lazzarinoa, M. (2012). "Triple coupled cantilever systems for mass detection and localization", *Sensors and Actuators A*, 175(1), 127– 131.
10   Yoon, J. H., Kim, B. J., & Kim, J.S. (2012). "Design and fabrication of micro hydrogen gas sensors using palladium thin film", *Materials Chemistry and Physics*, 133(2-3), 987– 991.
11   Gerlich, V., Sulovská, K., & Zálešák, M. (2013). "COMSOL Multiphysics validation as simulation software for heat transfer calculation in buildings: Building simulation software validation", *Measurement*, 46(6), 2003–2012.
12   Korsunsky, A. M., Cherian, S. R. R., & Berger, R. (2007). "On the micromechanics of microcantilever sensors: Property analysis and eigenstrain modeling", *Sensors and Actuators A*, 139(1-2), 70–77.
13   Firdaus, S. M., Azid, I. A., Sidek, O., Ibrahim, K., & Hussien, M. (2010). "Enhancing the sensitivity of a mass-based piezoresistive micro-electro-mechanical systems cantilever sensor", *Micro & Nano Letters, IET*, 5(2), 85–90.

14 Bashir, R., Gupta, A., Neudeck, G. W., McElfresh, M., & Gomez, R., (2000). "On the design of piezoresistive silicon cantilevers with stress concentration regions for scanning probe microscopy applications", *Journal of Micromechanics and Microengineering,* 10(4), 483–491.

Adithya H K Upadhya [1], Avinash K [2] and K Chandrasekaran [3]

Secure Data Management – Secret Sharing Principles Applied To Data Or Password Protection

Abstract: This paper deals with the introduction of an information security mechanism to deal with the increasing threats of password hijack attacks. The proposed work deals with application of secret sharing principles for the protection of critical credentials of the stakeholders such as login Id and password. It is based on the fact that individual segments of passwords or credentials are ineffective without the existence of the remaining segments. We try to establish this fact to our advantage by proposing an implementation where password segments are distributed across several databases to strengthen their security and integrity and thereby, eliminate the threat of credential hijacking or phishing. The distributed nature of the divided segments of data can ensure lower probability of a successful attack. To summarize, an alternative encryption mechanism based on secret sharing is proposed which could be computationally efficient and which could provide equivalent or greater security in comparison with the standard cryptographic algorithms.

Keywords: Secure Data Management, Augmented Password Protection, Secret Sharing Principles, and Cryptographic Algorithms.

1 Introduction

Data security is given prime importance in almost all modern day applications. This is very important because a large amount of damage can be caused if the malicious hackers obtain unauthorised access to the confidential data. As a result, many schemes of encrypting data have been proposed. Owing to the

1 Department of Computer Science and Engineering, National Institute of Technology Karnataka, Surathkal, Karnataka, India
2 Department of Computer Science and Engineering, National Institute of Technology Karnataka, Surathkal, Karnataka, India
3 Department of Computer Science and Engineering, National Institute of Technology Karnataka, Surathkal, Karnataka, India

growing number of threats, the login, password and other user related credentials are encrypted using certain standard encryption algorithm.

However, in the recent years, several methods have been determined to exploit the vulnerabilities in the encryption algorithm. Disclosure of encryption or decryption key could result in divulgence of confidential data. Therefore, it's evident that there is a scope for improvement in most of the algorithms utilised in the current scenario. There is a pressing need for an additional layer of security over the standard encryption algorithms. In this paper, a new approach is proposed so as to ensure data security through application of secret sharing based security mechanisms and distributed databases.

The crucial information such as user login name and password could be segregated into several segments and be stored in various different locations. The main motive behind this is that the probability of an attack being successful declines drastically as the number of databases used to store the credentials' segments increases. Hence, the encrypted information could be divided into several segments and be stored across different locations to impart an additional layer of security.

In some applications, the use of sophisticated encryption algorithms may not be required or feasible. Hence, computationally efficient and secure mechanisms will be the need of the hour. Succinctly speaking, the concept of secret sharing mechanisms where confidential information is segregated into segments and stored across distributed databases could be the solution to the need of secure, reliable and computationally efficient encryption and data storage mechanisms. The protection system proposed shall have a vast number of applications and could also contribute to the cloud computing security.

2 MOTIVATION

Several cryptographic algorithms such as Blowfish, MD5, DES, TripleDES and so on have been implemented in the past for secure data and password storage. However, the downside with these algorithms is that some of these are computationally intensive and algorithms such as DES or TripleDES require higher number of bits for their respective encryption keys in order to create a secure cipher text.

Besides, several algorithms mentioned above are Symmetric encryption and decryption algorithms. Hence disclosure of their encryption keys could in turn result in revelation of the passwords and credentials. Drawing inspiration from the problem stated above, our proposed work strives to provide greater or equivalent data security through application of secret sharing mechanisms on the targeted data, password and also the encryption key.

Furthermore, emphasis is also given on the time complexity and computational efficiency of the encryption and decryption algorithms. For research purposes, a simple and effective XOR encryption is chosen. Hence, the computational efficiency and the security advantages are the main motivating factors behind the application and adoption of this secret sharing based mechanisms for secure password and credential management.

3 OVERVIEW

An implementation of the proposed idea was made through utilisation of phpMyAdmin and XOR encryption. The password entered by the user will be encrypted through a simple XOR encryption mechanism and the cipher text was symmetrically divided into a fixed number of segments based on the total number of MySQL databases intended for storage.

XOR encryption is a computationally efficient algorithm where the same key will be used for both encryption and decryption. In order to ensure further security, the key shall also be divided into several segments and stored across other databases exclusively meant for storing key segments. Finally, the segments would be stored in the intended databases in a particular order. The cognition of the order in which the password is distributed and the order in which the keys are distributed are vital for the success of decryption mechanisms employed later.

When a valid or authentic user wishes to access his protected data or information, he will be prompted to enter his valid credentials. Next the cipher text segments are retrieved from the databases and reordered appropriately. In our implementation, the cipher text next gets decrypted through the key used earlier for encryption. The decrypted cipher text is compared with the password entered. If at all they match, the user is granted the permission to access the application or else he is denied the access.

Hence, we try to demonstrate that the mechanism proposed above uses an extremely efficient and light weight encryption mechanism while maintaining strong security and integrity through secret sharing mechanisms. This simple mechanism could be used to replace the computationally intensive and insecure encryption mechanisms. The typical disadvantage rendered by the symmetric cryptographic algorithms consists of security aspects of the key used for encryption. Asymmetric encryption could offer greater advantage in terms of security since different key is used for encryption and decryption.

4 RELATED WORK

This section deals with a brief overview of the previous work and research conducted in the area of secret sharing principles. Several concepts of the proposed work are inspired from the secret sharing principles stated below.

4.1 Shamir's secret sharing

Secret sharing is applicable in cases where the key would not be stored as a complete unit and all the various segments of the key are required to recover the secret. Suppose that the key is stored as a single complete unit, the chances of security breach will be very high as the attacker needs to obtain access to the single location containing the complete key. Due to this reason, the key needs to be segregated into several segments and be stored across several databases. The secret could become unrecoverable if any constituent part or segment of the key gets lost. Taking these points into consideration, it is rational to generate a method the secret could be accessible even if all the parts of the key are not available.

This idea has been used in Shamir's secret sharing principles. In this method, the key is divided into 'n' segments and any 'k' segments out of the 'n' parts are sufficient to retrieve the secret. The concept used here has been borrowed from graph theory and algebra where a particular polynomial of a given degree 'k' can be determined given the existence of 'k+1' points. For example, a straight line requires minimum two distinct points and a parabola requires minimum three points to fit a parabolic curve. Same principles of curve fitting are applicable to the polynomials of higher degree (Shamir (1979)).

The procedure followed in this algorithm is to take the secret stored as the co-efficient of the zero degree variable. All the other coefficients could be randomly chosen for the polynomial. The second stage is to get a fixed number of points 'n' which are distributed amongst all the stakeholders involved in the secret sharing mechanism. When a minimum of k stakeholders share their secrets, a polynomial of degree k+1 could be determined (Shamir (1979)).

The concept could be illustrated through a mathematical elaboration. A polynomial of degree M requires M+1 points to uniquely identify it from other polynomials. Let the threshold placed in order to share the secret S be (M, N).

The polynomial is built in such a way that all the coefficients are less than P, a
prime number. Hence the secret C_0 and all other coefficients C_1, C_2, C_3... C_M are
less than P.

The polynomial that we arrive at is (Shamir (1979)):

$$F(x) = C_0 + C_1 x + C_2 x^2 + C_3 x^3 + \cdots + C_M x^M \qquad (1)$$

Here, C_0 is equal to the secret that is intended for protection. Next, N distinct
points are randomly drawn from the equation to obtain the pairs [a, F(a)] for
different values of 'a'. The secret holders store the pair of values consisting of
input to the polynomial and the corresponding polynomial output (Shamir
(1979)). The entire equation and the coefficients can be obtained through the
method of curve fitting applied on any M subsets out of these N pairs. This
secret sharing principle was developed independently by A. Shamir and
Blakeley in 1979. This method provides a secure and reliable mechanism
intended for protection of secret.

4.2 Blakeley's secret sharing scheme

The concept of planar geometry is used in the Blakeley's secret sharing scheme.
The number of points of intersection possible for any two straight lines that are
not parallel to each other is one. In order to reduce the amount of specificity, let
us consider k planes that are not parallel to each other. These k planes have k-1
dimensions and each intersect at a specific point (Blakeley (1979)). It is possible
to take the secret and store it as the point where the planes intersect. However,
there is a major drawback with the allocation of a plane to the stakeholder is.
The secret holder will realise that the secret exists on the target plane allocated
to him. For a strong and secure system, any person involved in the secret sharing should not have any more cognition of the secret than any person outside
the system (Blakeley (1979)). After the planes have been allocated to the
stakeholders, the secret can be obtained only after the point of intersection of
all the different hyper planes is achieved.

5 ANALYSIS OF THE PROPOSED WORK

In this section, we strive to analyse the time complexity of the mechanisms
proposed for the credential management. The objective is to depict the estab-
lishment of reliable security while this password protection mechanism is utilised. In this proposed work, the storage of secret segments takes place across

several distributed databases. Let's consider a system hosting a total of N databases. Our intention is to distribute the secret segments across the M distributed databases.

For the simplification of the threat analysis, let's assume that the information pertaining to the total number of databases utilised for the maintenance of the cipher segments i.e. the value of 'M' is known beforehand to the malicious attacker. Under such circumstances, the malicious attacker has the problem of selecting 'M' databases out of the 'N' databases. The number of ways of selecting exactly 'M' databases out of 'N' databases is equal to ^N_MC .Out of these ^N_MC possible ways, only one set of the databases is valid which is responsible for the storage of cipher segments. Since there is just one legitimate configuration, the probability of arriving at the exact same 'M' valid databases out of the total 'N' distributed databases would be probabilistically equal to

$$P(selecting\ M\ valid\ databases\ out\ of\ N) = 1/^N_MC \quad (2)$$

Let the probability of a given database being hacked leading to information disclosure be α. In an Ideal and well protected database (established security against common attacks such as SQL injection), the value of α→0. Henceforth, the probability of occurrence of concurrent and simultaneous attacks on the protected databases leading to information disclosure of the secret segments would be

$$P\ (A_1 \cap A_2 \cap \ ...\ ...\cap A_m) = P(A_1) \times P(A_2) \times\ ....\ P(A_M) \quad (3)$$

As a result, this would be equal to $\alpha_1 \times \alpha_2 \times \alpha_3 ... \times \alpha_m$. If we make an assumption that all the databases storing the secret segments are equally protected, we can fairly assume that the value will converge to the result $(\alpha)^M$. Therefore, it's understood that the probability of information disclosure of all the well protected databases responsible for secret segment storage declines drastically as the number of databases involved in secret management increase.

$$lim\ (\alpha)^M \to 0\ as\ M \to \infty \quad (4)$$

Now let us consider another arena where this method ensures comprehensive security. The encrypted cipher text or secret segments are stored across M databases on a system consisting of total N databases. If the malicious attacker has the access to the encryption key, all he requires is the reordered cipher text consisting of reordered secret segments stored across the M databases. The total number of permutations of the M secret segments would be the factorial of M. Hence the probability of obtaining the accurate cipher text or complete arrangement of secret segments would be the inverse of factorial of M.

$$P(acquiring\ accurate\ order\ of\ cipher\ text) = {}^{1}\!/_{M!} \quad (5)$$

The probability of concurrent information disclosure of M databases holding secret cipher segments has been shown previously. This result can be utilized for further analysis along with the Eq. (5).

$$P(information\ disclosure\ of\ all\ M\ databases) = (\alpha^{M}) \quad (6)$$

Consequently, we arrive at the probability of concurrent disclosure of M databases and acquiring the precise order of cipher text or secret segments. The probability of successfully acquiring the cipher text would be the intersection of two events namely, the probability of disclosure of the M databases involved in secret storage out of N databases and the probability of reordering the M secret segments to obtain the legitimate configuration. Thereby, the resultant probability of the successful attack would be:

$$P(successful\ attack) = P(selecting\ M\ valid\ databases\ out\ of\ N)\ \times\ (\alpha^{M}) \times 1/M!$$

Therefore, the result would be,

$$P(successful\ attack) = 1/{}_{M}^{N}C \times (\alpha^{M})/M! = (\alpha^{M})/{}_{M}^{N}P \quad (7)$$

Here, ${}_{M}^{N}P$ represents the permutation of selecting M items out of N while considering the order and arrangement of the M items. In case the malicious user has the access to the encryption key, he could next decrypt the cipher text to obtain the secret password or code. Hence, it would be a rational solution to design similar secret sharing protection mechanisms in order to protect the encryption key. Similarly, let's assume that the encryption key is also distributed in an order among 'K' databases out to 'N'. Let the susceptibility of every database holding the key segments be β. Inferring details from the results mentioned above, the probability of obtaining the complete encryption key would be:

$$P(Acquiring\ the\ encryption\ key) = (\beta^{K})/{}_{K}^{N}P \quad (8)$$

Finally the probability of waging a successful attack and decryption would be the intersection of the two events, namely complete disclosure and reassembly of the key and disclosure and reassembly of the cipher text. Therefore,

$$P(Successful\ attack\ and\ decryption\ of\ cipher\ text) = (\alpha^{M})(\beta^{K})/({}_{M}^{N}P \times {}_{K}^{N}P)$$
$$(9)$$

Thereby, we can determine that the probability of the successful attack declines at a rapid rate and approaches zero for large number of databases involved in the key or the data segment storage.

$$\lim_{M\to\infty \ or \ K\to\infty} (\alpha^M)(\beta^K)/(^N_MP \times \ ^N_KP) = 0 \qquad (10)$$

For the analysis of the result, In the Fig.1, we considered an example where both M and K would be equal to 5 and also where the variables α and β attained the value 0.75. The probability of the successful attack would decline at a rapid rate with the increasing number of distributed databases.

Fig.2 describes the drastic increase in the permutation of the password or key segments with the increase in the number of databases storing those segments.

6 COMPLEXITY ANALYSIS OF ALGORITHMS

Most of the Block Cipher algorithms such as DES, Triple DES, AES, Blowfish work with blocks of fixed size. Since the block sizes are fixed, their time complexity will be independent of the length of the input and thus their complexity will be of the order of O(1). Depending on the mode of operation such as ECB, CBC or CFB and the length of the message or password, we would arrive at a time complexity of O(K) where 'K' signifies the number of data blocks utilized to encrypt the message or password.

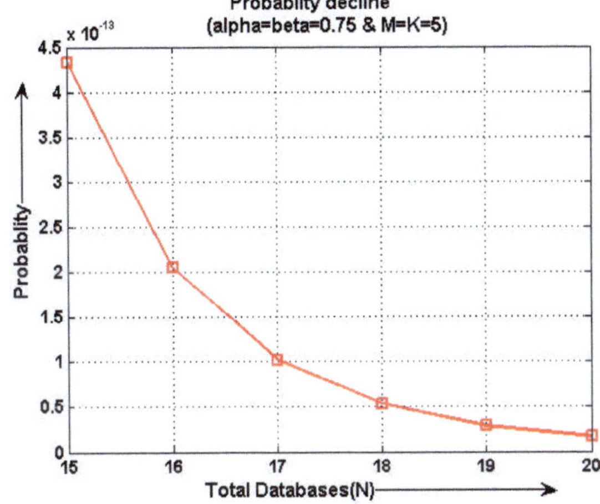

Figure. 1: Probablilty decline with increasing N(total databases)

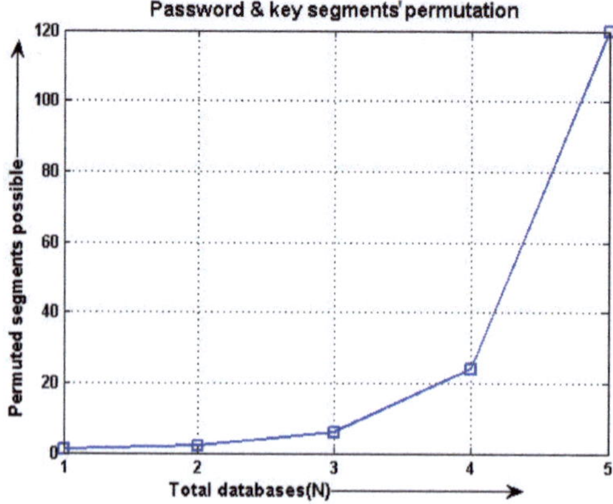

Figure. 2: The possible number of permutations of the segments of the key or the password increases with the databases involved in storage of segments

Encryption algorithms such as TripleDES usually have thrice the amount of computation in comparison with the DES counterpart. However, the time complexity in case of TripleDES will be restricted to the order of O(1) or O(K) depending upon the total data blocks utilized. Although, Blowfish algorithm on the other hand is slower owing to the key scheduling, the time complexity will still be restricted to O(1).

Modern day cryptographic algorithms utilize either MD5, SHA-1 or SHA-2 algorithms for encryptions. Recently MD5 has been proved to be insecure and is proven to be non-collision free. The block cipher algorithms utilize additional space for encryption but compensate for their requirement for space through efficient time complexity.

The prime reason for choosing the XOR encryption for the experiments in this proposed work are as follows; XOR encryption is symmetric and time efficient. The time complexity is of the order O(N) where N is the length of the target string. XOR encryption has been proven to be secure and reliable and requires no additional buffer or space for computation. Therefore, we propose to utilize lightweight encryption such as XOR encryption mechanism along with the secret sharing concepts for key and cipher text management instead of MD5, SHA-1 or SHA-2 since algorithms such as MD5 have been proved to cause collision and are slowly being phased out to introduce more robust hash algorithms such as SHA-256 or SHA-512.

7 PROMINENT CLOUD COMPUTING VULNERABILITIES

The major application of the password protection would obviously be the cloud computing infrastructure since the Cloud computing service providers identify and authenticate their customers through their credentials such as login and password. However, in the recent years, the cloud environment has been one among the favourite prey for the malicious hijackers owing to a humongous quantity of confidential data stored in the cloud infrastructure. Having asserted the importance of a password protection mechanism previously, this section deals with some other major challenges and drawbacks of the cloud environment apart from the necessity of a strong and reliable password and credential security management system in cloud infrastructure.

7.1 Resource pooling and rapid elasticity issues

Heterogeneity and Interoperability issues: Owing to the diversity in computational resources, security breach could occur if object's access to resources can't be handled in standard way. Hence, certain settings which are secure for a given resource could prove to be loophole for other resources and could cause certain to undergo halt

Host susceptibility: In the modern cloud computing era, user objects are vastly spread and distributed over various interconnected multi user hosts. In a multitenant environment, both the hosts and the user objects need to be protected from possible attacks. Evaluation of resource access request on every single user object on host needs to be made. Breach of host security could lead to inadvertent access to the host's resources and the tenant objects.

Vulnerability of objects: Security of objects could be violated in several ways. The service provider could access any user object currently residing at the hosts. Sometimes the mutual attack on each other's objects could take place which are the tenants of the same host. A malicious third party could also directly attack the user objects. A service provider is sometimes granted the privilege to access and modify the user objects for performing basic functions in the cloud environment. These could lead to an attack by service provider through modification of user objects. To prevent these, trust based secure systems could be implemented. Second form of attack could take place due to synchronous sharing of same resources by tenant objects. Third attack could consist of malfunction on the side of client due to bugs or faults.

7.2 Issues with broad network access and measured service security

Communication confidentiality: Confidential communication is the key factor of any networked computer system and PaaS clouds are not an exception. Objects must communicate in order to interoperate and to be accessed by their owners. The communication channel can be eavesdropped to extract valuable information as objects reside on remote hosts. Therefore, communication confidentiality must be obtained at all costs (Sandikkaya & Harmanc (2012)).
Requirement for authentication: Authentication is the first element of access control. It requires parties to prove the authenticity of their identities during an interaction. Unprivileged entities may access objects if authentication fails. The standards for the authentication mechanisms must be clear (Sandikkaya & Harmanc (2012)).

Requirement for authorization: Authorization mechanisms determine who can access to the objects based on predefined policies. Lack of authorization mechanisms may lead to unprivileged access. Role-based access control and federated access control help administrators manage authorization. Even then, they may not cover all problems in a PaaS cloud where objects migrate. Moreover, the hosts may be reconfigured in time and it may be difficult to keep up with the policies during these reconfiguration periods. A host that hosts numerous objects at a given time must know the policies of each object to apply them. The dynamic nature of PaaS clouds does not let the host keep these policies up-to-date locally.

Alternatively, the host may try to conduct online queries to fetch the policies of each object. This centralized approach may be a burden for the cloud user. Each object is likely to have its own access control requirements as an object may reside on any host at a given time. For example, when an object moves to a new host, the policies that are effective on the previous host must be effective on the current host too (Sandikkaya & Harmanc (2012)).

Traceability security issues: Traceability is majorly achieved through maintenance of records of the events occurred in a system. In addition to access control needs, measured service characteristic requires fairly kept event records. Event records are important for each cloud stakeholder. Service providers bill the users according to the amount of use and users monitor their applications' state and audit access to their data and jurisdiction investigates logs in case of conflicts in between parties. However, in a PaaS cloud, the logging system must be secure, protected against all interacting parties including the system administrators (Sandikkaya & Harmanc (2012)).

CONCLUSION

The objective of this proposed work is to explore new mechanisms in which encryption keys, passwords, debit card numbers, nuclear launch codes and equivalent vital entities could be protected through the secret sharing mechanisms. In case of password protection, several encryption mechanisms have been implemented so far. Some of these are extremely computationally intensive and are not suitable for small scale applications or applications which require comprehensive security. Some of the leading encryption mechanisms such as MD-5 have been found to cause collision and are proven to be insecure. There is a pressing need for computationally inexpensive, space efficient and reliable encryption mechanisms. Some of the contemporary encryption mechanisms also require encryption and decryption keys in order to maintain the security and consistency of the vital data. Nowadays, several cyber-terror attacks are also aimed at the disclosure and hijacking of the encryption or decryption keys.

The proposed solution ensures that in a large and distributed database system, the chances of acquiring the secret would be negligible. As shown in the previous section, the probability of the successful attack would tend to zero if we consider utilization of large number of databases involved in the secret storage.

The proposed work could offer solution to the large scale password and credential hijacking attacks that have been taking place recently. The document also describes the decline in the probability as the number of distributed databases increase. Another graph depicts the total number of permutations of the password segments or the key segments which increase at a rapid rate as the number of databases storing password or key segments keep increasing.

Henceforth, through algorithmic analysis and mathematical evaluation, we try to propose a password protection mechanism based on the secret sharing principles.

REFERENCES

1   Benaloh, J. & Leichter, J. (1990). "Generalized secret sharing and monotone functions", *Proceeding CRYPTO '88 Proceedings on Advances in cryptology*, Springer-Verlag New York, 27-35.

2   Bessani, A., Correia, M., Quaresma, B., Andre, F. & Sousa, P. (2013). "DepSky: Dependable and secure storage in a cloud-of-clouds", *ACM Transactions on Storage*, 9(4), article 12.

3   Blakley, G. R. (1979). "Safeguarding cryptographic keys". *Proc. AFIPS NCC*, 48, 313-317.

4   Catteddu, D. & Hogben, G. (2009). "Cloud computing: Benefits, risks and recommendations for information security". *Technical report, ENISA.*

5   Feldman, P. (1987). "A practical scheme for non-interactive verifiable secret sharing", *Foundations of Computer Science 28th Annual Symposium*, Los Angeles, USA, 427-438.

6   Jensen, M., Schwenk, J., Gruschka, N. & Iacono, L.L. (2009). "On Technical Security Issues in Cloud Computing", *IEEE International Conference in Cloud Computing*, 109-116.

7   Kandukuri, B. R., Paturi, V. R. & Rakshit, A. (2009). "Cloud security issues", *SCC '09. IEEE International Conference on Services Computing*, 517 – 520.

8   Kaufman L. M. (2009). "Data security in the world of cloud computing", *IEEE security and Privacy*, 7(4), 61–64.

9   Lombardi, F. & Pietro, R.D. (2010). "Transparent security for cloud", *SAC '10 Proceedings of the 2010 ACM Symposium on Applied Computing*, ACM New York, 414-415.

10   Morris, R., Thcompson, K. (1979). "Password security: a case history", *Communications of the ACM*, Volume 22, Issue 11, ACM New York, 594-597.

11   Pearson, S., Shen, Y., & Mowbray, M. (2009). "A Privacy Manager for Cloud Computing", *First International Conference, CloudCom, Beijing, China*, 90-106.

12   Pedersen, T. P. (1992). "Non-interactive and information-theoretic secure verifiable secret sharing", *Advances in cryptology —CRYPTO '91, Lecture Notes in Computer Science*, 576, 129-140.

13   Rabin, T., Ben-Or M. (1989). "Verifiable secret sharing and multiparty protocols with honest majority". *Proceeding STOC '89 Proceedings of the twenty-first annual ACM symposium on Theory of computing.* ACM New York, 73-85.

14   Rivest, R., Shamir, A. & Adleman, L. (1978). "A method for obtaining digital signatures and public-key cryptosystems", *Comm. ACM*, 120-126.

15   Ryan, M. D. (2013). "Cloud computing security: The scientific challenge, and a survey of solutions", *Journal of Systems and Software*, 86(9), 2263-2268.

16   Sandikkaya, M. T. & Harmanc, A.E. (2012). "Security problems of platform-as-a-service (PaaS) clouds and Practical Solutions to the Problems", *31st International Symposium on Reliable Distributed Systems*, 463-468.

17   Schneier, B. (1996). "Applied Cryptography", *Wiley, New York*, 70.

18   Shaikh, F.B. & Haider, S. (2011). "Security threats in cloud computing", *International Conference for Internet Technology and Secured Transactions (ICITST)*, 214-219.

19   Shamir, A. (1979). "How to share a secret", *Communications of the ACM*, 22(11), 612-613.

Jithendra.K.B [1] and Shahana.T.K [2]

Enhancing the Uncertainty Of Hardware Efficient Substitution Box based on Linear Cryptanalysis

Abstract: The industry and individuals are in search of compact devices with improved performance. This is true for secure communication systems also. Reduced hardware complexity usually reduces the security of a system. It is a real challenge to design a crypto system with high security with lesser hardware complexity. In this paper linear cryptanalysis of a part of hardware efficient Substitution box is carried out. The analysis proves that the security of the S Box is better than that of conventional one. Moreover, this paper gives some clues to enhance the uncertainty further, based on linear cryptanalysis.

Keywords: Encryption, block cipher, security, substitution box, linear cryptanalysis.

1 Introduction

The Substitution box introduced by C.Shannon (1949) is a significant element for the block ciphers since it contributes greatly to security through its nonlinear input-output characteristics. Normally the strength of the S box is decided by the complexity of the algorithm which determines the hardware complexity. The strength of the S box can only be proved by the cryptanalysis. Mainly there are two types of cryptanalysis for block ciphers: Linear cryptanalysis and differential cryptanalysis explained by Heys.H (2002), Musa et al. (2003) and Sinkov.A (1966). The paper by Jithendra.K.B and Shahana.T.K (2014) introduces a new S box which claims to have better security with reduced hardware complexity. Here the security is measured through the eye of linear

1 Department of Electronics and Communication, College of Engineering, Vadakara,Calicut (Dt), Kerala, INDIA
jithendrakb@gmail.com
2 Division of Electronics Engineering, School of engineering, Cochin University of Science and Technology, Kerala, INDIA
shahanatk@cusat.ac.in

cryptanalysis and compared with that of the conventional systems. The analysis opens new methods which can enhance the uncertainty to greater levels.

2 Linear Cryptanalysis Of A Conventional S Box

The linear cryptanalysis examines the extent to which linear relations exist between input vector and output vector, Matsui (1993). The existence of simple linear relations definitely unveils the key bits easily and thus defeats the very purpose of S box.

The Figure 1 shows a well defined 4 bit S box with Boolean variable set { x_3, x_2, x_1, x_0 } as input vector and Boolean variable set { y_3, y_2, y_1, y_0 } as output vector.

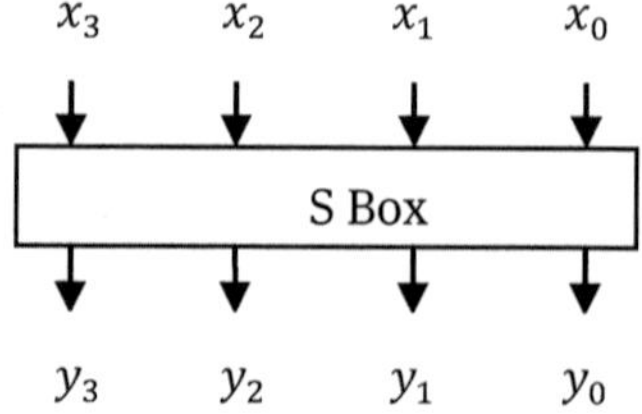

Figure. 1: S Box block diagram

Suppose the S box-1 is defined as in Table 1

Table 1. S box – 1

x_3x_2 / x_1x_0	0	1	0	1
00	C	D	5	2
01	B	A	6	3
10	0	8	E	9
11	7	1	F	4

This shows the input vector 0000 will be substituted by 1100 (C Hex), 0001 by 1011(B Hex) and so on.

Suppose a linear relation

$$x_3 \oplus x_2 \oplus y_1 \oplus y_0 = 0 \qquad (1)$$

is suspected. The probability of the existence of this linear relation can be found out by examining all possible 16 cases

Table 2. Linear approximation of Eq. (1) with S box-1 data

x_3	x_2	x_1	x_0	y_3	y_2	y_1	y_0	If Eq. (1) valid?	
								Yes	No
	0	0	0	1	1	0	0	√	
	0	0	1	1	0	1	1	√	
	0	1	0	0	0	0	0	√	
	0	1	1	0	1	1	1	√	
	1	0	0	1	1	0	1	√	
	1	0	1	1	0	1	0	√	
	1	1	0	1	0	0	0		√
	1	1	1	0	0	0	1	√	
	0	0	0	0	1	0	1	√	
	0	0	1	0	1	1	0	√	
	0	1	0	1	1	1	0	√	
	0	1	1	1	1	1	1		√
	1	0	0	0	0	1	0		√
	1	0	1	0	0	1	1	√	
	1	1	0	1	0	0	1		√
	1	1	1	0	1	0	0	√	

The Eq. (1) is true for 12 cases and false for 4 cases. So the probability bias of happening the condition given in Eq. (1) is 4/16 or 1/4.

The biases for all function pairs for S box – 1 is calculated as in C.K Shyamala et al. (2011) as in Table 3.

Table 3. Bias for function pairs of S box -1

	0	y_0	y_1	y_0y_1	y_2	y_2y_0	y_2y_1	$y_2y_1y_0$	y_3	y_3y_0	y_3y_1	$y_3y_1y_0$	y_3y_2	$y_3y_2y_0$	$y_3y_2y_1$	$y_4y_3y_2y_1$
0	8	0	0	0	0	0	0	0	0	0	0	0	0	0	0	0
x_0	0	2	4	-2	0	2	0	2	-2	0	-2	0	2	4	-2	0
x_1	0	0	-2	-2	0	0	-6	2	0	0	-2	-2	0	0	2	2
$x_0 x_1$	0	-2	2	0	-4	2	-2	-4	2	0	0	-2	2	0	0	-2
x_2	0	0	-2	2	-4	0	2	2	0	0	2	-2	0	4	2	2
$x_2 x_0$	0	2	2	0	0	-2	-2	0	2	-4	-4	2	2	0	0	2
$x_2 x_1$	0	0	4	0	0	-4	0	0	0	0	0	-4	-4	0	0	0
$x_2 x_1 x_0$	0	-2	0	2	0	2	0	-2	-2	-4	-2	0	-2	0	-2	4
x_3	0	0	2	2	2	2	0	0	-2	-2	0	0	0	0	6	-2
$x_3 x_0$	0	2	2	4	-2	0	0	2	0	2	-2	0	2	-4	0	2
$x_3 x_1$	0	0	0	0	-2	-2	-2	-2	-6	2	2	2	0	0	0	0
$x_3 x_1 x_0$	0	-2	0	-2	-2	-4	2	0	0	-2	-4	2	2	0	2	0
$x_3 x_2$	0	0	0	4	2	-2	-2	-2	2	2	-2	2	0	4	0	0
$x_3 x_2 x_0$	0	2	0	-2	2	0	2	-4	0	2	0	-2	2	0	2	4
$x_3 x_2 x_1$	0	0	-2	2	2	-2	0	0	-2	-2	0	-4	4	0	-2	-2
$x_4 x_3 x_2 x_1$	0	6	-2	0	-2	0	0	-2	0	-2	-2	0	-2	0	0	-2

From the above shown S Box, some weak trails are identified. The Figure 2 shows the Substitution Permutation Network (SPN) in block diagram form. Here the analysis is being carried out for a single stage.

b_i p[m,n] represent the ith bit of the nibble p[m,n].

p[m,n] represents input to n^{th} nibble (S- box) of mth round

q[m,n] represents output to n^{th} nibble (S- box) of mth round

w[m,n] represents input (n^{th} nibble of mth round) to SPN

k[m,n] represents key bits (n^{th} nibble of mth round)

Here

$$b_3\, \mathrm{p}[0,1] = b_3\, \mathrm{w}[0,1] \oplus b_3\, \mathrm{k}[0,1] \tag{2}$$

$$b_3\, \mathrm{p}[0,1] \oplus b_1\, \mathrm{p}[0,1] \oplus b_0\, \mathrm{p}[0,1] = b_3\, \mathrm{p}[0,1] \oplus b_1\, \mathrm{p}[0,1] \oplus b_0\, \mathrm{p}[0,1] \oplus b_3\, \mathrm{k}[0,1] \oplus b_1\, \mathrm{k}[0,1] \oplus b_0\, \mathrm{k}[0,1] \tag{3}$$

It can be seen from the Table that, the bit combination b_3, b_1 and b_0 at the input of Substitution Box S[0,1] decides the bit b_1 at its output side with a probability bias of -2/16 = -1/8. Replacing the input bits by output bit

$$b_1\, q[0,1] = b_3\, p[0,1] \oplus b_1\, p[0,1] \oplus b_0\, p[0,1] \oplus b_3\, k[0,1] \oplus b_1\, k[0,1] \oplus b_0\, k[0,1] \tag{4}$$

Further

$b_1 w[1,1] = b_1 q[0,1]$ $\qquad$ (5)

Combining Eq. (4) and Eq. (5)

$b_1 w[1,1] = b_3\ p[0,1] \oplus b_1\ p[0,1] \oplus b_0\ p[0,1] \oplus b_3\ k[0,1] \oplus b_1\ k[0,1] \oplus b_0\ k[0,1]$ $\qquad$ (6)

Equation 5 is valid with a bias of -2/16.

Further $\qquad b_1 p[1,1] = b_1 w[1,1] \oplus k_1 w[1,1]$

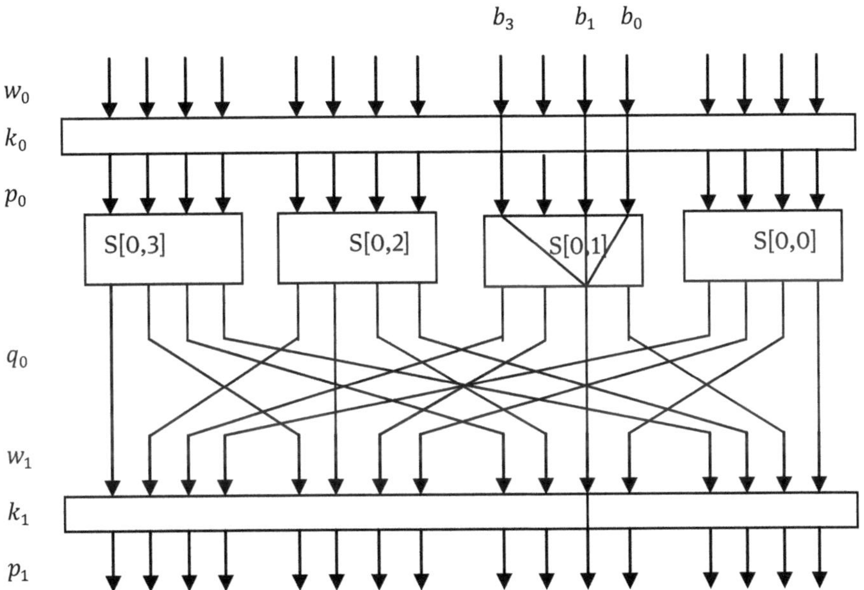

Figure. 2: SPN – single round

The same analysis can be carried out in coming stages and finally the relation between the output vector and the input vector can be find out, from which the keys can be obtained.

2.1 Analysis of the hardware efficient S Box

The paper above mentioned discuss about a hardware efficient S box, differing from conventional approaches given in Mister, S.. and Adoms (1996) and Heys, H.M, Tavares (1996). This concept divides the input vector into two vectors with half the number of bits of the parent vector. This effectively reduces the number of data required for substitution as shown in Figure 3. The additional number of LUTs provides the facility to select the S box at run time. This enhances the security. A diffusion system is also provided to attain increased uncertainty and to satisfy Strict Avalanche Criteria (SAC). Cryptanalysis of the diffusion system is not carried out here.

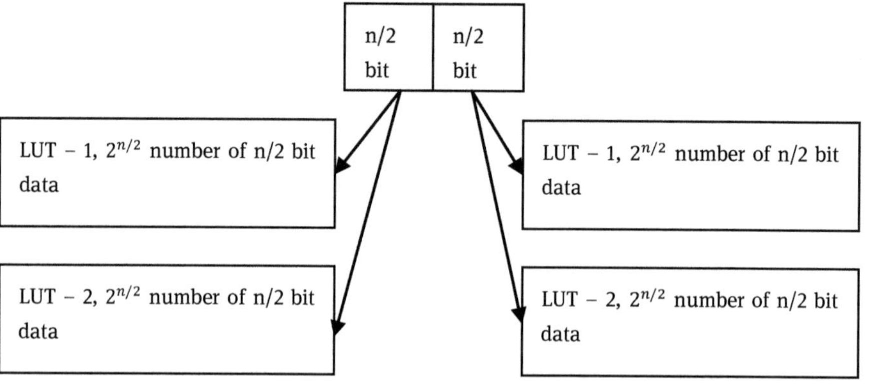

Figure. 3: Hardware efficient substitution approach

The security analysis of the LUT part is carried out below. Let the S box − 1 be the LUT-1 and S box − 2 be LUT -2 which is given in Table. 4.

Table 4 . S box -2

x_3x_2 / x_1x_0	00	01	10	11
00	7	6	3	9
01	C	4	5	E
10	F	D	1	0
11	2	B	A	8

Let n = 8 for the Figure 2. Now for the most significant 4 bits, 2 S Boxes are assigned in which only one will be used at a time depends on the control bit that is generated in a random fashion in Jithendra.K.B and Shahana.T.K (2014). Let LUT - 1 be S Box – 1(Table. 1) and LUT – 2 be S Box – 2 (Table. 2). Carrying out the linear cryptanalysis using the Table 3 and 5, for the S box shown in Figure 3 with the linear approximation taken in Figure 2, for a single round,

$b_1 q[0,1] = \overline{K} [b_3 p[0,1] \oplus b_1 p[0,1] \oplus b_0 p[0,1] \oplus b_3 k[0,1] \oplus b_1 k[0,1] \oplus b_0 k[0,]$
with a probability bias of -2/16

$+$ (7)

$K [b_3 p[0,1] \oplus b_1 p[0,1] \oplus b_0 p[0,1] \oplus b_3 k[0,1] \oplus b_1 k[0,1] \oplus b_0 k[0,1]]$
with a probability bias of 0. Where

K is the binary random bit, which selects either S box – 1 or S box – 2, for substitution, based on its status. Clearly the Eq.(7) indicates the cryptanalysis became more complicated. The probability biases in two conditions are different. This factor increases the security.

Table 5. Biases for different function pairs of S box – 2

	0	y_0	y_1	$y_0 y_1$	y_2	$y_2 y_0$	$y_2 y_1$	$y_2 y_1 y_0$	y_3	$y_3 y_0$	$y_3 y_1$	$y_3 y_1 y_0$	$y_3 y_2$	$y_3 y_2 y_0$	$y_3 y_2 y_1$	$y_4 y_3 y_2 y_1$
0	8	0	0	0	0	0	0	0	0	0	0	0	0	0	0	0
x_0	0	-4	0	0	0	0	4	0	2	2	-2	2	2	-2	2	2
x_1	0	0	0	0	-4	-4	0	0	2	-2	-2	2	-2	2	-2	2
$x_0 x_1$	0	0	-4	0	4	0	0	0	0	0	-4	0	-4	0	0	0
x_2	0	-2	-2	0	0	2	-2	-4	2	-4	0	-2	2	0	0	2
$x_2 x_0$	0	-2	-2	4	0	-2	2	0	0	-2	2	0	0	-2	-2	-4
$x_2 x_1$	0	-2	2	4	0	2	-2	4	0	-2	-2	0	0	2	2	0
$x_2 x_1 x_0$	0	2	-2	0	0	2	-2	0	2	0	0	6	2	0	0	-2
x_3	0	0	-2	2	-4	0	-2	-2	0	4	-2	-2	0	0	2	-2
$x_3 x_0$	0	0	-2	-2	-4	4	2	2	-2	-2	0	0	-2	-2	0	0
$x_3 x_1$	0	4	2	2	0	0	2	-2	2	-2	0	0	-2	-2	4	0
$x_3 x_1 x_0$	0	0	-2	-2	0	0	2	2	4	0	2	-2	0	4	2	-2
$x_3 x_2$	0	2	0	2	0	2	4	-2	-2	0	-2	0	2	4	-2	0
$x_3 x_2 x_0$	0	-2	0	2	0	2	0	-2	0	2	4	2	-4	2	0	2
$x_3 x_2 x_1$	0	-2	0	-2	0	-2	0	-2	-4	-2	0	2	0	2	4	-2
$x_4 x_3 x_2 x_1$	0	-2	4	-2	0	2	0	-2	2	0	-2	0	-2	0	-2	-4

3 SECURITY ENHANCEMENT OF DESIGN MENTIONED IN PREVIOUS SECTION

Based on the analysis done in the previous section, further uncertainty enhancement is possible. Let us assume all the S boxes in the fig. 2 are different

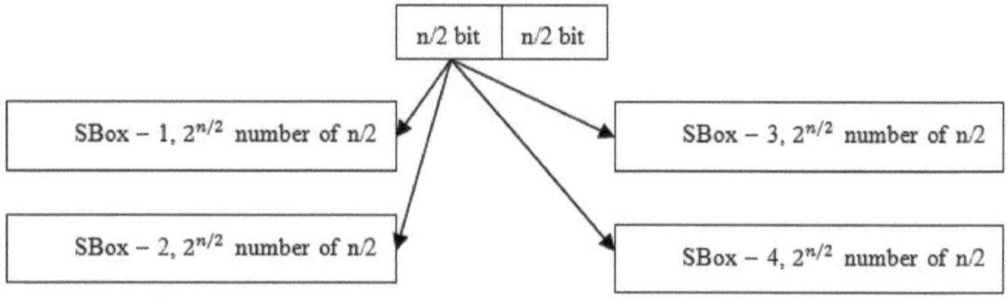

Figure. 4: Dynamic substitution with different S boxes

Let the S box – 3 be as shown in Table 6

The probability biases for all possible functions of S Box-3 shown in Table 6 are given in Table 7.

Table 7. Biases for function pairs of S box – 3

	0	y_0	y_1	$y_0 y_1$	y_2	$y_2 y_0$	$y_2 y_1$	$y_2 y_1 y_0$	y_3	$y_3 y_0$	$y_3 y_1$	$y_3 y_1 y_0$	$y_3 y_2$	$y_3 y_2 y_0$	$y_3 y_2 y_1$	$y_4 y_3 y_2 y_1$
0	8	0	0	0	0	0	0	0	0	0	0	0	0	0	0	0
x_0	0	2	0	-2	-2	0	2	0	2	0	-2	0	-4	2	-4	-2
x_1	0	0	-2	-2	-2	2	-4	0	0	0	-2	-2	-2	2	4	0
$x_0 x_1$	0	-2	-2	0	0	2	2	0	6	0	0	-2	2	0	0	2
x_2	0	-2	0	2	-2	0	-2	4	0	2	0	-2	-2	-4	-2	0
$x_2 x_0$	0	4	4	0	0	0	0	0	2	-2	2	-2	-2	-2	2	2
$x_2 x_1$	0	-2	2	-4	0	-2	2	4	0	2	2	0	0	2	2	0
$x_2 x_1 x_0$	0	0	-2	-2	-2	-2	0	0	-2	-2	0	0	0	0	-2	6
x_3	0	0	-4	0	0	-4	0	0	2	-2	2	2	-2	-2	2	-2
$x_3 x_0$	0	2	0	2	-6	0	2	0	0	2	0	2	2	0	2	0
$x_3 x_1$	0	0	-2	2	2	2	4	0	-2	2	0	0	-4	0	2	2
$x_3 x_1 x_0$	0	-2	2	0	0	-2	2	0	0	-2	-6	0	0	-2	2	0
$x_3 x_2$	0	2	0	2	2	-4	-2	0	2	4	-2	0	0	2	0	2
$x_3 x_2 x_0$	0	0	0	-4	0	0	0	-4	0	-4	0	0	0	-4	0	0
$x_3 x_2 x_1$	0	2	-2	0	0	-2	2	0	-2	0	0	-6	2	0	0	-2
$x_4 x_3 x_2 x_1$	0	4	-2	-2	2	2	0	4	0	0	-2	2	2	-2	0	0

Let the S Box – 4 be as shown in Table 8

Table 8. S box – 4

x_3x_2 / x_1x_0	00	01	10	11
00	9	4	8	3
01	2	A	C	1
10	6	E	5	0
11	D	F	B	7

The probability biases for all possible functions of S Box – 4 shown in Table 8 are given in Table 9

Table 9. Biases for function pairs of S box – 4

	0	y_0	y_1	y_0y_1	y_2	y_2y_0	y_2y_1	$y_2y_1y_0$	y_3	y_3y_0	y_3y_1	$y_3y_1y_0$	y_3y_2	$y_3y_2y_0$	$y_3y_2y_1$	$y_4y_3y_2y_1$
0	8	0	0	0	0	0	0	0	0	0	0	0	0	0	0	0
x_0	0	2	2	0	0	-2	2	4	2	0	0	2	-2	0	-4	2
x_1	0	2	2	0	4	-2	-2	-4	0	-2	-2	0	0	-2	-2	0
$x_0 x_1$	0	-4	0	4	0	0	0	0	-2	2	-2	2	-2	-2	-2	-2
x_2	0	0	2	-2	0	0	-2	2	-2	2	-4	-4	-2	2	0	0
$x_2 x_0$	0	-2	0	-2	0	2	4	-2	0	-2	0	-2	-4	-2	0	2
$x_2 x_1$	0	2	0	2	0	2	4	-2	-2	0	-2	0	2	4	-2	0
$x_2x_1x_0$	0	0	2	-2	4	0	2	2	-4	0	2	2	0	0	2	-2
x_3	0	2	-2	-4	-2	0	0	-2	-2	4	0	2	0	-2	-2	0
x_3x_0	0	0	0	0	-2	-6	2	-2	0	0	0	0	-2	2	2	-2
x_3x_1	0	0	0	0	2	2	-2	-2	2	2	2	2	-4	4	0	0
$x_3x_1x_0$	0	-2	6	0	-2	0	0	-2	0	2	2	0	2	0	0	2
x_3x_2	0	-2	0	-2	2	0	2	0	4	2	-4	2	2	0	2	0
$x_3x_2x_0$	0	0	-2	2	2	-2	0	0	-2	2	0	0	0	0	2	6
$x_3x_2x_1$	0	-4	-2	-2	2	-2	0	0	0	0	2	-2	2	2	-4	0
$x_4x_3x_2x_1$	0	-2	0	-2	-2	0	-2	0	-2	-4	-2	4	0	2	0	2

In Fig. 4, an additional key bit will provide the facility by which any n/2 bits (most significant or least significant) can be substituted by any S boxes (1, 2, 3 or 4). Instead of the key bit K used in the analysis in section III, here K_1 and K_2 are used.

For the upper n/2 bits,

$K_1 K_2$ = 00 selects SBox – 1.

$K_1 K_2$ = 01 selects SBox – 2

$K_1 K_2$ = 10 selects SBox – 3

$K_1 K_2$ = 11 selects SBox – 4

Carrying out the linear cryptanalysis for the S box shown in Figure 4, using the Table 3,5,7 and 9 with the linear approximation taken in Figure 2, for a single round,

$b_1\, q[0,1]$ = $\overline{K_1}\, \overline{K_2}\, [\, b_3\, p[0,1] \oplus b_1\, p[0,1] \oplus b_0\, p[0,1] \oplus b_3\, k[0,1] \oplus b_1\, k[0,1] \oplus b_0\, k[0,1]\,]$ with a probability bias of -2/16

+

$K_1\, \overline{K_2}\quad [\, b_3\, p[0,1] \oplus b_1\, p[0,1] \oplus b_0\, p[0,1] \oplus b_3\, k[0,1] \oplus b_1\, k[0,1] \oplus b_0\, k[0,1]\,]$

With a probability bias of 0.

+

$\overline{K_1}\, K_2\, [\, b_3\, p[0,1] \oplus b_1\, p[0,1] \oplus b_0\, p[0,1] \oplus b_3\, k[0,1] \oplus b_1\, k[0,1] \oplus b_0\, k[0,1]\,]$ with a probability bias of 2/16

+

$K_1\, K_2\, [\, b_3\, p[0,1] \oplus b_1\, p[0,1] \oplus b_0\, p[0,1] \oplus b_3\, k[0,1] \oplus b_1\, k[0,1] \oplus b_0\, k[0,1]\,]$

With a probability bias of 6/16.   (8)

Eq. (8) shows that the uncertainty increased many times that of the existing system. When the cryptanalysis is done for multiple rounds it becomes too complex. The 'piling up lemma' given in C.K.Shyamala et al. calculated using piling up principle will not be a constant because of the dynamic substitution. It is clearly evident from Eq. (8) that different combinations of key lead to different probability bias.

CONCLUSION AND FUTURESCOPE

Linear cryptanalysis is carried out for a hardware efficient Substitution box and compared it with that of a Conventional S box. It is proved that the security of the hardware efficient S box with static as well as dynamic properties is better than that of the conventional system. The technique to enhance uncertainty further is also elaborated. It can be easily seen from the analysis that increasing the number of key bits increases the security. Further research can be done to increase the security using differential cryptanalysis also.

REFERENCES

1   Heys, H. "A Tutorial on Linear and Differential Cryptanalysis", *Cryptologia*, July 2002.
2   Heys, H.M and Tavares, S.E. Substitution Permutation Networks resistant to differential and linear cryptanalysis, Journal of Cryptology, pp 1-19, 1996
3   Jithendra.K.B, Shahana. T.K, "Hardware Efficient Parallel Substitution Box for Block Ciphers with Static and Dynamic Properties", *Procedia Computer Science,Elsevier,Vol.46,pp.540-547*, 2015
4   Matsui, "Linear Cryptanalysis Method for DES cipher." *Proceedings, EUROCRYPT '93*, 1993; published by Springer-Verlag
5   Mister, S.. and Adoms, C. "Practical S Box Design." *Proceedings, Workshop in selected area of Cryptography, SAC'96*, 1996
6   Musa, M.;Schaefer.;and Wedig, S. "A simplified AES Algorithm and Its Linear and Differential Cryptanalysises." Cryptologia, April 2003.Phoon, K. K. & Kulhawy, F. H. (1999). "Characterization of geotechnical variability", *Canadian Geotechnical Journal*, 36(4), 612-624.
7   Shannon.C, Communication theory of secrecy systems, *Bell Systems Technical Journal*, vol.28, 1949
8   Shyamala C K, N Harini, T R Padmanabhan, Cryptography and Security, Wiley India Pvt ltd, New Delhi, 2011
9   Sinkov,A. "Elementary cryptanalysis: A Mathematical Approach". Washington,D.C.: The Mathematical Association of America, 1966

Nishtha [1] and Manu Sood [2]

Comparative Study Of Software Defined Network Controller Platforms

Abstract: Rapid evolution and deployment of Software Defined Network (SDN), the latest network architecture is because it has the ability to operate and manage the network through programs. Therefore, the users may alter, prioritize or shape network traffic dynamically according to their demands and requirements. This is mainly possible because of software based logically centralized control program – the controller. In SDN, the control program that decides packet forwarding in the network is removed from the network devices. It resides in network servers while the data plane remains in the network devices only. Therefore, SDN is a split architecture that separates control and the data plane and also provides open programming interfaces between control and data planes. OpenFlow (OF) is regarded as a standardized protocol for SDN that provides interfaces for communication among the data and the control planes. The controller may dynamically install rules in the flow table of every network devices and according to these rules network traffic is carried out in the network. So, it is possible for the controller to program flow-table of every device attached in the network, thereby programmatically control network traffic in the network. The controller in SDN is often regarded as a Network Operating System (NOS). As a result, various commercial vendors as well as universities have developed SDN controllers. This paper summarizes most of the common SDN controllers available till date.

Keywords: Software-Defined Network, data plane, control plane, OpenFlow

1 Introduction

Software Defined Network apart from providing centralized monitoring (Levin et al., 2012) vendor independence (Foster et al., 2013), less energy consumption (Prete et al., 2012), low cost (Sood & Nishtha, 2014) easy implementation of net-

1 Computer Science Department, Himachal Pradesh University, Shimla, Himachal Pradesh, India
2 Computer Science Department, Himachal Pradesh University, Shimla, Himachal Pradesh, India

work policies (Kang, 2012), increased flexibility and reliability (Sood & Nishtha, 2014) in network, also enables each device in the network to be configured programmatically (Foster et al., 2013). This is only possible because this architecture separates control and the data plane. The control plane is decoupled from the network device and is based in network servers as software based logically centralized control program- the controller Heller et al. (2012), Shirazipour et al. (2012), Jeong et al. (2012), Levin et al. (2012). While the data plane is the infrastructure formed in which devices in the network are interconnected with one another. Therefore, the network intelligence is situated in the control plane and data plane resides in the network device thereby, making the network device to be a traffic carrying agent only. Simultaneously, this architecture also offers open Application Programming Interfaces (APIs) from both control and the data planes Heller et al. (2012), Shirazipour et al. (2012), Kreutz et al. (2014).

For communication among the controller and the data plane a number of protocols have been developed such as OpenFlow developed by ONF (Open Networking Foundation), ForCES (Forwarding & Control Element Separation) developed by IETF (Internet Engineering Task Force) or any other proprietary protocol Shirazipour et al. (2012), Jeong et al. (2012). The controller also provides open programmable interfaces for development of management applications. Therefore, distinct applications that perform network management tasks may be easily developed Monsanto et al. (2013), Gude et al. (2008). The controller as per the directions of these management applications directs the underlying network devices to carry network traffic Gude et al. (2008), Prete et al. (2012).

OpenFlow (OF) is regarded as a standardized protocol for SDN which is a flow-based architecture. But there is no standardized API for communication among controller and the network applications Gude et al. (2008), Shahreza et al. (2013), Foster et al. (2013), Prete et al. (2012), Jeong et al. (2012), Open Networking Foundation (2012). In 2011, group of people including researchers, network administrators, network operators and vendors; who were concerned about the development of standards on SDN formed Open Networking Foundation (ONF) Open Networking Foundation (2012).

The following part of the paper briefs about Software Defined Network by conceptually dividing this architecture into a number of layers, which is continued by SDN controllers. Subsequently, various controller platforms are compared in tabular form, succeeded by conclusions in the end.

2 THREE TIERS IN SDN ARCHITECTURE

SDN may broadly be represented as three tier architecture:

Infrastructure Layer (Tier-1): This layer comprises of network devices such as switches and includes all software as well as hardware components in the network devices.

Control Layer (Tier-2): Network intelligence resides in software-based logically centralized SDN controllers. The control layer regulates and manages network devices i.e. Tier-1.

Application Layer (Tier-3): Applications and services take advantage of infrastructures in the above layers i.e.Tier-1, Tier-2. Conceptually placed above the control layer, these applications accomplish management tasks Shirazipour et al. (2012), Open Networking Foundation (2012).

As a result, through open programming interfaces, communication between the controller and switches, as well as, the controller and applications is possible.

Therefore, controller is the major component in SDN architecture which is logically centralized in the network, have open interfaces for communication with the network devices as well as for development of management applications. Consequently, the controller imparts centralized monitoring of the entire network, increases reliability, better performance, change in network traffic as per needs and requirements, easy implementation of network- wide policies etc.; making deployment or management of such networks easy.

2.1 SDN CONTROLLER

The methodology behind SDNs is that the network elements rely on the logically centralized controller for all decision making capabilities Heller et al. (2012), Shirazipour et al. (2012). It is regarded as network operating system and enables all control and management related functions (Gude et al., 2008).

In SDN architecture open interfaces used to communicate between different layers of the SDN stack, as per their functions, are categorized as:

Northbound APIs: For communication among controller & applications (Monsanto et al., 2013).

Southbound APIs: For communication among controller & underlying hardware.

East/Westbound APIs: For communication among group of controllers (Sood & Nishtha, 2014).

An OpenFlow switch consists of three main components: (i) a flow table, (ii) secure channel and (iii) OpenFlow protocol. In an OpenFlow compliant switch flow entries resides in Flow Table. These entries are further divided into three parts: each entry has a (i) packet header to match incoming packets (ii) list of action and (iii) counters. Incoming packets are compared with the header field of each entry and if matched, the packets are processed according to the action specified in that entry. Therefore, the controller may dynamically install rules in the flow table of every network devices and according to these rules network traffic forwards. Therefore, it is possible for a controller to program flow-table of every device attached in the network and control network traffic programmatically. Counter are used to keep packet statistics that keeps information related to the flow as well as time of every flow Jarschel et al. (2011), Shahreza et al. (2013), Shin et al. (2013).

3 COMPARISON BETWEEN VARIOUS SDN CONTROLLER PLATFORMS

In SDN, the controller is regarded as Network Operating System (NOS) that provides programmable interfaces, and the network is managed by network applications executed on the top of the network operating system. Maximum benefits in SDN are provided because of centralized controller that performs network management functions such as routing, access control, network virtualization, energy management etc. Erickson (2013), Gude et al. (2008), Jeong et al. (2012), Levin et al. (2012), Mogul et al. (2012)].

A number of controller platforms have been developed and these according to their use may be divided into two categories: one for research purpose mostly developed by universities and others for commercial uses which are mostly developed by commercial vendors. The controllers for research purposes are open source, single-instance controllers and are developed using different programming languages that may vary from one controller to another such as Beacon is developed using Java language and McNettle using Haskell language. Whereas, the controllers for commercial purposes are closed-source, distributed controllers as these controllers distribute their state across multiple running instances for fault tolerance (Erickson, 2013).

The first generation controller platforms such as NOX and POX provide a low-level programming interfaces while next generation controllers provide additional features such as composition, isolation and virtualization and most of these controllers support multi-core technology (Foster et al., 2013). The features of scalability and fault tolerance are provided by closed-source, distributed controllers such as Onix (Koponen et al., 2010), Maestro (Cai et al., 2010).

Centralized controller is an intrinsic requirement in SDN. But, as the size of network increases, the number of data plane elements as well as flows increases. Therefore, instead of a single controller, a number of controllers are needed to manage the entire network (Cai et al., 2010). To provide global network view in such networks, a logically centralized control plane operates on distributed control plane (Levin et al., 2012). Large SDN networks may be divided in multiple domains where in a domain a single controller takes decisions regarding its domain and communicates with other controllers for executing global policies. A model for distributed computing called "Supported locality model" is the best for SDN networks where by using existing local algorithms, coordination protocols may be developed and each controller respond to events that take place in its local neighborhood only (Schmid et al., 2013).

Significant characteristics that are mainly responsible for evaluating performance of these controller are scalability, security, centralized monitoring, virtualization; reliability (Levin et al., 2012) and may include some SDN centric characteristics such as OpenFlow support and network programmability.

The existing controllers may be distinguished broadly on the characteristics of open interfaces provided, programming languages, use in production or research, single core or multi core that is whether centralized or distributed, open source or not etc.

Table 1 compares most of the existing controller platforms.

Table 1. Common SDN Controllers

Controller Name, Vendor/ Developer	Language used, APIs provided	Distinguished characteristics
Application Policy Infrastructure Controller (APIC). Vendor: Cisco	OpFlex, Representational State Transfer (REST) APIs, CLI, XML, JSON (Java Script Object Notation).	– Supports physical, virtual as well as cloud. – Fast & simple application development. – Easy integration with VMware, Microsoft, OpenStack (Cisco).
Beacon	Language: Java.	– Multithreaded, high perfor-

Developer: Stanford University	OpenFlowJ, ad-hoc API.	mance, centralized. – Developer friendly, cross- platform Open Source: Yes Astuto et al. (2013), Erickson (2013), Kreutz et al. (2014).
Big Network Controller Vendor: Big Switch	Language: Java. (Same as used by Floodlight). OF, RESTful API.	- Commercial, distributed controller. - Resilient & scalable (Big Switch).
Distributed Multi-domain SDN Controllers (DISCO)	Language: Java. OpenFlow, REST APIs.	- Distributed, multi domain controller. - Each DISCO controller manages its own network domain Kreutz et al. (2014), Phemius et al. (2014).
Extensible Operating System (EOS) Vendor: Arista	Linux based. OpenFlow, Directflow, XMPP, Python, SQLite Databases Advanced Event Manager (AEM), eAPI & SDK.	- Offers pre-built applications. - Separates information & packet forwarding from protocol processing & application logic. - Easy integration with third party applications (Arista).
Fleet	Ad–hoc API, OpenFlow	- Deals with malicious administrator problem. - Distributed Kreutz et al. (2013), Matsumoto et al. (2014)
Floodlight Developers :Open community (Supported by Big Switch)	Language: Java. OpenFlow, RESTful API.	- Centralized, multi threaded. - Base of commercial product from BigSwitch Network. - Support OpenStack cloud orchestration platform. Open Source: Yes Astuto et al. (2013), (Floodlight), Kreutz et al. (2013).
FlowN Developers: Princeton University & University of Pennsylvania	Language: Python. As an extension of NOX controller.	- Built as a NOX application it provides each tenant working on the same network, an illusion of having their own resources e.g. own address space, controller. - Instead of running a separate controller for each tenant, it performs a lightweight container based virtualization Drutskoy et al. (2013).
		- Acts as a transparent virtualization

FlowVisor Developers: Stanford/ Nicira	Language: C. OpenFlow	layer between controller and OpenFlow switches. - Allows multiple researchers to run independent experiments on the same production network. - Act as a hypervisor, it separately manages traffic for each researcher on the same production network, and also manages shared resources such as controller, bandwidth Astuto et al. (2013), Sherwood et al. (2010).
FortNOX Independent Developers	Language: C++.	- NOX extension. - Enables role-based authorization & enforces security constraint that checks automatically about violation of security policies by the new flow rules. - Each application in FortNOX has separate compartments Porras et al. (2012).
Helios Developers: NEC	Language: C. OpenFlow	- Distributed, scalable. -Used for research and enable different researchers to coordinate with one another. -Without stopping data plane operation, module maintenance may be performed. Open Source: No Astuto et al. (2013), (NEC).
HyperFlow Developers: University of Toronto	Language:C++. OpenFlow	- The first distributed event control plane. - Processing of a particular flow request is localized to an individual controller. - Built as a NOX application. Open Source: Yes Kreutz et al. (2014), Tootoonchian et al. (2010).
HP VAN (Virtual Application Network) Controller Vendor: HP	Language: Java. OpenFlow, RESTful API.	- Suitable for data center, campus & branch. - Distributed, provides full integration with HP Intelligent Management Center (IMC). - Scalable, robust (HP), Kreutz et al.

		(2014).
Jaxon Independent Developers.	Language: Java.	- NOX-based, flexible, provides thin interfaces with NOX controller. - Robust. Open Source: Yes Astuto et al. (2013), (Jaxon).
Kandoo Developers: University of Torento.	Language: Combination of C, C++, Python. RCP (Remote Procedure Call) API, OpenFlow.	- Hierarchically distributed, two layers of controllers: (i) bottom layer locally manage most of the events. (ii) upper layer is a logically centralized controller that provides global network view. - This structure decreases load on centralized controller by having replicated local controllers which carries maximum load and only rare events are processed in the central location. - Modular Kreutz et al. (2014), Yeganeh et al. (2012).
Maestro Developers: Rice University.	Language: Java. OpenFlow, ad-hoc API.	- The first controller that uses parallelism to have linear performance by using multi-core processors. - Centralized, multi-threaded. - Robust, secure, flexible. Open Source: Yes Astuto et al. (2013), Cai et al. (2010), Kreutz et al. (2014).
McNettle Developers: Yale University.	Language: Haskell. McNettle API.	- Use multi core servers to achieve low latency and high throughput. - Scalable, Extensible (McNettle), Voellmy et al. (2012).
Meridian	Language: Java. Extensible API Layer, REST API, OpenFlow.	- Cloud networking platform that may be easily integrated with OpenStack and other commercial cloud platforms. - Built on the open source Floodlight controller platform. - Centralized, multithreaded Banikazemi et al. (2013) Kreutz et al. (2014).
MobileFlow	SDMN API, OpenFlow	- Flexible architecture for mobile networks. Kreutz et al. (2014), Pentikousis et al. (2013).
MUL	Language: C.	- Centralized multithreaded, supports

Developer: Kulcloud	OpenFlow, OVSDB as well as Netconf., CLI, GUI, RESTful API.	modular applications. - Provides multi-level multi-language northbound interfaces for developing network application. Open Source: Yes Astuto et al. (2013), Kreutz et al. (2014), Saikia & Malik (2013).
NodeFlow Developers: Independent Developers	Language: Pure Java Script (JS).	- Node.JS provides a dynamic platform for fast, simple applications, server side scripting. - High performance system. Open Source: Yes Astuto et al. (2013), (NodeFlow).
NOX Developer: Nicira	Language: Python/ C++ OpenFlow, ad-hoc API.	- Centralized, single threaded, provides event driven interface. - A number of network applications are built on the top of NOX. Open Source: Yes Astuto et al. (2013), Cai et al. (2010), Gude et al. (2008), Kreutz et al. (2014).
NOX MT Developer:	Language: C++. OpenFlow, ad-hoc API	- NOX based multi threaded controller Kreutz et al. (2014) Tootoonchian et al. (2012).
Network Virtualization Platform (NVP) Vendor: Vmware	OpenFlow, NVP API	Commercial, distributed, network virtualization solution for multi-tenant datacenters Koponen et al. (2014), Kreutz et al. (2014).
ONIX	Language: C++. Onix's API. Onix NIB (Network Information Base) API.	- Commercial control plane. - A distributed system that may run one or more servers, each of which may run multiple Onix instances. Open Source: No Kreutz et al. (2014), Koponen et al. (2010).
ONOS	Language: Java. RESTful API, OpenFlow	- A controller designed to increase scalability, performance. - Open & distributed control plane Berde et al. (2014), Kreutz et al. (2014).
OpenContrail Controller Vendor: Juniper	Interface to Metadata Access Point (IF-MAP), REST API	- Distributed, support clouds. - Interoperable (Juniper), Kreutz et al. (2014).
OpenDayLight		- Distributed.

(ODL).Developer: Collaborative project under The Linux Foundation.	Language: Java. REST API, OpenFlow.	- Project for SDN development & forming basis for NFV (Network Functions Virtualization). Open Source: Yes Kreutz et al. (2014), OpenDaylight (2013).
OVS Controller Developer: Pica 8.	Language: C. OpenFlow	- Using OpenFlow protocol, manages any number of distant switches. Open Source: Yes (Open Networking Foundation), (Pica 8).
PANE	Language: Haskell and Nettle library. OF, PANE API	-Event driven, multi core, distributed. Ferguson et al. (2013), Kreutz et al. (2014).
POX Developer: Nicira	Language: Python. ad_hoc API	- Derived from NOX. - Performs better than NOX for Python based applications. - Reusable sample components for path selection, topology discovery, centralized. - Used in research and education, centralized. Open Source: Yes Kreutz et al. (2014), (POX).
ProgrammableFlow Vendor: NEC	Language: C. OpenFlow	- Network virtualization. - Allows data centers and service providers to easily deploy, control, secure, monitor and manage multi tenant network infrastructure. - Centralized. Open Source: No Kreutz et al. (2014), (NEC).
Programmable Network Controller Vendor: IBM	OpenFlow, Web GUI, Web API	- Provides Virtual Tenant Network (VTN) (IBM Programmable Network Controller).
RouteFlow Developed by: CPQD		- By separating the control and the data plane allows remote IP routing services in a centralized way. This enables to provide network as a Service (IPNaaS).

	Language: C++.	- New protocols and algorithms may be easily added, making these IP networks to be more flexible. - Virtualized IP routing services over OpenFlow enabled hardware. Open Source: Yes Nasciment et al. (2011).
Rosemary	Language: C. Ad_hoc API, OpenFlow	- Centralized, robust, secure and high-performance Kreutz et al. (2014), Shin et al. (2014).
Ryu Vendor: NTT, OSRG group.	Language: Python. OF, Nicira Extensions, OF-config, Netconf, ad-hoc API.	- Centralized multi-threaded. - Component based & acts as a Network Operating System for OpenStack (Apache licensed 2.0). Open Source: Yes Astuto et al. (2013), Kreutz et al. (2014), (Ryu).
SMaRT Light	Language: Java. OpenFlow, RESTful API.	- Fault-Tolerant. - Distributed Botelho et al. (2014), Kreutz et al. (2014).
SNAC (Simple Network Access Control). Funded by Stanford clean state program (Nicira).	Language: C++. Provides web based GUI for network management, web based API, OF.	- Centralized, multithreaded, provide flow level traffic information which increases visibility. - Campus networks, requires authentication. Open Source: Yes Astuto et al. (2013), Kreutz et al. (2014), (SNAC).
Trema Vendor: NEC	Language: Ruby, C. Ad-hoc API	- Developer friendly. Open Source: Yes Astuto et al. (2013), Kreutz et al. (2014), (Trema).
VellOS Vendor: Vellos	Linux based OS System. VellOS RESTful APIs, OpenFlow.	- Commercial, - Users' have graphical view of network, have automatic network discovery feature & dynamic access control (Vello System).
VMware NSX Vendor: VMware	RESTful API, NSX API	- Distributed controller. - Virtual networks and overlays (VMwareNSX).
Yanc Developer: University of Colorado at	Linux extension. For application developmnts supports	- Distributed control that exposes the network configuration and state as a file system.

Boulder	any language supported by modern operating systems.	- Each application is like a process. - Network operating system that may be used in a variety of ways Kreutz et al. (2014), Monaco et al. (2013).

First generation controllers including NOX and POX are centralized, single-threaded system and provide low-level programming interfaces. As a result, only a single flow request may be processed at any time. Whereas, most of the recently developed controllers use network virtualization to provide network abstraction and automated orchestration. Most of these controllers support network slices. All these network slices may concurrently execute different programs, and maintain traffic isolation and share resources. One such controller, FlowVisor, allows different researchers to simultaneously work on the same production network. Similarly, FlowN is another architecture that allows large number of tenants to share the same network by providing each tenant the illusion of its own space, topology and controller.

The growth of the network after a specific limit becomes a bottleneck in the network operation as the network may not be able to process all events/requests. This even limits scalability. Therefore, use of parallel core processors is essential for handling these frequent events as the size of the network increases. These may even be lowered by reducing flow arrivals on controller by proactively pushing the network state. Kandoo takes a different approach of distributed control plane by maintaining scalability, without changing the switches.

The significant differences that exist in these SDN controllers are listed below:

- Most of the controllers developed by universities for research purposes are open source and are developed using different programming languages whereas others developed by various vendors for commercial use are closed source controllers.
- Most of these controllers use OpenFlow as a southbound API only, whereas others may   use different set of southbound APIs that may or may not include OpenFlow.
- Some of these controllers support single northbound interface whereas, others use a set of multilevel and multi language northbound APIs such as offered by mul and yanc.
- To provide services such as overlay networking, virtual private networks, cloud computing & dynamic resource allocation etc. most of the commercial controller vendors combine virtualization with SDN.

- Large network require number of controller. Therefore, most of the commercial controllers and even some controllers developed for research purposes are distributed in state in order to provide global network view.

CONCLUSION

The software based logically centralized controller allows easy deployment and management of SDNs. These existing SDN controllers mainly differ on parameters such as open interfaces provided, programming languages, use in production or research, whether centralized or distributed, open source or not and scalability etc. OpenFlow is regarded as southbound standardized API is offered by most of the controllers whereas few other controller platforms offer a set of APIs which may not include OpenFlow. The sole purpose behind SDNs is to provide flexibility and easy innovations in networks. By using different southbound APIs, the network may again be restricted to use proprietary hardware and custom software. This vertical integration of software and hardware may curtail rapid flexibility and innovations in the networks. Therefore, to make OpenFlow a southbound standard for SDNs in real sense it may be enhanced to cope with the requirements all networks. In that case only all controller platforms may offer OpenFlow as a southbound API and it may become standardized API in real sense.

Even the open interfaces for communication between the controller and applications which is used for development of network management applications have not been standardized. As a result, the management application built for a particular controller platform may not work for other platforms. Therefore, these controller platforms should provide all or most of the languages supported by existing operating systems for application development. So that applications developed may not remain specific to a particular platform. This would increase network applications development and simultaneously it would also enhance deployment of SDNs.

REFERENCES

1   Arista. Software Driven Cloud Networking. Retrieved from
     http://www.arista.com/en/products/software-driven-cloud-networking#datatab213

2   Astuto, B.N., Mendonca, M., Nguyen, X.N., Obraczka, K. & Turletti, T. (2013). "A survey of software-defined networking: past, present, and future of programmable networks", In IEEE Communications Surveys & Tutorials, doi: 10.1109/SURV.2014.012214.00180

3   Banikazemi, M., Olshefski, D., Shaikh, A., Tracey, J. &. Wang G. (2013). "Meridian: an SDN platform for cloud network services", Communications Magazine, IEEE, vol. 51, no. 2, 120–127 doi: 10.1109/MCOM.2013.6461196

4   Berde, P., Gerola, M.,Hart, J. et al. (2014) "ONOS: towards an open, distributed SDN OS", HotSDN'14, ACM, 1-6.

5   Big Switch. Big Network controller. Retrieved from http://bigswitch.com/products/SDN-Controller

6   Botelho, F., Bessani, A., Ramos, F. M. V. & Ferreira, P. (2014). "On the design of practical fault-tolerant SDN controllers", in Third European Workshop on SDN, doi: 10.1109/EWSDN.2014.25

7   Cai, Z., Cox, A. L. & Ng, T.S.E. (2010). "Maestro: a system for scalable openflow control", Technical Report TR10-08, Rice University.

8   Cisco. Cisco Application Policy Infrastructure Controller (APIC). Retrieved from http://www.cisco.com/c/en/us/products/cloud-systems-management/application-policy-infrastructure-controller-apic/index.html

9   Drutskoy, D., Keller, E., Rexford J. (2013). "Scalable network virtualization in software-defined networks", Internet Computing, IEEE (Volume: 17, Issue: 2), 20 - 27, doi = 10.1109/MIC.2012.144, 2013.

10   Erickson, D. (2013), "The Beacon openflow controller". HotSDN'13, ACM. doi:10.1145/2491185.2491189

11   Ferguson, A. D., Guha, A., Liang, C., Fonseca, R. & Krishnamurthi, S. (2013). "Participatory networking: an API for application control of SDNs", In Proceedings of the ACM SIGCOMM 2013 conference on SIGCOMM, SIGCOMM '13. ACM, Volume 43 Issue 4, 327–338. doi:10.1145/2486001.2486003.

12   Floodlight (2012). Project Floodlight. Retrieved from http://www.projectfloodlight.org/floodlight/

13   Foster, N., Freedman, M. J., Guha, A., Harrison, R., Katta, N.P., Monsanto, C., Reich J., Reitblatt, M., Rexford, J., Schlesinger, C., Story, A. & Walker D. (2013). "Languages for software-defined networks", In IEEE (Volume: 51, Issue: 2) Communications Magazine, 128-134.

14   Gude, N., Koponen, T., Pettit J., Pfaff, B., Casado, M., McKeown, N. & Shenker S. (2008). "NOX: towards an operating system for networks", In the Computer Communication Review Newsletter, ACM SIGCOMM, Volume 38 Issue 3, ACM, 105-110. doi:10.1145/1384609.1384625

15   Heller, B., Sherwood, R. & McKeown N. (2012). "The controller placement problem", Proceeding of HotSDN '12, ACM, 7-12. doi:10.1145/2342441.2342444

16   HP. HP Virtual Application Networks SDN Controller: Retrieved from http://h17007.www1.hp.com/docs/networking/solutions/sdn/4AA4-8807ENW.PDF

17   IBM. Programmable Network Controller. Retrieved from http://www-01.ibm.com/common/ssi/ShowDoc.wss?docURL=/common/ssi/rep_ca/0/897/ENUS212-310/index.html&lang=en&request_locale=en

18   Jarschel, M., Oechsner, S., Schlosser, D., Pries, R., Goll, S. & Tran-Gia, P. (2011). "Modeling and performance evaluation of an openflow architecture", Proceedings of the 23rd International Teletraffic congress ITC'11, 1-7.

19 Jaxon. Jaxon: Java Bindings for NOX. Retrieved from
http://www.noxrepo.org/2012/03/335/

20 Jeong, K., Kim, J. & Kim, Y-T. (2012). "QOS-aware network operating system for software
defined networking with generalized openflows", In NOMS, 2012 IEEE, 1167 – 1174.
doi:10.1109/NOMS.2012.6212044

21 Juniper Networks Contrail. White paper Juniper network Contrail. Retrieved from
http://www.juniper.net/us/en/local/pdf/whitepapers/2000535-en.pdf

22 Kang, N., Reich, J., Rexford, J. & Walker, D. (2012). "Policy transformation in software
defined networks ", In the proc. of the ACM SIGCOMM 2012 conference on Applications,
technologies, architectures, and protocols for computer communication SIGCOMM '12,
ACM, 309-310.

23 Koponen, T., Casado, M., Gude, N., Stribling, J., Poutievski, L., Zhu, M., Ramanathan, R.,
Iwata, Y., Inoue, H., Hama, T. & Shenker, S. (2010). "Onix : a distributed control platform
for large-scale production networks", Proc. of the 9th USENIX conference OSDI'10,
USENIX, Article No. 1-6.

24 Koponen, T., Amidon, K., Balland, P., Casado, M., Chanda, A., Fulton, B, Ganichev, I
Gross,, J., Gude, N., Ingram, P., Jackson, E., Lambeth, A., Lenglet, R., Li, S.H.,
Padmanabhan, A., Pettit, J., Pfaff, B., Ramanathan, R., Shenker, S., Shieh, A., Stribling, J.,
Thakkar, P., Wendlandt, D., Yip, A. & Zhang, R. (2014). "Network virtualization in multi-
tenant datacenters", In 11th USENIX Symposium NSDI 14, 203–216.

25 Kreutz, D., Ramos, F .M. V., Verissimo, P., Rothenberg, C.E., Azodolmolky, S. & Uhlig, S.,
"Software-defined networking: a comprehensive survey", Proceedings of IEEE. doi:
10.1109/JPROC.2014.2371999

26 Levin, D., Wundsam, A., Heller, B., Handigol, N. & Feldmann, A. (2012). "Logically
centralized? state distribution trade-offs in software defined networks", In HotSDN '12,
ACM,1-6.

27 Matsumoto, S., Hitz, S. & Perrig, A. (2014). "Fleet: defending SDNs from malicious administrators", In Proceedings of the Third Workshop HotSDN '14 ACM, 103–108.doi:
10.1145/2620728.2620750

28 McNettle. Retrieved from http://haskell.cs.yale.edu/wp-
content/uploads/2012/08/mcnettle.pdf

29 Mogul, J. C. & Congdon, P. (2012). "Hey, you darned counters!: get off my ASIC!",
Proceedings of the first workshop on Hot topics in software defined networks HotSDN '12,
ACM, 25-30 , doi = 10.1145/2342441.2342447, 2012.

30 Monaco, M., Michel, O. & Keller, E. (2013). "Applying operating system principles to SDN
controller design", In Twelfth ACM Workshop HotNets-XII. doi:10.1145/2535771.2535789

31 Monsanto C., Reich J., Rexford J., Walker D. & Foster N. (2013). "Composing software-
defined networks", Proceedings of the 10th USENIX conference, nsdi'13, 1-14.

32 Nasciment, M.R., Rothenberg, C.E., de Lucena S.P. & Magalhaes M.F. (2011), "Virtual rou-
ters as a service: the RouteFlow approach leveraging software-defined networks",
Proceedings of the 6th International Conference on Future Internet Technologies CFI '11,
ACM, pp 34-37, doi =10.1145/2002396.2002405

33 NEC Helios. Global Environment for Network Innovation (GENI) Helios OpenFlow controller by NEC. Retrieved from google.co.in//
groups.geni.net/geni/.../Helios_OpenFlow_Controller_GEC9_Demo.pdf

34 NEC. NEC Programmableflow Networking. Retrieved from http://www.necam.com/SDN/

35 NodeFlow. An OpenFlow Controller Node Style. Retrieved from
http://garyberger.net/?p=537

36 OpenDaylight (2013) OpenDaylight: A Linux Foundation Collaborative Project. Retrieved
from http://www.opendaylight.org

37 Open Networking Foundation (2012). Software-Defined Networking: The New Norm for
Networks. Retrieved from
https://www.opennetworking.org/images/stories/downloads/sdn-resources/white-
papers/wp-sdn-newnorm.pdf

38 Pentikousis, K., Wang, Y. & Hu, W. (2013). "MobileFlow: toward software defined mobile
networks", Communications Magazine, IEEE, vol. 51, no. 7, 44–53. doi:
10.1109/MCOM.2013.6553677

39 Phemius, K., Bouet, M. & Leguay J. (2014). "DISCO: distributed multi-domain SDN controllers", Published in NOMS, IEEE, 1-4. doi: 10.1109/NOMS.2014.6838330

40 Pica8 controller. Pica 8 Integrated Open OVS Switch & Controller. Retrieved from
http://www.pica8.com/sdn-technology/integrated-open-ovs-switch-controller.php

41 Porras, P., Shin, S., Yegneswaran, V., Fong, M., Tyson, M. & Gu, G. (2012). "A security
enforcement kernel for openflow network", Proceedings. of HotSDN '12, ACM, 121-126.
doi:10.1145/2342441.2342466

42 POX. About POX. Retrieved from http://www.noxrepo.org/pox/about-pox

43 Prete, L., Farina, F., Campanella, M. & Biancini, A. (2012). "Energy efficient minimum
spanning tree in openflow networks", In EWSDN, 36-41. doi:10.1109/EWSDN.2012.9

44 Ryu SDN Framework. Build SDN Agility. Retrieved from http://osrg.github.io/ryu/

45 Saikia, D. & Malik, N. (2013). White paper An Introduction to Open MUL SDN Suite
Retrieved from http://www.openmul.org/uploads/1/3/2/6/13260234/openmul-sdn-
platform.pdf

46 Schmid, S. & Suomela, J. (2013). "Exploiting locality in distributed SDN control",
HotSDN'13, ACM.

47 Shahreza, S.S. & Ganjali, Y. (2013). "FleXam: flexible sampling extension for monitoring
and security applications in OpenFlow", In the proc. of the second ACM SIGCOMM workshop HotSDN'13, ACM, 167-168.

48 Sherwood, R., Chan M., Covington, A., Gibb, G., Flajslik, M., Handigol, N., Huang, T. ,
Kazemian, P., Kobayashi, M., Naous, Seetharaman, S. , Underhill, D., Yabe,T., Yap,
Yiakoumis, Y., Guido, H., Johri, A. R., McKeown N., Parulkar U. (2010). " Carving research
slices out of your production networks with openflow", Computer Communication Review
Newsletter ACM SIGCOMM, ACM, Volume 40 Issue 1, January 2010, pp 129-130, doi =
10.1145/1672308.1672333, 2010.

49 Shin, S., Porras, P., Yegneswaran, V., Fong, M., Gu, G. & Tyson, M. (2013). "FRESCO: modular composable security services for software-defined networks", In NDSS Symposium.

50 Shin, S., Song, Y., Lee, T, Lee, S., Chung, J., Porras, P., Yegneswaran, V., Noh, J. & Kang, B.
B. (2014). "Rosemary: a robust, secure, and high performance network operating system",
In Proceedings of the 21st ACM Conference on CCS, ACM, 78-89.
doi:10.1145/2660267.2660353

51 Shirazipour, M., Zhang, Y., Beheshti, N., Lefebvre, G. & Tatipamula, M. (2012). "OpenFlow
and multi-layer extensions: overview and next steps", In EWSDN, 7-12. doi:
10.1109/EWSDN.2012.22

52 SNAC. Stanford University/ Big switch networks SNAC OpenFlow controller. Retrieved from
google.co.in// SNAC-poster-gec9-final(1).pdf-Reader

53 Sood, M. & Nishtha (2014). "Traditional verses software defined networks: a review paper", Published in IJCEA, Vol VII issue 1.

54 Tootoonchian, A. & Ganjali, Y. (2010). "Hyperflow: a distributed control plane for openflow", Proceedings of the 2010 INM/WREN'10 , USENIX, pp 3-3.

55 Tootoonchian, A., Gorbunov, S., Ganjali, Y., Casado, M. & Sherwood R. (2012). "On controller performance in software-defined networks", In Proceedings of the 2nd USENIX conference Hot-ICE'12, 10–10.

56 Trema.Retrieved from trema: http://trema.github.io/trema/

57 Vellos. Retrieved from http://www.vellosystems.com/why-vello/

58 Voellmy, A., Ford, B, Hudak, P. & Yang, Y.R. (2012). "Scaling software-defined network controllers on multicore servers".

59 VMwareNSX. Retrieved through https://www.vmware.com/files/pdf/products/nsx/VMware-NSX-Datasheet.pdf

60 Yeganeh, S.H. & Ganjali, Y. (2012). "Kandoo: a framework for efficient and scalable offloading of control applications", Proceedings of the first workshop on HotSDN'12, ACM, 19-24.doi: 10.1145/2342441.2342446

Arya G S [1], Oormila R Varma [2], Sooryalakshmi S [3],
Vani Hariharan [4] and Siji Rani S [5]

Home Automation using Android Application

Abstract: Home automation can be defined as automating the ability to control the electronic devices in our home. In recent years, the home environment is getting more acquainted with network enabled devices which gives rise to the need of such a system. This paper presents a prototype of an energy efficient home automation system where the user can control the light and fan in his room from anywhere around the world using an android application, through a website. The system predominantly works on android and arduino platform with the support of Bluetooth. A user-friendly interface is developed in android through which the user can check the status and control the light and fan by providing the domain name. The server passes the data to the controller, programmed in android, which in turn passes the information to the arduino platform. End user can query the status and control the electronic device through his mobile. The information from the mobile is is passed to controller through cloud and control will take corresponding actions. The development in mobile technology has made the realization of such a system simpler.

Keywords: Home Automation, Android, Arduino, Bluetooth

1 Introduction

Home automation can be defined as automating the ability to control the electronic devices in our home. In recent years, the home environment is getting more acquainted with network enabled devices which gives rise to the need of such a system. This paper presents a prototype of an energy efficient home au-

1 Dept. of Computer Science and Engineering, Amrita Vishwa Vidyapeetham.
E-mail: hvani4@gmail.com
2 Dept. of Computer Science and Engineering, Amrita Vishwa Vidyapeetham.
3 Dept. of Computer Science and Engineering, Amrita Vishwa Vidyapeetham.
4 Dept. of Computer Science and Engineering, Amrita Vishwa Vidyapeetham.
5 Dept. of Computer Science and Engineering, Amrita Vishwa Vidyapeetham, Kollam.
E-mail: sijiranis@am.amrita.edu

tomation system where the user can control the light and fan in his room from anywhere around the world using an android application, through a website. The system predominantly works on android and arduino platform with the support of Bluetooth. A user-friendly interface is developed in android through which the user can check the status and control the light and fan by providing the domain name. The server passes the data to the controller, programmed in android, which in turn passes the information to the arduino platform. End user can query the status and control the electronic device through his mobile. The information from the mobile is is passed to controller through cloud and control will take corresponding actions. The development in mobile technology has made the realization of such a system simpler.

2 Related Work

Various home automation systems have been developed till date with PC as the controller or Bluetooth being the overall technology used making it quite common and not accessible all the time. The systems that have been proposed so far:

In [1], by R.A.Ramlee, M.H.Leong, the system discussed is implemented using Bluetooth technology for the communication between the PC/laptop and the smart phone. This holds back the system from communicating with each other outside the Bluetooth range and is also an energy consuming one.

In [2-3], the controls can be performed only through the PC which implies a dedicated user should be monitoring it all the time.

Al-Ali and Al-Rousan [4] presented a design and implementation of a Java-based automation system through World Wide Web. It had a standalone embedded system board integrated into a PC-based server at home. They did not highlight the low level details of the type of peripherals that can be attached.

The design and implementation of a microcontroller based voice activated wireless automation system is presented in [5]. The user speaks the voice commands through a microphone, which is processed and sent wirelessly via radio frequency (RF) link to the main control receiver unit. Voice recognition module is used to extract the features of the voice command. This extracted signal is then processed by the microcontroller to perform the desired action. The drawback is that the system can only be controlled from within the RF range.

Paper [6] proposed mobile IP based architecture and its potential applications in Smart homes security and automation without any actual deployment and testing.

The studies in [7, 8] have presented Bluetooth based home automation systems using Android Smart phones without the Internet controllability. The devices are physically connected to a Bluetooth sub-controller which is then accessed and controlled by the Smart phone using built-in Bluetooth connectivity

Recently, more and more Smart Living applications based on Android and Bluetooth have been developed. Android system equips with SDK and APIs for developers to build new applications. With Bluetooth already integrated into Android system, many Smart Living systems are constructed under Android system.

3 Proposed Work

3.1 Proposed Architecture

Fig.1 is the block diagram of the system we are proposing which is an energy efficient, extendable and globally available one. It can be divided into four different modules namely:

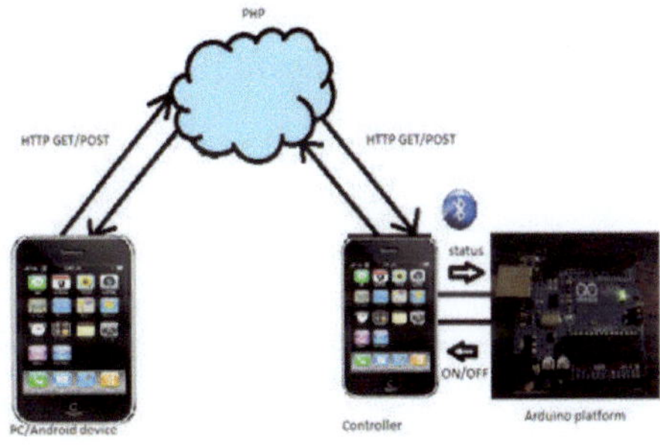

Fig. 1. Block Diagram

- Client-side
communication
- Server-side hosting
- Data fetching
- Device controlling

In the first module, at the user end, we have an application developed using android through which the user can control the electronic devices as well as check the status. This is done by logging onto the specified domain name so that the hosted web page can be accessed. In this system, 'studentsinnovations.com/Home_Automation' is the specified domain name. Once the user enters into this web page, he/she can view and control the above mentioned parameters.

The second module deals with the request given by the user through his or her android mobile device which is updated in the database at the server. There is a field named 'status' which is 'clear' by default. When the user presses a button in his android mobile device to turn on/off the light or fan, the field value of the status changes accordingly. For example,if the user requests to switch on the light,'L' is updated in the database. This change in field is captured by the controller which is continuously checking for any user requests. The request is then forwarded to the device controlling part.

The third module, data fetching operation is done by the controller developed using Android, thus eliminating the use of a PC which is more energy consuming one. It provided checks for any data request in the cloud from the user after every few seconds. If there is such a request, it is caught by the controller using HTTP get/post methods.

The final module, device controlling, is performed by the arduino platform which interfaces with the controller via Bluetooth. The arduino platform consists of an arduino nano microcontroller, a Bluetooth module, power supply, voltage regulator, relays, relay driver, a tester LED, an LED to represent light and a motor to represent fan; all soldered onto a single board (Fig.2).
The specifications of the components are:
- Arduino nano
- ULN2803 relay driver
- 3 LEDs
- NSK Bluetooth module
- Capacitors of 10microFarad and 100 microFarad
- 3 resistors of 330ohm
- 2 relays
- Power supply

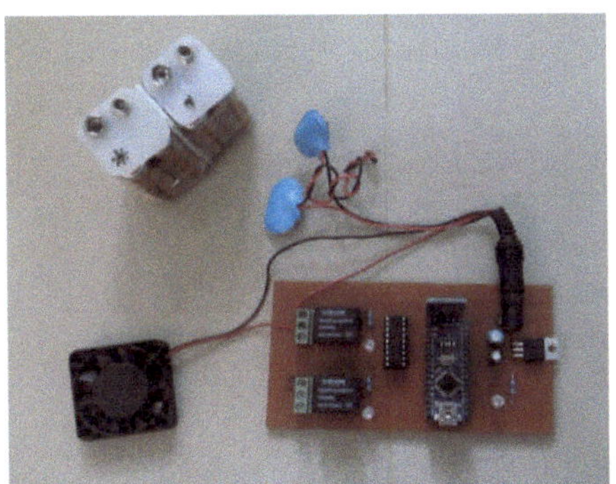

Fig. 2. Arduino Platform

The system was implemented in three stages as follows:

Software development using android: We have chosen android as the working platform due to its widespread poularity and availability in the market. The android SDK provides different tools and APIs for developing an application in Java language. 'ControlHM.apk' is the programmed java file created and incorporated into the mobile device. It is an interface between the server and the user (client). Using this interface, the user can control the light and fan from anywhere around and check the status.

Bluetoothchat.apk (available in the SDK sample code kit) file, modified as per the system requirements, is incorporated into another mobile device which acts as the controller. The program checks for any Bluetooth-activated device and pairs. So, the arduino platform which contains a Bluetooth module gets paired with the controller and the Bluetooth chat takes place. It is through this Bluetooth chat that the data is sent and received between the controller and the arduino microcontroller.

Server side Web hosting: Fig.3 is the screenshot of the application where the user can check the status and turn On/Off the light/fan.

This page can be viewed by typing in the domain name through the android application. Once the user presses the 'Light ON' button, the data is updated in the database in the server and the controller captures the state change. This information is passed onto the device controlling part explained below.

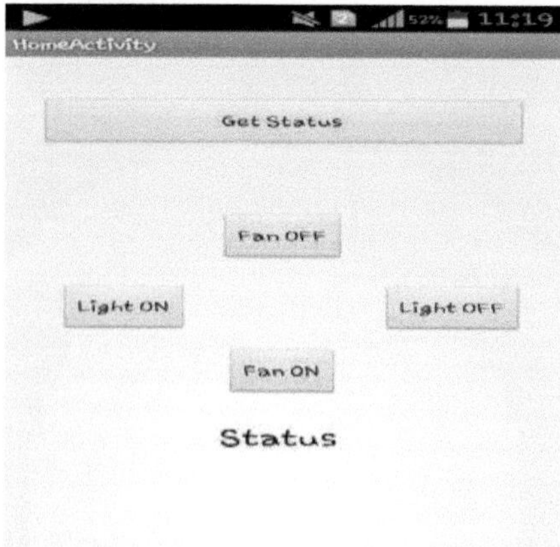

Fig. 3. User Interface

Device Controlling: Fig.4 is the screenshot of the controller where the Bluetooth chat takes place. Initially, we have to pair the arduino device controlling platform and the android developed controller via bluetooth. So, we install the Bluetoothchat.apk file in a mobile device and open the application. As soon as we enter the domain name mentioned in the previous section, it asks for permission for switching on the bluetooth of the device. When this permission is granted, select 'connect to a device' option and since the arduino is switched on at the other end, it detects an open bluetooth connection named here as 'HC-05' and pairs with it. The Embedded C code required by the arduino to control the devices is uploaded into it via serial port. The circuit diagram of the arduino platform is shown in Fig.5. Resistors of 330ohm are used to protect the LED as it runs on 2.2V and the power supply is 9V.

Programming the microcontroller: The arduino microcontroller was serially loaded with Embedded C code as follows:

- Set the communication parameters like the baud rate (9600 bits/s) and the input/output ports.
- Initially, the relays are set to LOW.
- Fan relay is set to the 3rd pin and light relay to the 4th.
- A tester LED is kept to know whether the board is working or not.
- if (Serial.available () 0)

This function checks for any message from the controller and if so, the cases are tested against each of them.

- recVal = Serial.read ();

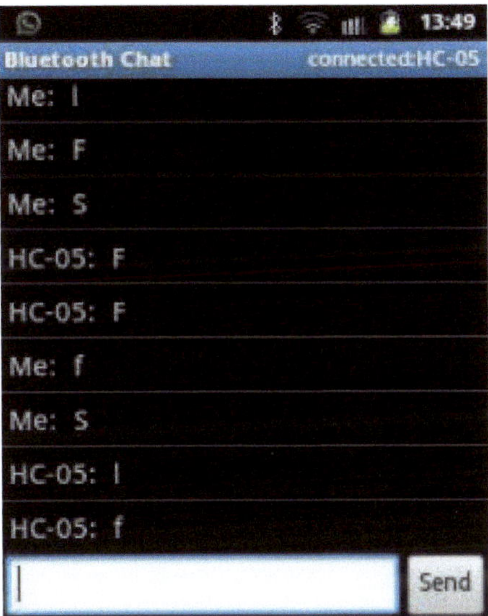

Fig. 4. Bluetooth chat

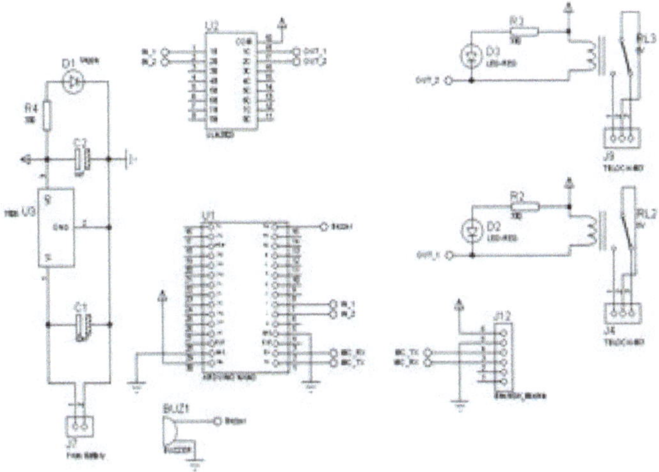

Fig. 5. Home Automation Circuit

The received message is read and processed.

    - The message can be any of the following letters: S,L,l,F,f.

4 Results and Analysis

After the completion of the hardware coupling, several tests were done on the design. Some observations and corrections were made as follows:

(a)   It was observed that the android app hung up every time the phone was filled with recent apps used. So, after clearing it up it worked well.

(b)   The request and response takes time according to the data rate/speed of the Internet connection used.

(c)   Domain for hosting the web page was not easily available. So, initially an IP address was used for testing. Later, we purchased the domain from the 'studentsinnovations.com'.

Other than the above situations, the system worked well and reliably.

Conclusion

The home automation system discussed in this paper is fully developed and tested to analysis its effectiveness. The system aims at controlling the electronic devices at home from anywhere around using an android mobile application by means of a Wi-Fi connection. It is dominantly working on android and arduino open source. The system is expandable to more devices according to which the program has to be modified. The speed of the data sent and received depends on the data rate of the GSM/Wi-Fi connection used.

It is evident from this project work that an individual control home automation system can be assembled and implemented easily using low cost, easily available components and can be used to control different home appliances ranging from the refrigerator, the television to the air conditioning system and even the entire house security system. And the better part is that the components required are so small and few that the space required by them is negligible. Finally, this home automation system can be also implemented over Bluetooth and Infrared connectivity without much change to the design and yet still be able to control a variety of home appliances. Hence, this system is scalable and expandable.

Future Work

In consonance with the project work and in view of the researched methods and undertakings in the project design, the following are recommended:
Our system can be further developed to control more than one home appliance at once through the use of short message service texts or voice dial. Also, to cut the cost of mobile phone, the project may be implemented using standalone GSM modems that only perform specialised functions like text messaging and/or phone calls. This GSM modems often are cheaper and more reliable than GSM mobile phones. This project can also include more specialized properties like automatic detection of faults in the control appliances. Similarly, sensors can be attached to the system so that the appliance can sense the presence of a person and automatically set the parameters in the room accordingly.

References

1   R.A.Ramlee, M.H.Leong, R..S.S.Singh, M.M.Ismail, H.A.Sulaiman, M.H.Misran, M.A.Meor Said,(2013) "Bluetooth Remote Home Automation System Using Android Application" Vol. 2, pp 149-153.
2   N. Sriskanthan and Tan Karande, "Bluetooth Based Home Automation Systems," Journal of Microprocessors and Microsystems, 2002, Vol. 26, pp. 281-289
3   Kwang Yeol Lee & Jae Weon Choi, "RemoteControlled Home Automation System via Bluetooth Home Network" in SICE Annual Conference in Fukui, 2003, Vol. 3, pp. 2824- 2829
4   Al-Ali, Member, IEEE & M. AL-Rousan,"Java-Based Home Automation System R." IEEE Transactions on Consumer Electronics, Vol. 50, No. 2, MAY 2004
5   K. P. Dutta, P. Rai, and V. Shekher, "Microcontroller Based Voice Activated Wireless Automation System," *VSRD Internation Journal of Electrocal, Electronics & Communication Engineering*,vol. 2, pp. 642-649, 2012.
6   Park, B.: Mobile IP-based architecture for smart homes. International Journal of Smart Home, vol. 6, no.1, 29--36 (2012)
7   Piyare, R., Tazil, M.: Bluetooth based home automation system using cell phone. In Consumer Electronics (ISCE), 2011 IEEE 15th International Symposium on, pp. 192- 195.(2011)
8   Yan, M., Shi, H.: Smart Living Using Bluetooth-Based Android Smartphone. International Journal of Wireless & Mobile Networks (IJWMN), vol. 5, no.5, pp. 65--72 (2013)

Sujay Saha [1], Praveen Kumar Singh [2] and Kashi Nath Dey [3]

Missing Value Estimation in DNA Microarrays using Linear Regression and Fuzzy Approach

Abstract: High-throughput microarray experiments usually generate gene expression profiles in the form of matrices with high dimensionality. Unfortunately, it may happen that microarray experiments generate data sets with multiple missing values, which significantly affect the performance of subsequent statistical and machine learning experiments. There exist numerous imputation algorithms to estimate those missing values. In this paper, we propose a new approach, LRFDVImpute, for time series gene expression analysis that estimates multiple missing observations by first find the most similar genes of the target gene and then applying the linear regression on those similar genes. The new approach works in two stages. At the first stage, we estimate the real missing cells of SPELLMAN_COMBINED dataset and at the later stage, we make some cells miss forcefully of the same dataset and then using the estimated results from the first step, we estimate those missed cells using the same approach used earlier. Absolute Error has been calculated from the difference between the original value and the estimated value. Root Mean Square Error (RMSE) of those absolute errors are then determined. It has been seen that our approach shows significant improvements over the current state-of-the-art methods in use as far as the RMSE value is concerned.

Keywords: DNA microarray, Fuzzy Difference Vector, Linear Regression, Missing value estimation, SPELLMAN_COMBINED dataset

1 Introduction

A gene expression microarray is a collection of microscopic DNA spots attached to a solid surface, which is used to study the expression levels of thousands of genes under various conditions simultaneously [8]. Gene expression microarray

1 Heritage Institute of Technology, Kolkata, West Bengal, India, sujay.saha@heritageit.edu
2 Heritage Institute of Technology, Kolkata, West Bengal, India, pcubesingh@gmail.com
3 University of Calcutta, Kolkata, West Bengal, India, kndey55@gmail.com

experiments generate datasets of massive order which are in the form of matrices of gene expression levels under various experimental conditions. Each row of a gene expression matrix is basically a gene of the organism used in the gene expression matrices that contain missing values. These missing values occur due to errors in the experimental process that lead to corruption or absence of expression measurements. Various statistical methods used for gene expression analysis requires the complete gene expression matrix for providing accurate results. Methods such as hierarchical clustering, K-Means clustering are not robust to missing values. Hence, it is necessary to devise proper and accurate methods which impute data values when they are missing. Time series data are a sequence of data points sampled at regular intervals of time. Gene expression time series data is a special class of microarray data where gene expression levels are sampled at regular intervals of time. Data sets measuring temporal behavior of thousands of genes offer rich opportunities for computational biologists [1]. A time-series gene expression data set is very sparse in nature as it contains a handful of data points. So a very accurate prediction method must be used for estimation.

2 Linear Regression

Regression can be used to approximate the given data. In (simple) linear regression, data are modeled to fit a straight line. For example, a random variable y, also called as response variable, can be represented as a linear function of another random variable, x, referred to as predictor variable, by the equation $y = h_\theta(x) = \theta_0 + \theta_1 x$, where regression coefficients θ_0 & θ_1 specify the Y-intercept and the slope of the line respectively and $h_\theta(x)$ is a hypothesis [9]. Multiple linear regression is an extension of linear regression, that allows y to be modeled as a linear function of two/ more predictor variables, where the response variable y can be modeled as a linear function of other predictor variables x_1, x_2,...,x_n by the following equation:

$$y = h_\theta(x) = \theta_0 + \theta_1 x_1 + \theta_2 x_2 + ... + \theta_n x_n \tag{1}$$

The coefficients mentioned above can be solved for by using a cost function that minimizes the error between the actual line separating the data and the estimate of the line [9]. The objective of the linear regression is to minimize the cost function, which is defined as follows:

$$J(\theta) = \frac{1}{2m} \sum_{i=1}^{m} (h_\theta(x^{(i)}) - y^i)^2 \tag{2}$$

where m denotes number of training samples present. $h_\theta(x^{(i)})$ represents estimated output of ith training sample and y^i represents actual output of the same sample.

3 Literature Review

Gene expression microarray experiments can generate data sets with multiple missing expression values. Unfortunately, many algorithms for gene expression analysis require a complete matrix of gene array values as input. For example, methods such as hierarchical clustering and K-means clustering are not robust to missing data, and may lose effectiveness even with a few missing values. Methods for imputing missing data are needed, therefore, to minimize the effect of incomplete data sets on analyses, and to increase the range of data sets to which these algorithms can be applied. Rest of this section briefly describes some of those widely used, existing methods for estimation of the missing values from DNA microarray.

The earliest method, named as Row averaging or filling with zeroes, used to fill in the gaps for the missing values in gene data set with zeroes or with the row average.

Troyanskaya et al. [2] proposed KNNImpute method to select genes with expression profiles similar to the gene of interest to impute missing values. After experimenting with a number of metrics to calculate the gene similarity, such as Pearson correlation, Euclidian distance, variance minimization, it was found that Euclidian distance was a sufficiently accurate norm.

The SVDImpute method, proposed by Troyanskaya et al. [2] uses Singular Value Decomposition of matrices to estimate the missing values of a DNA micro array. This method works by decomposing the Gene data matrix into a set of mutually orthogonal expression patterns that can be linearly combined to approximate the expression of all genes in the data set. These patterns, which in this case are identical to the principle components of the gene expression matrix, are further referred to as eigengenes. [3], [4]

Golub et al. [5] proposes another method named as LLSImpute that represents a target gene with missing values as a linear combination of similar genes. The similar genes are chosen by k-nearest neighbors or k coherent genes that have large absolute values of correlation coefficients followed by least square regression and estimation.

BPCAImpute method, proposed by Oba et al. [6] uses a Bayesian estimation algorithm to predict missing values. BPCA suggests using the number of samples minus 1 as the number of principal axes. Since BPCA uses an EM-like repetitive algorithm to estimate missing values, it needs intensive computations to impute missing values.

Ziv Bar-Joseph et al. [1] presents algorithms for time-series gene expression analysis that permit the principled estimation of unobserved time-points, clustering, and dataset alignment. Each expression profile is modeled as a cubic spline (piecewise polynomial) that is estimated from the observed data and every time point influences the overall smooth expression curve. The alignment algorithm uses the same spline representation of expression profiles to continuously time warp series.

FDVImpute method, proposed by S. Chakraborty, S. Saha, K. N. Dey [7], incorporates some fuzziness to estimate the missing value of a DNA microarray. The first step selects nearest (most similar) genes of the target gene (whose some component is missing) using Fuzzy Difference Vector algorithm. Then the missing cell is estimated by using least square fit on the selected genes in the second step.

Another method, FDVSplineImpute, proposed by S.Saha et al. [8], consists of two distinct steps. In the first step, we select the most similar genes to the target gene using the fuzzy difference vector (FDV) algorithm with a suitable membership threshold. In the second step, a representative gene is chosen from the set of k similar genes, and a smoothing spline is fitted to the representative gene to reconstruct the target gene and estimate the missing value.

CFBRST, proposed by Bo-Wen Wang and Vincent S. Tseng [11], imputes multiple missing values using collaborative filtering (CF) and rough set theory. User based CF method is used to find the similar genes of the target gene and then rough set theory is used to estimate the missing values. Experimental results show that the proposed approach has better accuracy than that of a k-NN approach for yeast cDNA microarray datasets, especially when the percentage of missing values is high.

A Decision Tree based Missing value Imputation technique (DMI) proposed by Md. Geaur Rahman and Md. Zahidul Islam [12] makes use of an decision tree algorithm such as C4.5 for datasets having categorical attributes and an Expectation Maximisation (EM) based imputation technique (EMI) for numerical values. The proposed method use various evaluation criteria such as coefficient of determination, Index of agreement and root mean squared error (RMSE). Experimental results show that DMI performs significantly better than EMI.

4 Proposed Work

Our proposed algorithm, LRFDVImpute, works in two distinct phases. In the first phase, the algorithm estimates the real missing cells present in the SPELL-MAN_COMBINED dataset. It has been observed that in the said dataset, there are majority of rows containing single missing values, two missing values and three missing values. So we first estimate those cells one by one. There may be some small number of rows with more than three missing cells in the said dataset. Since this number is very much insignificant we'll not consider them. Whereas, in the second phase, the algorithm makes some known cells in different rows missing and then estimates those missing values on the basis of the estimation done in the first phase. So the second phase can be treated as the justification of the first phase.

Let us consider a time series gene expression microarray dataset Data of m genes $Data_1$, $Data_2$,..., $Data_m$ each of which has n observations. Let's now assume that out of these m genes, for m_1 genes there are no missing values. So these m_1 genes are treated as the training data, denoted by say $Data_{train}$. Now let's assume that the dataset Data contain s number of rows with single missing values, d number of rows with two missing values and t number of rows with three missing values. The proposed algorithm starts the first phase by estimating single missing cells first, then two missing cells and then three missing cells. On the way of doing this, the algorithm first choose one row, r, with single missing value randomly. Then it finds the most similar genes of r among the set $Data_{train}$ using fuzzy difference vector (FDV) algorithm proposed in [7] and used in [8]. Before applying this FDV algorithm all the columns from $Data_{train}$ is removed that correspond to missing values in target row, i.e. row r. So $Data_{train}$ now contains (n – 1) observations. Now, the difference vector V_i of the ith gene y_i is calculated as follows:

$$DifferenceTable_{i,k} = y_i(k) - y_i(k+1), 1 \leq k \leq (n-1) \qquad (3)$$

We then calculate $Membership_i$ to obtain the number of matches between difference vectors $DifferenceTable_i$ and $DifferenceTable_j$ of the respective genes. $Membership_i$ defines how much differentially each gene g_i in $Data_{train}$ is expressed with respect to the target row. We then define a fuzzy membership function Memgrade() for the ith gene as follows:

$$Memgrade(i) = \frac{Membership(i)}{q-1} \tag{4}$$

If this Memgrade(i) is greater than the threshold θ, then the corresponding i^{th} gene will be selected as the similar gene of the target gene. Once we have found all the similar genes of the target gene r then linear regression is applied on those similar genes by using a built in function LM() of R programming language to estimate the missing cell at that position. Once this cell is estimated then this row will be included in $Data_{train}$ for the next estimation. That means for the estimation of next cell we can have more enriched training dataset $Data_{train}$. In this way we have estimated all the rows with single, two and three missing values. Every time the training set is larger than the previous one. That's how the first phase is ended. At the start of the second phase of this algorithm, all the missing cells of SPELLMAN_COMBINED dataset contain the estimated values. Now to justify the estimation of the done at the first phase, this phase of the proposed method first makes some cells of the complete dataset missed forcefully one at a time. Now the values of those cells are estimated using the same method that has been employed at the first phase. This estimated value is in some way will be influenced by the estimation done at the first phase. If the estimation at the first phase is good then we hope to get better estimation at the later phase also. We have done this experiment couple of times and take the RMSE value of the errors got at each case. The complete pseudo code algorithm for the proposed method is shown below:

LRFDVImpute (Data, MaximumMissing , FeatureType, θ)
{
    //FeatureType can be any one of alpha, cdc15, cdc28, elu
Create ProcessedData from Data using FeatureType.
Select row and column positions of nearly 50 cells in ProcessedData from rows which have no missing values. Put it in 2-D matrix Row_Col.
/*Missing values for only those rows will be estimated that have a maximum number of missing cells, less than or equal to "MaximumMissing"*/
for (i =1 to MaximumMissing){
//FIRST PHASE
While(1){
TargetRow $\leftarrow$ select randomly a row with the least number of missing values.
if (number of missing values in TargetRow >i) Break .
/*SimilarGene provides a Boolean vector which is true for the index corresponding to the rows that are similar to the TargetRow and with no missing cells*/
// θ is the membership parameter

```
        SimilarVector ← SimilarGene(ProcessedData,TargetRow, θ)
        MissingCell ← randomly select one missing cell from the TargetRow.
/*creating a linear model object for the estimation of the MissingCell*/
/*left side of '~' indicates the columns that constitute features and right side
indicates the target or output column*/
Formula ← Columns in ProcessedData corresponding to
    which there is value in TargetRow ~ MissingCell_Col.
Lmobj ← lm (formula, Data= ProcessedData, Subset = SimilarVector)
MissingCell ← Estimation (Lmobj, TargetRow)
}
//SECOND PHASE
for (j=1 to length(Row_Col))
{
//selecting  the jth row and column entry from Row_Col.
                        MissingCell←ProcessedData[Row_Col(1, j), Row_Col(2, j
]
        SimilarVector ←SimilarGene (ProcessedData, MissingCell_Row, θ)
Formula ← Columns in ProcessedData corresponding to which there is value in
MissingCell_Row ~ MissingCell_Col.
Lmobj ← lm(formula, Data = ProcessedData, Subset =SimilarVector)
MissingCell ← Estimation (Lmobj, MissingCell_Row)
}
All the results generated from the above loop are stored in the ith column of
 RESULT_TABLE.
 }
   Return RESULT_TABLE
}
```

5 Experimental Results

The proposed method, LRFDVImpute, is applied on the yeast cell cycle time series dataset from Spellman et al. The dataset consists of a total of 6178 genes and 82 experiments performed on each gene. The dataset was pre-processed by decomposing the complete gene expression matrix into four specific time series gene expression matrices – alpha, cdc15, cdc28, elu and do the experiments on each of these components separately.

Table 1: Summary of gene expression time series analyzed

Dataset	Start	End	Sampling
alpha	0m	119m	Every 7m
cdc15	10m	290m	Every 20m for 1 hr., 10 m for 3hr, 20 m for final hr.
cdc28	0m	160m	Every 10m
elu	0m	390m	Every 30m

After doing the required pre-processing on the said dataset, we choose random gene expression levels from each experiment and applied our proposed method. We compute the error as e = |EstimatedValue – OriginalValue|. We then compute the root mean square error (RMSE) as follows:

$$RMSE = \sqrt{\frac{1}{n}\sum_{i=1}^{n} e_i^{2}}$$

(5)

We have taken threshold value θ as 0.6. TABLE II below shows the complete result of the experiments done on four gene expression data sets at the second phase of the algorithm, where RMS error is shown for both linear features and quadratic features and from the results it is clear that estimation is much more accurate for the added quadratic features.

Table 2. Experimental Result of the Experiment on ALPHA, CDC_15, CDC_28 & ELU dataset (End of the paper).

The following figure shows a comparative study of the results of our approach LRFDVImpute with the earlier two methods, like FDV-LLSImpute and FDVSplineImpute. Comparison is done on the basis of RMSE metric and with ALPHA, CDC_15 and ELU dataset of SPELLMAN_COMBINED dataset.

From the above plot it is very clear that as far as RMSE metric is concerned our proposed approached is far better than compared to existing state-of-the-art approaches like FDV-LLSImpute and FDVSplineImpute. That means the estimation that our proposed method does is very close to the original one.

We have done another set of experiments on the same dataset where we vary the value of θ, membership degree from 0.5 to 0.7. Following table shows that result.

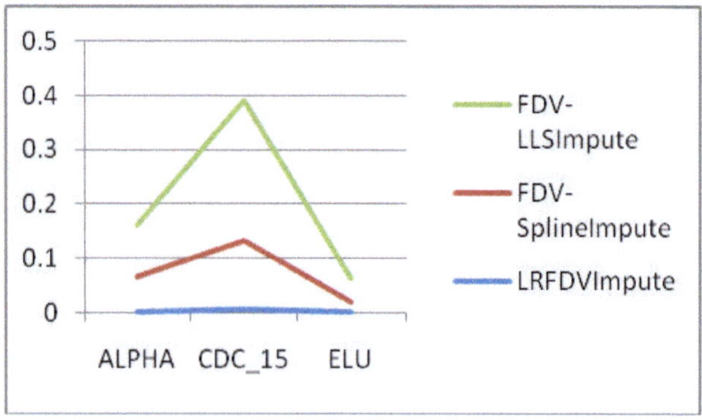

Figure. 1: Comparative Study w.r.t RMSE on ALPHA, CDC_15 & ELU for LRFDVImpute, FDVSplineImpute and FDV-LLS methods

Table 3. RMSE value of various values of θ

Dataset	θ = 0.5	θ = 0.55	θ = 0.6	θ = 0.65	θ = 0.7
RMSE					
Alpha	0.000599	0.001192	0.001624	0.00117	0.001034
cdc15	0.002806	0.00072	0.001003	0.001352	0.001526
cdc28	0.000965	0.001264	0.004256	0.001256	0.00125
elu	0.000622	0.000431	0.000713	0.000885	0.00102

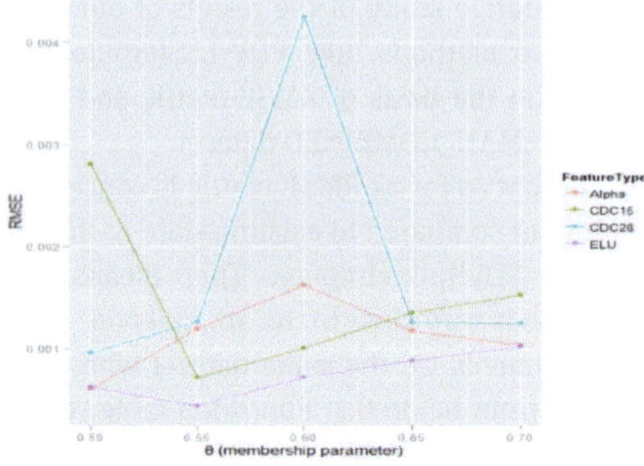

Figure. 2: Plot of RMSE w.r.t. θ

Conclusion & Future Scope

Results show that our proposed approach, LRFDVImpute, performs significantly better than the existing methods, like FDVLLSImpute, FDVSplineImpute. Fuzzy Difference Vector (FDV) is used to find the most similar genes of the target gene. A package of R, named as LM is used for linear regression. Performance is measured on RMSE metric. Moreover, the proposed approach can estimate multiple missing values occurred simultaneously and by experiments, we have considered those rows having one, two and three missing cells. There are some rows having more than three missing cells, but since those are insignificant numbers we have not considered those cells.

One drawback that we observe for this approach is that at the first phase, when we finish estimation of the missing cells in a particular row, that row is included in the training set. Now if the estimation is very close then there is no problem. But when estimation is not very close then by including that row in the training set will propagate the error further. That may affect the estimation of the following cells.

We have only measured the error in estimation of missing cells using one statistical metric like RMSE. It'll sometimes necessary to validate the results biologically too. Gene ranking methods can be used as an extension of this work where biological validation can be done using ranking of the genes.

References

1 Z. Bar-joseph , G. K. Gerber , D. K. Gifford , T. S. Jaakkola , and Itamar Simon, Continuous Representations of Time-Series Gene Expression Data, *Journal of Computational Biology*, 10 (3-4):341-56 (2003)

2 O. Troyanskaya, M. Cantor, G. Sherlock, P. Brown, T. Hastie, R. Tibshirani, D. Botstein, and R. B. Altman, Missing value estimation methods for DNA microarray, *Bioinformatics*, 17, 520–525 (2001)

3 O. Alter, P. O. Brown, and D. Botstein, Singular value decomposition for genome-wide expression data processing and Modelling, *Proc. Natl. Acad. Sci. USA*, 97, 10101–10106 (2000)

4 G. H. Golub and V. Loan, *Matrix Computations*, 3rd edn. (Johns Hopkins University Press, Baltimore, CA, 1996)

5 H. Kim, G. H. Golub, and H. Park, Missing value estimation for DNA microarray gene expression data: local least squares imputation, *Bioinformatics*, 21, 187 – 198 (2005)

6 S. Oba, M. Sato, I. Takemasa, M. Monden, K. Matsubara and S. Ishii, A Bayesian missing value estimation method for gene expression profile data, *Bioinformatics*; 19(16):2088-96, (2003)

7   S. Chakraborty, S. Saha, and K.N. Dey, Missing Value Estimation in DNA Microarray – A Fuzzy Approach, *IJAINN*, Volume 2, Issue 1 (2012)

8   S. Saha, K. N. Dey, R. Dasgupta, A. Ghose, and K. Mullick, Missing Value Estimation In DNA Microarrays Using B-Splines, *Journal of Medical and Bioengineering*, Volume 2, Number 2, pp. 88 – 91, (2013)

9   J. Han and M. Kamber, *Data Mining Concepts and Techniques*, 2ndEdition, ELSEVIER

10 Z. Kai, M. Heydari, and G. Lin, Iterated local least squares microarray missing value imputation, *Journal of Bioinformatics and Computational Biology*, 4(5), 935-57 (2006)

11 B. W. Wang and V. S. Tseng, Improving Missing-Value Estimation In Microarray Data With Collaborative Filtering Based On Rough-Set Theory, *International Journal of Innovative Computing, Information and Control*, Volume 8, Number 3(B), (2012)

12 Md. G. Rahman and Md. Z. Islam, A Decision Tree-based Missing Value Imputation Technique for Data Pre-processing, Proc. 9-th Australasian Data Mining Conference (AusDM), Ballarat, Australia (2011)

Experiment	Row	Column	Original Value	After estimating all rows with single missing cells at 1st phase			After estimating all rows with two missing cells at 1st phase			After estimating all rows with three missing cells at 1st phase		
				Estimated Value	Error	RMSE	Estimated Value	Error	RMSE	Estimated Value	Error	RMSE
ALPHA	81	13	-0.02	-0.02013	0.00013	0.0023	-0.0201	0.00010	0.0022	-0.02014	0.00014	0.0022
	3066	18	-0.02	-0.0197	0.00030	.	-0.01969	0.00030	.	-0.0197	0.00030	.
	4852	1	0.02	0.019691	0.00030	.	0.019733	0.00026	.	0.019771	0.00023	.
	4492	17	-0.16	-0.1605	0.00049	.	-0.16041	0.00041	.	-0.16038	0.00038	.
	4528	10	-0.02	-0.01929	0.00071	.	-0.0193	0.00070	.	-0.01934	0.00066	.
	4982	1	0.07	0.071176	0.00117	.	0.071178	0.00117	.	0.07117	0.00117	.
	3357	12	0.06	0.058798	0.00120	.	0.058766	0.00123	.	0.058752	0.00124	.
	1277	5	0.04	0.038785	0.00122	.	0.038766	0.00123	.	0.038782	0.00121	.
	4332	16	0.1	0.101385	0.00139	.	0.101384	0.00138	.	0.101393	0.00139	.
	1817	14	-0.32	-0.31858	0.00142	.	-0.3186	0.00140	.	-0.31861	0.00139	.
	4063	18	0.01	0.011595	0.00159	.	0.011557	0.00155	.	0.011533	0.00153	.
	4784	17	-0.07	-0.06757	0.00243	.	-0.06757	0.00243	.	-0.06758	0.00242	.
	324	8	0.87	0.867396	0.00260	.	0.867487	0.00251	.	0.867467	0.00253	.
	1338	7	0.36	0.356491	0.00350	.	0.356582	0.00341	.	0.356585	0.00341	.
	266	14	0.84	0.83342	0.00658	.	0.833494	0.00651	.	0.833442	0.00655	.

Experiment	Row	Column	Original Value	After estimating all rows with single missing cells at 1st phase			After estimating all rows with two missing cells at 1st phase			After estimating all rows with three missing cells at 1st phase		
				Estimated Value	Error	RMSE	Estimated Value	Error	RMSE	Estimated Value	Error	RMSE
CDC_15	2088	1	-0.31	-0.31057	0.00057	0.0060	-0.31047	0.00047	0.0060	-0.31045	0.00045	0.0060
	5867	14	-0.24	-0.2406	0.00060	.	-0.24063	0.00062	.	-0.24055	0.00054	.
	5101	17	-0.01	-0.01076	0.00076	.	-0.01071	0.00071	.	-0.01071	0.00070	.
	2383	3	-0.03	-0.03086	0.00086	.	-0.03079	0.00079	.	-0.0308	0.00080	.
	4760	23	0.04	0.038857	0.00114	.	0.038908	0.00109	.	0.038895	0.00110	.
	4097	19	-0.39	-0.3883	0.00170	.	-0.38827	0.00173	.	-0.38828	0.00172	.
	809	9	-0.52	-0.52257	0.00257	.	-0.52249	0.00248	.	-0.52247	0.00246	.
	5910	7	0.15	0.153584	0.00358	.	0.15353	0.00353	.	0.153521	0.00352	.
	1642	10	0.37	0.363079	0.00692	.	0.363208	0.00679	.	0.363201	0.00679	.
	1604	16	-0.56	-0.56823	0.00822	.	-0.56785	0.00784	.	-0.56768	0.00767	.
	2998	20	-0.53	-0.53827	0.00827	.	-0.53833	0.00832	.	-0.53834	0.00834	.
	2019	24	-0.14	-0.13122	0.00877	.	-0.13116	0.00883	.	-0.13117	0.00883	.
	3388	16	0.02	0.029095	0.00909	.	0.029134	0.00913	.	0.02913	0.00913	.
	868	15	0	-0.00921	0.00920	.	-0.00954	0.00953	.	-0.00957	0.00957	.
	1096	3	0.34	0.349233	0.00923	.	0.349299	0.00929	.	0.349306	0.00930	.

Experiment	Row	Column	Original Value	After estimating all rows with single missing cells at 1st phase			After estimating all rows with two missing cells at 1st phase			After estimating all rows with three missing cells at 1st phase		
				Estimated Value	Error	RMSE	Estimated Value	Error	RMSE	Estimated Value	Error	RMSE
CDC_28	3639	4	-0.33	-0.32934	0.00066	0.0031	-0.32949	0.00051	0.0034	-0.32945	0.00054	0.0034
	2272	8	-0.11	-0.11067	0.00067	.	-0.11092	0.00091	.	-0.11089	0.00089	.
	4820	2	0.7	0.700963	0.00096	.	0.701017	0.00101	.	0.701009	0.00100	.
	1774	6	-0.21	-0.20851	0.00149	.	-0.20868	0.00131	.	-0.20873	0.00127	.
	1610	11	0.2	0.198468	0.00153	.	0.198339	0.00166	.	0.198314	0.00168	.
	1980	6	0.38	0.378329	0.00167	.	0.378257	0.00174	.	0.378133	0.00186	.
	2712	1	0	-0.00183	0.00182	.	-0.00281	0.00280	.	-0.00287	0.00286	.
	5524	2	-0.04	-0.03794	0.00206	.	-0.03812	0.00187	.	-0.03817	0.00183	.
	1742	6	0.15	0.152935	0.00293	.	0.15286	0.00286	.	0.153007	0.00300	.
	1859	6	0.22	0.216977	0.00302	.	0.216983	0.00301	.	0.216892	0.00310	.
	2044	16	0.02	0.0159	0.0041	.	0.015634	0.00436	.	0.01569	0.00431	.
	5968	13	-0.17	-0.17435	0.00434	.	-0.17589	0.00588	.	-0.17588	0.00587	.
	594	8	-0.32	-0.31534	0.00466	.	-0.31481	0.00518	.	-0.31457	0.00542	.
	2966	3	-0.27	-0.2747	0.00469	.	-0.27446	0.00446	.	-0.27429	0.00428	.
	4181	3	0.81	0.804245	0.00575	.	0.803856	0.00614	.	0.803862	0.00613	.

Experiment	Row	Column	Original Value	After estimating all rows with single missing cells at 1st phase			After estimating all rows with two missing cells at 1st phase			After estimating all rows with three missing cells at 1st phase		
				Estimated Value	Error	RMSE	Estimated Value	Error	RMSE	Estimated Value	Error	RMSE
ELU	2783	7	-0.04	-0.03988	0.00011	0.0009	-0.03989	0.00011	0.0009	-0.03989	0.00011	0.0009
	2326	1	0.24	0.240196	0.00019	.	0.240192	0.00019	.	0.240182	0.00018	.
	2211	9	0	0.000209	0.00020	.	0.00021	0.00021	.	0.00021	0.00021	.
	1947	4	0.26	0.259723	0.00027	.	0.259731	0.00026	.	0.259731	0.00026	.
	1856	14	0.04	0.039688	0.00031	.	0.039687	0.00031	.	0.039683	0.00031	.
	2705	2	-0.28	-0.28033	0.00032	.	-0.28033	0.00032	.	-0.28033	0.00032	.
	4630	14	0.11	0.110363	0.00036	.	0.110362	0.00036	.	0.110362	0.00036	.
	3093	2	-0.15	-0.15038	0.00037	.	-0.15037	0.00037	.	-0.15037	0.00037	.
	4579	1	0.12	0.120473	0.00047	.	0.120475	0.00047	.	0.120475	0.00047	.
	1803	8	-0.15	-0.15048	0.00048	.	-0.15048	0.00047	.	-0.15048	0.00047	.
	931	2	-0.39	-0.39101	0.00100	.	-0.39099	0.00098	.	-0.39101	0.00100	.
	4472	10	-0.26	-0.25863	0.00137	.	-0.25863	0.00137	.	-0.25863	0.00137	.
	260	12	0.39	0.388364	0.00163	.	0.388364	0.00163	.	0.388364	0.00163	.
	4846	6	0.67	0.668131	0.00186	.	0.668131	0.00186	.	0.668131	0.00186	.
	4862	8	-0.03	-0.02799	0.00201	.	-0.028	0.00200	.	-0.028	0.00200	.

Amrita Roy Chowdhury[1], Rituparna Saha[2] and
Sreeparna Banerjee[3]

Detection of Different Types of Diabetic Retinopathy and Age Related Macular Degeneration

Abstract: Diabetic Retinopathy (DR) and Age related Macular Degeneration (AMD) are quite similar as they appear in retina. Varity of DR and AMD are detected using frequency, color, area and shape of abnormal objects in retina as characteristics. A training image set is used to detect the disease for a test set using Naïve Bayes classifier.

Keywords: Diabetic Retinopathy (DR); Age related Macular Degeneration (AMD); Naïve Bayes Classifier

1 Introduction

DR and AMD both are becoming the biggest threats for the ophthalmologists. Diabetic Retinopathy (DR) is a serious complication in retina. It causes clouding of vision as the blood vessel tree, supplying blood and nutrition to the retina, leaks blood and fluids like lipids and cholesterol. Age related Macular Degeneration(AMD) is the disease associated with elderly people as the shape of macula gets deformed with progressing age. Retina may contain small blood spots or tiny white fluid spots which leak due to damage of blood vessel. These are early symptoms of non-proliferative DR and can be arrested if treated early. Severe leakage of blood vessel tree causes hemorrhages and if untreated, can lead to blindness. Symptoms of AMD are very close to those of DR. In this paper, separation of AMD and DR images are done depending on some pathological

1 Department of Computer Science and Engineering, West Bengal University of Technology, Salt Lake, West Bengal 700064 India
E-mail: amrita.me.cse@gmail.com
www.wbut.ac.in
2 Department of Computer Science and Engineering, West Bengal University of Technology, Salt Lake, West Bengal 700064 India
3 Department of Computer Science and Engineering, West Bengal University of Technology, Salt Lake, West Bengal 700064 India

reports of the patients. DR has four types of symptoms: cotton wool spots (also known as soft exudates), hard exudates, micro aneurysms and hemorrhages. In cotton wool spots, a few whitish yellow, nearly circular objects appear in retina. In hard exudates, the frequency of circular objects increases than cotton wool spot and color is bright yellow. Micro aneurysms are very small red dots of blood with no specific shape whereas hemorrhages are large clots of blood covering a big area of retina. In this research work, several retinal images having different types of DR are considered as input image and the type of abnormality is detected automatically. In AMD, mainly two varieties are noticed: dry AMD and wet AMD. Dry AMD contains many yellow spots called drusen which are clustered near macula. Wet AMD contains drusen along with blood clots. An automatic system is developed to detect the damaged portion of retina and the type of abnormality. Early detection of DR or AMD can prevent permanent vision loss. The advanced stages of DR or AMD can be checked through proper medication. In this research work, several Matlab codes are developed to detect the abnormalities of retinal images due to DR and AMD. Naive Bayes Classifier available in Weka is used for machine learning purpose.

2 Related Works

Several research works have been carried out to detect abnormalities in retina. Naive Bayes classifier and Support Vector Machine classifier are used separately for feature selection and classification of exudates.[1] A large database of retinal images are used. Naive Bayes Classifier is used to select best features and these selected features are initially fed to Support Vector Machine. The best features selected by Naive Bayes and Support Vector Machine Classifier is given to Nearest Neighbor Classifier. It is noticed that both Naive Bayes and Support Vector Machine works good compared to Nearest Neighbor classifier. Retinal exudates are automatically detected using a support vector machine.[2] In this research work, localization of optic disc is also done. Eliminating optic disc, detection of yellowish exudates are found. An approach to detect drusen, exudates and cotton wool spots is developed depending on machine learning algorithm.[3] In this paper, each pixel of a retinal image is assigned a probability that indicates the pixel's chance to be bright lesion. Clustering of pixels is done on the basis of this probability value. In Kande et.al.[4], micro aneurysms and hemorrhages are detected using candidate detection scheme using SVM. In this approach, for preprocessing of the input image, both red channel and green channel are used. Using morphological top-hat transformation, separation of vasculature and red lesions are done. In Aravind et. al.[5] , automatic detection of micro aneurysms are done along with the detection of stage of diabetic retinopathy. SVM technique is used for this purpose. Extracting features from preprocessed retinal images, the condition of diabetic retinopathy is identified. A Random Forest based approach is developed to detect and classify hard exudates from retinal images.[6] In this research work, both K means clustering and Fuzzy C means clustering are used to compare the effectiveness. Features like standard deviation, mean are extracted from the segmented image and these are given as input to Random Forest for classification. Another approach using back propagation neural network is used for exudates detection.[7] In this paper, after eliminating optic disc from preprocessed image, features like hue, intensity, mean intensity are extracted. Significant features are identified using Decision Tree. Then classification is done using Artificial Neural Network. In B. Ramasubramanian et.al.[8], L*a*b color space is used to segment the preprocessed image. Texture and color based features are extracted using Gray Level Co-occurrence Matrix. K- Nearest Neighbor classifier is used to classify the segmented images into exudates and non-exudates. An FCM based exudates detection is done by Sopharak et.al.[9] After enhancing the contrast of the input images, FCM is applied to generate eight different clusters. This gives a unclear identification of exudates. To enhance this, morphological reconstruction with a marker image and a mask image is applied. Decision tree is used for detecting different types of abnormalities of Diabetic Retinopathy.[10] Differentiating AMD and DR is done using some contextual information as those can be combined with images in Decision Tree.

3 Motivation

Detection of retinal abnormality due to DR or AMD and the type of disease needs thorough observation and huge amount of time. Developing a automated system capable of detecting the abnormality would help the doctors to detect the disease accurately with less time.

4 Innovative and proposed method

4.1 Separation of DR and AMD

The patients suffering from diabetes for a long period are more prone to have DR. AMD is much related to aged persons. Postprandial blood glucose (PPBG) is an important feature to detect DR or AMD. Generally, if a patient's PPBG is more than 180, then DR is more probable, otherwise AMD. Depending on patients past medical records, it can be decided whether it is DR or AMD.

4.2 Algorithm for detecting the type of DR

Step 1: Retinal image affected by Diabetic Retinopathy (DR) is taken as input.
Step 2: The colored input image is converted to gray image.
Step 3: Image enhancement is done for prominence of the diseased area and other retinal objects.
Step 4: Fuzzy C Means clustering algorithm is applied to the image where number of cluster is specified as three.
Step 5: Three different clusters are assigned three different colors.
Step 6: Clusters related to darkest portion and brightest portions are extracted separately.
Step 7: Hard exudates and soft exudates along with optic disk are included in the cluster which is generated corresponding to the brightest portion of retina. After elimination of optic disk, hard exudates and soft exudates are identified.
Step 8: Hemorrhages and micro aneurysms are included in the cluster which is generated corresponding to the darkest portion of retina. After elimination of blood vessel tree, hemorrhages and micro aneurysms are found.
Step 9: The frequency, color, area and shape of the abnormal objects are calculated.

4.3 Algorithm for detecting the type of AMD

Step 1: Retinal image affected by Age related Macular Degeneration (AMD) is taken as input.
Step 2: The colored input image is converted to gray image.
Step 3: Image enhancement is done for prominence of the diseased area and other retinal objects.
Step 4: Fuzzy C Means clustering algorithm is applied to the image where number of cluster is specified as three.

Step 5: Three different clusters are assigned three different colors.

Step 6: The clusters of darkest portion and brightest portion are extracted separately.

Step 7: As the cluster of darkest portion contains blood vessel tree along with blood clots, eliminating blood vessel tree, region of blood clots are tracked.

Step 8: As the cluster of brightest portion contains optic disk along with drusen, elimination of optic disk generates image with drusen.

Step 9: The frequency, color, area and shape of the abnormal objects are calculated.

4.4 Discussion on proposed algorithm

At first a retinal image affected by either DR or AMD is considered as input image. Then the colored image is converted to gray scale for convenience in handling the image. Quality of the image is improved by adjusting the contrast and brightness. This step makes the significant regions more prominent. Three different clusters are generated by applying Fuzzy C Means algorithm. For each pixel, membership value for belonging to each cluster is generated by FCM. Here, number of cluster is specified as three because three different colors are noticed in any retinal image. The bright yellow color is related to optic disk as well as hard exudates and cotton wool spots in case of DR, drusen in case of AMD; the dark red color is associated with retinal blood vessel tree along with hemorrhages and micro aneurysms in case of DR and blood clots in case of AMD. The rest of the retina has reddish yellow color. After applying FCM, the three clusters are assigned white, gray and black colors. For each pixel, among the three membership values for three different clusters, the maximum membership value is selected and that pixel is assigned to the particular cluster for which it has the highest membership value. Now the cluster related to brightest portion of retina is extracted. Optic disk along with hard exudates, cotton wool spots or drusens, if present, gets extracted with this cluster. To eliminate optic disk, a mask is generated with the help of optic disk's position and size. As optic disc is the most prominent element of retina, by detecting the brightest pixel from retinal image, position of OD is determined. Elimination of optic disk leaves hard exudates and cotton wool spots in case of DR or drusens in case of AMD. The cluster related to darkest region is treated differently. As this cluster contains blood vessel tree along with hemorrhages, micro aneurysms or blood clots, to eliminate blood vessel tree, structures like holes are extracted from the image. Here hole like structures are selected as hemorrhages or micro aneurysms are nearly rounded whereas blood vessel tree is not. As a result blood vessel tree gets rejected as its structure is not matched with hole. Frequency of abnormal objects and the area in terms of pixel are now calculated automatically. The color and shape of the abnormal objects are observed from the main input image. If a normal retinal image is given as input, then elimination of optic disk or blood vessel tree gives an output image with no objects as any type of abnormality is not present in the main input image. In this research work, three normal retinal images are considered in each training dataset and test dataset. While removing OD or blood vessel tree from normal retina, some portions may retain in the processed image due to uneven shape of OD and its mismatch to the mask.

4.5 Training of Naïve Bayes classifier in WEKA

Nearly 34 retinal images, either affected by DR, AMD or normal, are taken as input with the knowledge of disease type. These images are passed through the above mentioned algorithm and the abnormal objects are detected automatically. Then some features like frequency, color, shape, area of those detected objects are recorded. This set is supplied to the Naïve Bayes Classifier as training data set. For each image, the frequency, color, shape and area of abnormal objects calculated by the proposed algorithm and the type of the disease is considered as distinct characteristics. Once the classifier is trained with the specified characteristics, a test image set can be given to it for identifying the disease.

4.6 Testing using Naïve Bayes classifier in WEKA

For each image of the test data set, all the feature values are mentioned except the name of the disease. The disease type is automatically detected by the classifier. Now by comparing the decision of Naive Bayes classifier with the prior knowledge of the disease for each image, the accuracy of the classifier is calculated.

5 Problem formulation

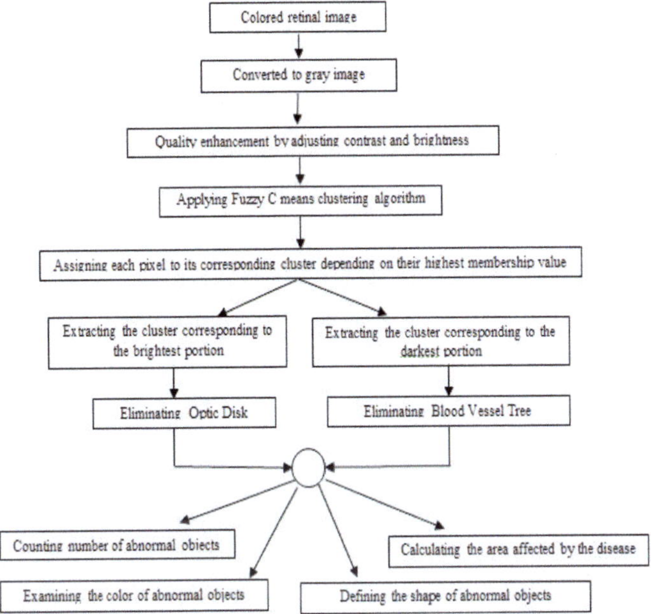

Fig. 1: Flow chart for detection of abnormalities in retina

6 Test results

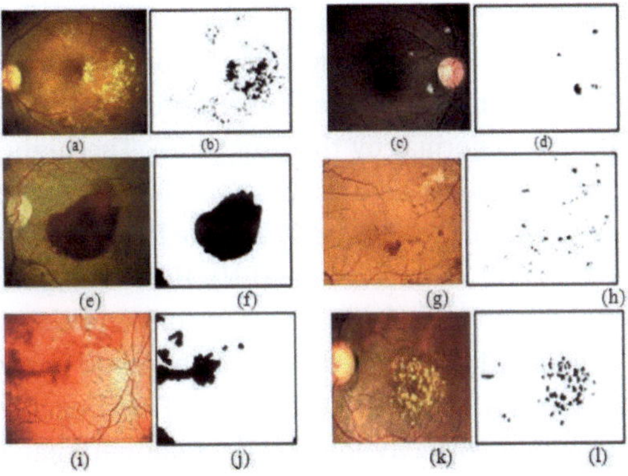

Fig. 2. Retinal images containing (a) hard exudates (b) detection of hard exudates (c) Cotton wool spots (d) detection of cotton wool spots (e) hemorrhages (f) detection of hemorrhages (g) micro aneurysms (h) detection of micro aneurysms (i) blood clots due to wet AMD (j) detection of blood clots due to wet AMD (k) drusens (l) detection of drusens

Developing different Matlab functions, detection of abnormal objects in retinal images are done. Features like number of abnormal objects (frequency), color, shape and affected area are calculated. Table 1 shows those features for the above mentioned images.

Table 1. Table for showing feature values of retinal images displayed above having different types of abnormalities

Image	Disease	Frequency	Color	Shape	Area
Fig.1. (a)	Hard Exudates	182	Bright yellow	circular	2.18
Fig.1. (c)	Cotton Wool Spot	5	Whitish yellow	bounded	0.08
Fig.1. (e)	Hemorrhages	2	Dark red	bounded	4.7
Fig.1. (g)	Micro Aneurysms	47	Bright red	circular	0.08
Fig.1. (i)	Wet AMD	8	Dark red	bounded	1.53
Fig.1. (k)	Dry AMD	41	Bright yellow	circular	0.98

7 Sensitivity analysis

The features like frequency, color, shape, percentage of area in terms of pixel are obtained from the processed images and provided as input to the Naïve Bayes Classifier. To evaluate the classifier performance the measures like sensitivity, specificity and accuracy are used. Sensitivity means the percentage of abnormal image classified as abnormal by the procedure. Specificity means percentage of normal image classified as normal by the procedure. The higher values of sensitivity and specificity implies the high accuracy of the procedure. All measures can be calculated based on four values namely True Positive (TP), True Negative (TN), False Positive (FP) and False Negative (FN). Here true positive (TP) means the no. of abnormal image of each class that is correctly classified with that abnormality, false negative (FN) means the no. of abnormal image of each class that are not correctly classified under that class, True negative (TN) means the no. of normal image that correctly identified as normal image, False positive (FP) means the no. of normal image that identified as abnormal image. The result of Naïve Bayes classification is shown in Table 2. Table 3 shows the sensitivity, specificity, accuracy value for the retinal images. The sensitivity, specificity, accuracy are calculated using the following formula [1].

$$Sensitivity = \frac{TP}{TP+FN} \qquad (1)$$

$$Specificity = \frac{TN}{TN+FP} \qquad (2)$$

$$Accuracy = \frac{TP+TN}{TP+TN+FP+FN} \qquad (3)$$

Table 2. Results of Naïve Bayes Classifier

Classes	No. of Training images	No. of Testing images	No. of correctly classified image	Classification (%)
Normal	3	3	2	66.67
Dry AMD	4	6	4	66.67
Wet AMD	5	5	5	100
Hemorrhage	6	9	7	77.78
Micro aneurysms	4	6	5	83.33
Hard Exudates	6	10	7	70.00
Soft Exudates	6	7	5	71.43

Table 3. Results of Sensitivity, Specificity, Accuracy (of Test Data)

TN	TP	FN	FP	Sensitivity (%)	Specificity (%)	Accuracy (%)
2	33	10	1	76.74	66.67	76.08

8 Justification of the results

Our proposed system can automatically detect all type of Retinopathy combining both AMD and DR and can classify the individual classes of AMD and DR with an average efficiency of 76.55% which is obtained from Table 2 by calculating the average of percentage of classification of all types of images. Then sensitivity and specificity is calculated using the formula mentioned above and the values are 76.74% and 66.67% respectively.

Conclusion

In this paper, verity of Diabetic Retinopathy and Age related Macular Degeneration are considered as input set. A system for automatic detection of abnormalities of retinal images is developed using Matlab. After detection of abnormal objects from an input image, a data set is constructed which serves as training dataset to Weka Naive Bayes Classifier. After training the classifier, a test data set is given where the type of the disease is not mentioned. The classifier generates its answer corresponding to a particular data set. Then its accuracy is checked. By widening the range of database of retinal images, the accuracy of the classifier to predict the disease can be increased. This will be helpful for the eye specialists to detect the type of a particular disease in a very short time.

Future work

The paper focuses on automatic detection of different types of abnormalities found in retina due to DR and AMD. The algorithm described in this paper holds good for hard exudates, cotton wool spots, hemorrhages and drusen but some false detection of micro aneurysms are observed for some output images. This is due to some portion of blood vessel tree which are very thin and hence could not be removed. This problem has to be modified in future. Scope of modification is present for processing normal retinal images also. In some cases, false detection of abnormal objects in normal retinal image is found for poor quality of image. Improved preprocessing steps using filters can avoid this problem. The size of database for training and test images should be increased to get more accurate results. In this research work, the retinal images used are available from free online sources. Collection of some real life images can improve the quality of analysis. For machine learning purpose, Naïve Bayes classifier is used. Other classifiers like Support Vector Machine can be used for the same.

Acknowledgments

The authors would like to acknowledge a grant from TEQIP and Department of Biotechnology, Government of India (No.BT/PR4256/BID/7/393/2012 dated 02.08.2012) for supporting this research.

References

1   A. Sopharak, M. N. Dailey, B. Uyyanonvara, S. Barman, T. Williamson, K.T. Nwe and Y. A. Moe, Machine learning approach to automatic exudate detection in retinal images from diabetic patients, Vol. 57(2), pp:124-135, 2009 (Journal of Modern Optics)
2   K. Wisaeng, N. Hiransakolwong, E. Pothiruk, Automatic Detection of Retinal Exudates using a Support Vector Machine, Vol. 32, No. 1/2013, pp: 33-42 (Applied Medical Informatics, 2013).
3   M. Niemeijer, B. V. Ginneken, S. R. Russell, M. S. A. Suttorp-Schulten, and M. D. Abramoff, Automated Detection and Differentiation of Drusen, Exudates, and Cotton-Wool Spots in Digital Color Fundus Photographs for Diabetic Retinopathy Diagnosis,Vol. 48, No. 5 (Invest Ophthalmol Visual Science., May 2007).
4   G. B. Kande, T. S. Savithri and P. V. Subbaiah, Automatic Detection of Microaneurysms and Hemorrhages in Digital Fundus Images, Vol 23, No 4 (August), pp 430-437 (Journal of Digital Imaging, 2010)
5   C.Aravind, M.PonniBala, S. Vijayachitra, Automatic Detection of Microaneurysms and Classification of Diabetic Retinopathy Images using SVM Technique, (International Journal of Computer Applications (0975 – 8887))
6   T. Akila, G. Kavitha, Detection and Classification of Hard Exudates in Human Retinal Fundus Images Using Clustering and Random Forest Methods, Vol. 4, Special Issue 2, April (International Journal of Emerging Technology and Advanced Engineering, 2014)
7   A. G. Karegowda, A. Nasiha, M.A. Jayaram, A.S .Manjunath, Exudates Detection in Retinal Images using Back Propagation Neural Network, Vol. 25– No.3, July 2011 (International Journal of Computer Applications (0975 – 8887))
8   B. Ramasubramanian, G. Prabhakar, An Early Screening System for the Detection of Diabetic Retinopathy using Image Processing, Vol. 61– No.15, January 2013(International Journal of Computer Applications (0975 – 8887))
9   A. Sopharak, B. Uyyanonvara and S. Barman, Automatic Exudate Detection from Non-dilated Diabetic Retinopathy Retinal Images Using Fuzzy C-means Clustering, Vol. 9, pp: 2148-2161, March 2009 (Sensors (1424-8220))
10 S. Banerjee, A. Roy Chowdhury, Case Based Reasoning in the Detection of Retinal Abnormalities using Decision Trees, International Conference on Information and Communication Technologies (ICICT 2014)

Inadyuti Dutt[1] and Samarjeet Borah[2]

Design and Implementation of Intrusion Detection System using Data Mining Techniques and Artificial Neural Networks

Abstract: Intrusion detection in a real-time network has been an issue for the past two and half decades. Much work has been done using computational techniques and machine intelligence. But due to voluminous data that invades a network such knowledge-based techniques become less efficient and accurate. And for this reason newer and more sophisticated methods are tried and implemented using Data Mining techniques as well as Artificial Neural Network. The beauty of these two knowledge-based approaches is that, they both are efficient decision makers once a proper set of inputs are fed to them. As they take lesser computational time thus a new hybrid approach based on Data Mining and Artificial Neural Network approaches have been taken into consideration. The results show that the hybrid approach works well in a real-time network and can be efficiently used as an Intrusion Detection System.

Keywords: Intrusion Detection System (IDS), Artificial Neural Network (ANN), Data Mining Techniques, Decision Support System, Decision Tree

1 Introduction

The security of our computer systems and data is at continual risk due to the extensive growth of the Internet and increasing availability of tools and tricks for intruding and attacking networks. For the past two decades, there has been increased reliance of government, military and commercial bodies on Internet technologies for their day to day business. This has created a myriad of new challenges for combating external attacks. Such attacks external to these bodies are deliberate in action against data, software or hardware and that can destroy, degrade, disrupt or deny access to a network computer system are called Cyber

1 B. P. Poddar Institute of Management & Technology, Dept. of Computer Applications, Kolkata, India
inadyuti@gmail.com
2 Sikkim Manipal Institute of Technology, Dept. of Computer Applications, Sikkim
samarjeetborah@gmail.com

attacks. In an attempt to guard against the unknown cyber attacks, much effort has been given in researching and developing Intrusion Detection System (IDS), which tries to filter out the cyber attacks from the network traffic. IDS are software tools meant specifically for strengthening the security of information and communication systems. An IDS dynamically monitors logs and network traffic, applies detection algorithms to identify intrusions in a network. Intrusion detection is based on the assumption that intrusive activities are noticeably different from normal system activities and thus are identifiable. Intrusion detection is not used to replace prevention-based techniques such as authentication and access control; instead, it is intended to complement existing security measures. Intrusion detection system is therefore considered as a second line of defence for computer network systems to detect actions that bypass the security monitoring and control component of the system. The majority of IDS can be classified into host-based or network based whilst the former collects data from a host (for example, system calls, system and application log files), whereas the latter collects the data directly from the network in the form of packets.

2 Types of Intrusion Detection Systems

There are two types of intrusion detection systems that employ one or both of the intrusion detection methods outlined above.

2.1 Host-based Intrusion Detection System

Host-based systems base their decisions on information obtained from a single host (usually audit trails). A generic intrusion detection model proposed by Denning (1987) is a rule-based pattern matching system in which the intrusion detection tasks are conducted by checking the similarity between the current audit record and the corresponding profiles. If the current audit record deviates from the normal patterns, it will be considered an anomaly. Several IDSs were developed using profile and rule-based approaches to identify intrusive activity (Lunt et al., 1988).

2.2 Network-based Intrusion Detection System

With the proliferation of computer networks, more and more individual hosts are connected into local area networks and/or wide area networks. However,

the hosts, as well as the networks, are exposed to intrusions due to the
vulnerabilities of network devices and network protocols. The TCP/ IP protocol
can be also exploited by network intrusions such as IP spoofing, port scanning,
and so on. Therefore, network-based intrusion detection has become important
and is designed to protect a computer network as well as all of its hosts. The
installation of a network-based intrusion detection system can also decrease the
burden of the intrusion detection task on every individual host.

3 Types of Intrusion Detection System based on Detection Methods

An ID falls into two categories according to the detection methods they employ
(i) Anomaly detection and (ii) Misuse detection.

3.1 Anomaly Detection

Anomaly detection considers that an intrusion will always reflect some deviations from normal patterns. Anomaly detection may be divided into static and
dynamic anomaly detection. In static, anomaly detector assumes that there is a
portion of the system being monitored that does not change and therefore these
detectors only address the software portion of a system and as the hardware
portion need not be checked. The static portion of a system is the code for the
system and the constant portion of data upon which the correct functioning of
the system depends. It is assumed that if the static portion of the system ever
deviates from its original form, an error has occurred or an intruder has altered
the static portion of the system. Thus static anomaly detectors focus on integrity
checking. Dynamic anomaly detectors typically operate on audit records or on
monitored networked traffic data. Audit records of operating systems do not
record all events; they only record events of interest and therefore only
behaviour that results in an event that is recorded in the audit will be observed
and these events may occur in a sequence. In distributed systems, partial
ordering of events is sufficient for detection and in other cases, the order is not
directly represented; only cumulative information, such as cumulative
processor resource used during a time interval, is maintained. In this case,
thresholds are defined to separate normal resource consumption from
anomalous resource consumption.

3.2 Misuse Detection

Misuse detection is based on the knowledge of system vulnerabilities and known attack patterns is concerned with finding intruders who are attempting to break into a system by exploiting some known vulnerability. Ideally, a system security administrator should be aware of all the known vulnerabilities and eliminate them. The term intrusion scenario is used as a description of a known kind of intrusion; it is a sequence of events that would result in an intrusion without some outside preventive intervention. An intrusion detection system continually compares recent activity to known intrusion scenarios to ensure that one or more attackers are not attempting to exploit known vulnerabilities. To perform this, each intrusion scenario must be described or modelled. The main difference between the misuse techniques is in how they describe or model the behaviour that constitutes an intrusion. The original misuse detection systems used rules to describe events indicative of intrusive actions that a security administrator looked for within the system. Large numbers of rules can be difficult to interpret. If-then rules are not grouped by intrusion scenarios and therefore making modifications to the rule set can be difficult as the affected rules are spread out across the rule set. Since it is rather difficult to enumerate the various intrusion scenarios and build different models for them respectively thus anomaly-based system is more concrete in detecting intrusions. Therefore, the proposed model is an anomaly-based intrusion detection system which tries to detect abnormality in the system's behaviour.

The proposed model not only uses decision support system but also biologically-inspired Artificial Neural Network (ANN) in order to detect abnormal behaviour in the system. It has been introduced in order to overcome these difficulties. The usage of ANN in culmination with decision support system provides a strong backbone for mitigating the problems. The proposed model uses decision support based system invokes a decision tree for different values of input parameters. Like, the first tree accumulates the login frequency input and if the value is on the higher side then this value would be taken into consideration by the Artificial Neural Network Processor (ANNP). Similarly the next tree would use the values for location frequency, followed by another tree for number of password failures, then for number of sessions and finally for execution frequency and usage of computing devices (CPU, Memory, I/O and Printer). For every input, a decision tree is maintained and values which are considered to be abnormally high would be treated as suspicious inputs for the next phase of Artificial Neural Network Processor. This is how the number of inputs being processed by the decision trees, are further considered by the ANNP to maximize the accuracy of detecting the intrusion in a system.

4 Related Work

The proposed model is based on the assumption that an abnormal behaviour in the system can be depicted by considering the resource consumption presently made by the system. This concept of categorizing the behaviour of intruder with respect to resource consumption was theorized by Dorothy Denning in [1]. According to the proposed work by Denning, security violations can be detected by monitoring a system's audit records for abnormal patterns of system usage [1]. The model includes profiles for representing the behaviour of subjects with respect to objects in terms of metrics and statistical models, and rules for acquiring knowledge about this behaviour from audit records and for detecting anomalous behaviour [1].

Carlo A. Catani et. al. in their paper [2] has also mentioned the importance of using resource based consumption as one of the metrics for measuring the intrusion.

In this paper [3], we adopt an anomaly detection approach by detecting possible intrusions based on program or user pro/les built from normal usage data. In particular, program pro/les based on Unix system calls and user pro/les based on Unix shell commands are modeled using two di1erent types of behavioral models for data mining. The dynamic modeling approach is based on hidden Markov models (HMM) and the principle of maximum likelihood, while the static modeling approach is based on event occurrence frequency distributions and the principle of minimum cross entropy.

The paper [4] describes the design of a genetic classifier-based intrusion detection system, which can provide active detection and automated responses during intrusions. It is designed to be a sense and response system that can monitor various activities on the network (i.e. looks for changes such as malfunctions, faults, abnormalities, misuse, deviations, intrusions, etc.). In particular, it simultaneously monitors networked computer's activities at different levels (such as user level, system level, process level and packet level) and use a genetic classifier system in order to determine a specific action in case of any security violation.

In the paper [5], the immunity-based agents roam around the machines (nodes or routers), and monitor the situation in the network (i.e. look for changes such as malfunctions, faults, abnormalities, misuse, deviations, intrusions, etc.). These agents can mutually recognize each other's activities and can take appropriate actions according to the underlying security policies. Specifically, their activities are coordinated in a hierarchical fashion while sensing, communicating and generating responses. Such an agent can learn and adapt to its environment dynamically and can detect both known and unknown intrusions.

In this [6] work, Holger et al set out to by understanding and predicting the CPU and memory consumption of such systems. They begin towards this goal by devising a general NIDS resource model to capture the ways in which CPU and memory usage scale with changes in network traffic. They have then used this model to predict the resource demands of different configurations in specific environments. Finally, they present an approach to derive site-specific NIDS configurations that maximize the depth of analysis given predefined resource constraints.

Dayu Yang, Alexander Usynin, and J. Wesley Hines [7] proposed traffic profiles in order to capture the resource usages. Traffic profiles are created using symptom-specific feature vectors, called system indicators, such as link utilization, CPU usage, and login failure. These profiles then are classified by time of day, day of week, and special days, such as weekends and holidays. When new traffic data fails to fit within a predetermined confidence interval of the stored profiles, then an alarm is triggered [7].

In this paper [8], they introduced a novel mechanism that identifies abnormal system-wide behaviors using the predictable nature of real-time embedded applications. They introduced, Memory Heat Map (MHM) to characterize the memory behavior of the operating system. Their machine learning algorithms automatically (a) summarize the information contained in the MHMs and then (b) detect deviations from the normal memory behavior patterns. These methods are implemented on top of multicore processor architecture to aid in the process of monitoring and detection.

5 Motivation

The model proposed is based on the theory of misuse detection and the hypothesis suggests that when a system is penetrated by external intruders there is an abnormal usage of the resource consumption. The attempted break-ins by these externals to the system and sometimes abuses by the internal users would always result into abnormal patterns of system usage. This system usage can be related to abnormal memory consumption, unnecessary CPU time usage or may be abnormal patterns of I/O usage. The basic idea is to monitor the standard operations on a target system such as logins, command, program executions, file and device accesses, etc., looking exactly for deviation in usage pattern. This proposed model would be based on Anomaly Detection whereby the main objective is to extract deviation from the normal day-to-day patterns of log-ins, command and program executions', file and device accesses in memory and CPU time consumption.

The following examples illustrate such abnormal behaviours:

1   Attempted break-in: An intruder trying to break into a system might generate an abnormally high rate of password failures w.r.t. a single account or a system as a whole.

2   Impersonate breaking-in: An intruder behaving like the actual, authorised user might have a different login time, location or network connection type from that of the legitimate user. Such intruder might have a different browsing history or may typically explore the files in the system without compiling and executing them.

3   Unusual data retrieval: A user, who is not a legitimate one, attempts to obtain unauthorised data from a database and might want o retrieve more records than usual.

4   Trojan Horse: A Trojan horse planted as a substitution for a program may exhibit abnormality in terms of CPU time usage and I/O activity.

5   Abnormal high activity: An intruder attempting to access a particular resource might exhibit abnormally high activity with respect to all other users at that specific time or location.

An important objective of the current proposed study is to determine what activities and statistical measures provide the best measures in order to discriminate between legitimate user and an intruder.

6 The Proposed Model or Method

The proposed model is an independent with respect to system, application environment, system vulnerability, or type of intrusion thereby providing a framework for general-purpose IDS. The proposed model has six main components:

A) Subjects: Subjects are the initiators of activity on target system specifically normal users.

B) Objects: Resources such as files, programs, CPU, memory or other devices etc.

C) Audit Records: These are generated by the subjects' behaviour or a specific object such as user login, command execution or file access etc.

D) Profiles: The characteristic behaviour of subjects with respect to objects in terms of statistical measure and models of observed activity.

E) Anomaly records: These are generated when an abnormal behaviour is detected.

F) Activity rules: Actions that can be taken when some condition related to behaviour of a subject on an object is satisfied, which would update the profiles on encountering a new pattern or behaviour, produce reports thereby.

The proposed model in Fig. 1 can be regarded as a hybrid of decision support-based system and artificial neural network methods. In this model, decision trees are used in order to match the parameters, which are generated from audit records of a particular system. The audit records are generated both by the legitimate users or an intruder and these are considered to be the input parameters to the decision trees.

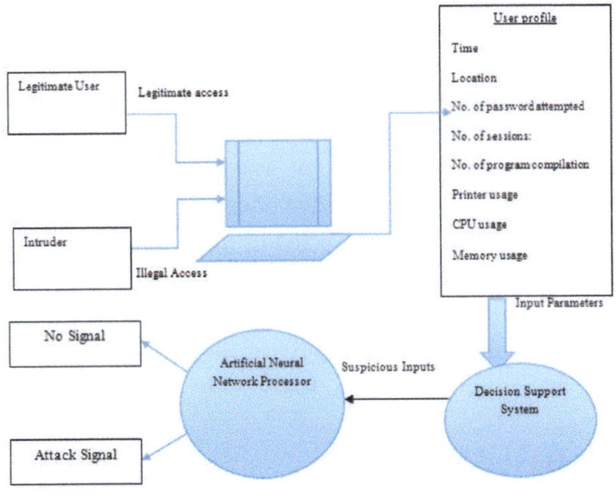

Figure. 1: An Overview of the Proposed Model

Following are the types of input parameters generated from a user or
candidate profile:

The inputs to the decision support system are the profiles that exhibit the
characteristic behaviour of a subject on an object. These profiles are called
candidate profiles. The different measures for candidate profile are as follows:

- Login Frequency: This measures login frequency by day and time.
  Since a user's login behaviour may vary considerably during a
  workweek, login occurrences may be used to detect intruders who log
  into an unauthorized account during off-hours when the legitimate
  user is not considered to be accessing the account.
- Location Frequency: A frequency that can be maintained in order to
  determine the frequency of log-ins at different locations. It may be
  prevalent that an intruder may be accessing an account of a legitimate
  user from different location which is never being previously used by
  the legitimate user.
- Number of password failures: This can be used to measure number of
  password attempts at login. This can reflect an attack involving many
  trail passwords on a particular account thereby showing and
  abnormally high number of password failures w.r.t. a user profile.
- Number of Sessions: This is used to measure number of log-in session
  in average a user maintains daily. An abnormally high frequency, or
  beyond working hours might suggest breach in security.
- Execution frequency: It is used to measure the number of times a program is being executed during some time period. It is evident that an
  intruder illegally accessing an account might execute a program
  frequently as compared to the legitimate user. Moreover, the intruder
  might use different commands as compared to the legitimate user for
  successful penetration into the system.
- CPU, Memory, I/O usage: The abnormal usage of CPU, memory and I/O
  devices might occur when a virus gets injected in the original program
  or computer. Usage of printer or other printing devices might detect the
  presence or absence of intruder in the system. In all of these cases, the
  usage of these resources shoot up as otherwise they tend to exhibit
  normal values.

6.1 Decision Support-Based System

The inputs are fed as audit records for further processing to the Decision Support-based System. The role of the system is to invoke a decision tree for diffe-

rent values of input parameters. The first tree accumulates the login frequency input and if the value is lies on the higher side then this value would be taken into consideration by the Artificial Neural Network Processor. Similarly the next tree would use the values for location frequency, followed by another tree for number of password failures, then for number of sessions and finally for execution frequency and usage of computing devices (CPU, Memory, I/O and Printer). For every input, a decision tree is maintained and values which are considered to be abnormally high would be treated as suspicious inputs for the next phase of Artificial Neural Network Processor. The values which are abnormally low would not be further processed by the Artificial Neural Network Processor. Fig. 2 illustrates an example of how the decision tree works for a set of input from user profile.

6.2 Artificial Neural Network Processor (ANNP)

At any given time, t, to detect whether an attack has been encountered by the network the ANNP assigns a weight, W_1, for the input, I_1 coming from the decision tree of the DSS. For each of the inputs, say I_n a weight say, W_n is being assigned. The value of W_n differs with the input values. A weight of 1 is assigned for higher input values as otherwise 0. Then the neural network equation is computer using the input function of the neural equation:

$$X = \sum_{i=1}^{n} W_i I_i + \mu_i$$

where I_i, is the input value from the Decision Tree
and
W_i, is 1 if input is abnormally high, or, W_i, is 0 if input is abnormally low and μ_i, is termed as negative, if W_i is 0, otherwise is termed as positive, if W_i is 1.

The value of input function, X, is evaluated and if it is seen that the computed value of X, is more than or equal to the fifty percent of the actual value of X being ideally computed, then it is stated that the attack is not encountered in the network. A mail alert is sent to the network to all the computing nodes so that they might become alert and aware of the intruder.

7 Test Results and Discussion

The results of the proposed model, its performance have been implemented on Pentium V machines using Java as the programming language. The output shows the efficiency with which the Java code interacts with the computing nodes of a network and checks whether an attack has actually been encountered by any other node or node. Once an attack has been detected the alert message is propagated to the network. The result in the Fig. 3 shows attack being detected for network having nodes 4, 6, 8, 10 and so on. The proposed model has been tested on the backdrop of a real-time network having 28 computing nodes working simultaneously for the network, Fig. 4 shows that attack can be successfully detected and managed by sending an alert message across the network.

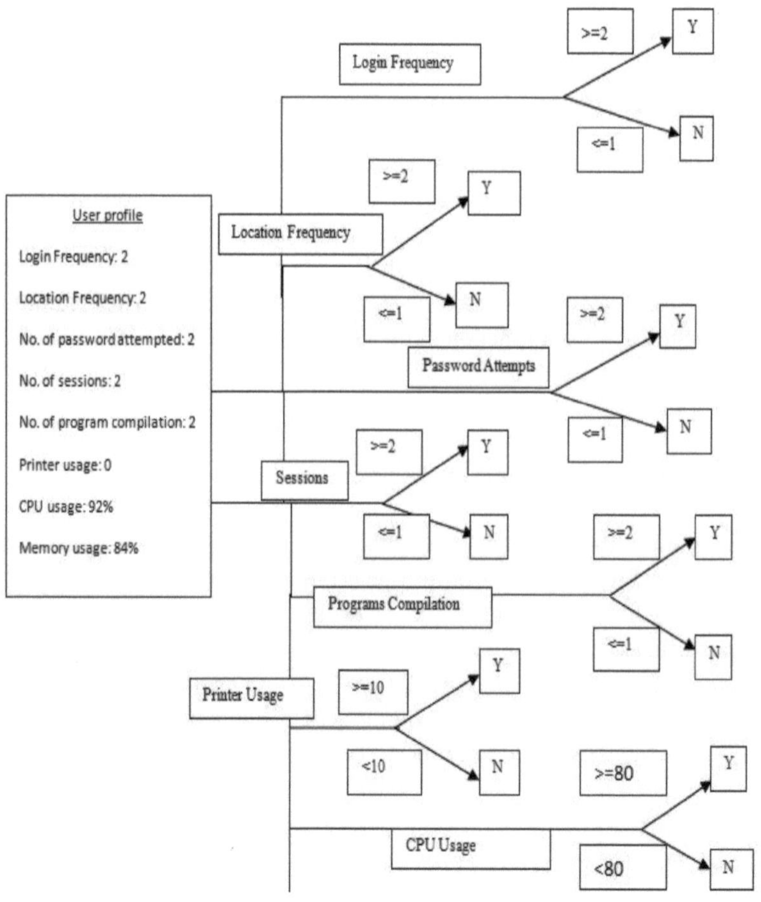

Figure. 2: Decision Tree for a User Profile

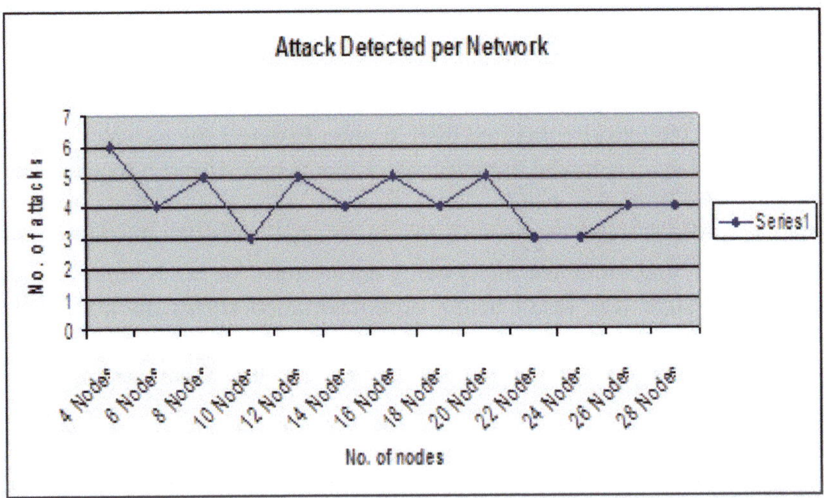

Figure. 3: Attacks Detected per Network

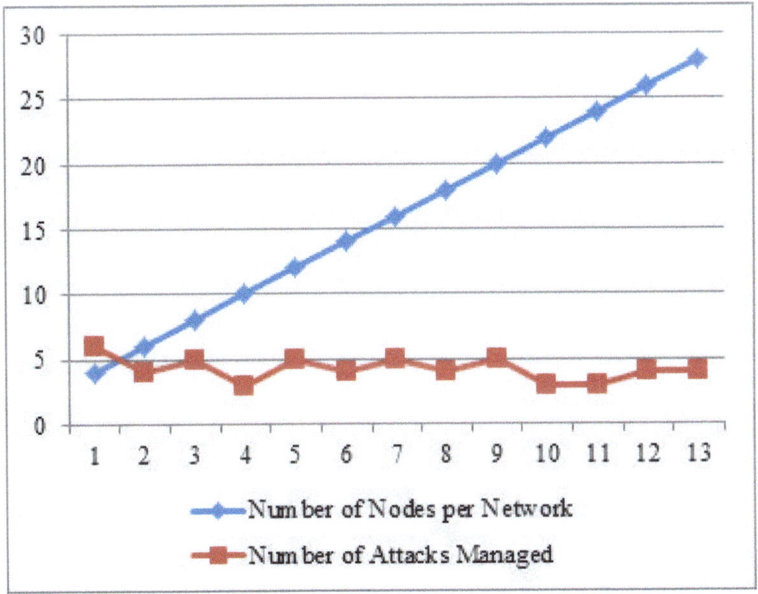

Figure. 4: Attacks Managed per Network

Conclusion

The proposed model combines the decision support system of the data mining
technique and biologically-inspired ANN in order to detect the intrusions in the
system accurately. The model uses decision support-based system to accept the
various inputs from the events. The abnormal behaviour of the system is
entrapped in the decision support-based system whereas the final computation
to find out whether an attack really being encountered or not is computed by
ANN based system. The result in the experimentation shows the ability and
accuracy with which intrusions per node have been detected and managed
which becomes difficult if being handled by only a decision support-based system.

References

1   Dorothy E. Denning, "An Intrusion -Detection Model", IEEE Transactions on Software
     Engineering, Vol. SE-13, No. 2, Feb 87.
2   Carlos A. Catania, Carlos Garcia Garino, "Automatic Network Intrusion Detection: Current
     Techniques and Open Issues", Computers and electrical Engineering, 38(2012) 1062-1072,
     2012.
3   Dit-Yan Yeung, Yuxin Ding "Host-based intrusion detection using dynamic and static
     behavioral models", Pattern Recognition 36 (2003) 229 – 243, Elsevier.
4   Dipankar Dasgupta and Fabio A. Gonzalez, "An Intelligent Decision Support System for
     Intrusion Detection and Response", LNCS 2052, pp. 1-14, 2001, Springer-Verlag Berlin Hei-
     delberg 2001.
5   Dipankar Dasgupta, "Immunity-Based Intrusion Detection System: A General Framework",
     In Proc. of the 22nd NISSC'99.
6   Holger Dreger, Anja Feldmann, Vern Paxson, Robin Sommer, "Predicting the Resource
     Consumption of Network Intrusion Detection Systems", RAID 2008, LNCS 5230, pp. 135-
     154, Springer, 2008.
7   Dayu Yang, Alexander Usynin, and J. Wesley Hines, "Anomaly-Based Intrusion Detection
     for SCADA Systems", IAEA Technical Meeting on Cyber Security of NPP I&C and Informati-
     on systems, Idaho Fall, ID, Oct. 2006.
8   Man-Ki Yoon*, Sibin Mohany, Jaesik Choiz, and Lui Sha, "Memory Heat Map: Anomaly
     Detection in Real-Time Embedded Systems Using Memory Behavior", DAC '15, June 07 - 11,
     2015, San Francisco, CA, USA, 2015, ACM 978-1-4503-3520-1/15/06.

S. Pranavanand[1] and A. Raghuram[2]

Development of Sliding mode and Feedback Linearization Controllers for the Single Fluid Heat Transfer System

Abstract: To optimize cost and space in the pharmaceutical industry, it is desirable to heat and cool the utility fluid from the same equipment. The heat transfer for both heating and cooling is done using a single liquid. The system implemented with this concept is called Single Fluid Heat Transfer System (SFHTS). In this paper an actual SFHTS used by a pharma major company is considered. The essential objective is to eliminate overshoot and achieve set point in the shortest time possible, as the process ramp rate is approximately 1°C/minute. The real time SFHTS was simulated using LabVIEW software. PI,PID, Sliding Mode Control (SMC) and Feedback linearization controllers were developed and applied to the heating and cooling module of the SFHTS. The results showed that Feedback linearization controller provided best control than the other controller considered here.

Keywords: Single Fluid Heat Transfer system, Virtual Instrumentation, LabVIEW, Sliding Mode Control, Feedback Linearization, Non linear control system.

1 Introduction

The paper is organized as detailed. Section I: Introduction, Section II: Block Diagram of SFHTS, Section III: Implementation of PID controller to Thermal Transfer Module of SFHTS, Section IV: Development and Implementation of Sliding Mode controller to Thermal Transfer Module of SFHTS, Section V: Development and implementation of Feedback linearization controller to thermal transfer module of SFHTS, Section VI: Virtual Instrument, Section VII: Results.

1 Dept of EIE, VNR Vignana Jyothi Institute of Engineering and Technology, Hyderabad, 500039, India
E-mail: pranavanand_s@vnrvjiet.in
www.vnrvjiet.ac.in
2 Dept of EEE, JNTUH College of Engineering, Hyderabad, Hyderabad, 500039, India

The objective of SFHTS is to use a single fluid for heating and cooling of
mass(substance to be heated) in a chemical process .The fluid chosen for the
process is Therminol-D12.Therminol-D12 also known as utility fluid was chosen
on account of its stability under both high and low temperatures. The order of
temperatures involved in the use of Therminol D-12 is -85°C to 250 °C. In addition, Therminol D-12 posses good heat transfer properties. The SFHTS is divided
into two modules: Thermal Transfer Module and Level Control Module. In the
Thermal Transfer Module, Therminol-D12 is heated or cooled to reach the set
temperature. In the level control module, the desired level of Therminol-D12 is
maintained in the concentric jacket tank and in the expansion tank. The work
reported has two objectives. One, to achieve the set temperature, two, to achieve
the set levels of both concentric jacket tank and expansion tank, at the shortest
time possible for Thermal Transfer Module and Level control Module
respectively. To this end, the heating /cooling   process is controlled via Proportional Integral Derivative Controller (PID) control, Sliding Mode Control
(SMC) and Feedback linearization controller. The responses of SFHTS's Thermal
Transfer Module to PID Control, SMC and Feedback linearization were
compared. The Thermal transfer process and level control process are non linear
systems. Investigation was carried out using LabVIEW software which lends
itself very closely to actual implementation. The Control System Design and
Simulation tool associated with LabVIEW were also used.

2 Block diagram of SFHTS

The Block diagram of the SFHTS is shown in the Figure 1. The Chemical/Mass to
be heated is stored in the inner tank of the Concentric Jacket Tank. The
Therminol-D12 is used as thermal agent to heat or cool the Chemical/Mass.
Therminol-D12 is heated or cooled so that it can transfer the thermal energy to
Chemical/Mass. It necessitates Therminol-D12 to be heated or cooled as per the
requirement.

2.1 Delineation of the Block Diagram

The Therminol-D12 is used as thermal agent to heat or cool the Chemical/Mass.
Therminol-D12 is heated or cooled so that it can transfer the thermal energy to
Chemical/Mass. It necessitates Therminol-D12 to be heated or cooled as per the
requirement.

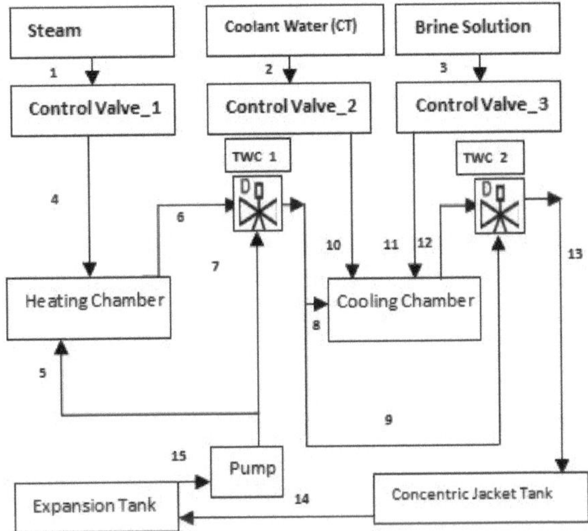

Fig. 1: Block diagram of the SFHTS

Heating Process: Therminol–D12 is drawn from the Expansion Tank (Storage Tank) by the pump via the pipe(15).Therminol-D12 is allowed to flow through heating chamber via pipe(5). During the heating process the cooling chamber is put in off position manually. Simultaneously, Therminol-D12 is allowed to flow into pipe(7), which is attached to the TWC_1. At any instant, the TWC_1 allows flow from either from terminal 1 or terminal 3 but not both. Hence the pipe (7) throughput, which is coupled to 3rd terminal of Three way control valve-TWC_1, is blocked and Therminol-D12 flows through pipe (5) alone. The heating chamber has steam supply via Control valve_1, pipe (1) and Therminol-D12 via pipe (5).The heat is transferred from Steam to Therminol-D12. Here the heat transfer takes place by non contact conduction method. The heated Therminol-D12 is delivered via pipe (6) to TWC_1 which allows the Therminol-D12 to flow through it. The output from the TWC_1 is allowed to flow through pipe (9) which has a parallel path, pipe(8).The pipe(8) is connected to the cooling chamber. In the SFHTS, at any instant, either heating or cooling can take place but not both. Hence the cooling chamber is in off condition during heating. Terminal 1 of TWC_2 blocks the flow of Therminol-D12. This arrangement makes Therminol-D12 to flow through pipe (9) to TWC_2's 3rd terminal. The 3rd terminal is allowed to access the TWC_2's output and hence Therminol-D12 flows to Concentric Jacket Tank. The level of Therminol-D12 is maintained at a preset level in the Concentric Jacket Tank by maintaining the flow through the control valve_4. The piping arrangement for Therminol-D12 for heating is via pipes 15-5-6-9-13-14.Therminol-D12 then flows back to the Expansion Tank. The process is repeated till the desired temperature of Therminol-D12 is reached.

Cooling Process: During the cooling process, the heating chamber is put in off condition manually. The TWC_1's 3rd terminal is open and Therminol-D12 flows via pipe (7) to pipe (8). The entire Therminol-D12 flows to the cooling chamber. The cooling chamber has two source of coolants: Coolant Water (CT) and Brine Solution. Depending upon the requirement either the Coolant Water or Brine Solution is allowed to flow via their respective control valves, 2 and 3. The output of the cooling chamber is connected to TWC_2's 1st terminal via pipe (12). The TWC_2 gives access of its output to 1st terminal. Theminol-D12 flows to Concentric Jacket Tank via pipe (14). The level of Therminol-D12 is maintained at a preset level in the Concentric Jacket Tank by maintaining the flow through the control valve_4. The piping arrangement for Therminol-D12 to be cooled is 15-7-8-12-13-14.The process is repeated till the desired set point is reached.

3 Implementation of PID controller to Thermal Transfer Module of SFHTS

The PID controller was tuned using Zieger-Nichols Method, also known as Ultimate cycle method.[3] The response of the simulated closed loop system with set point = 40˚C, Initial input temperature = 30˚C,Initial flowrate of 10 lph and Proportional gain, K_c =11 with Integral Time

(T_i) and Derivative time$(T_d) = 0$ is shown below.

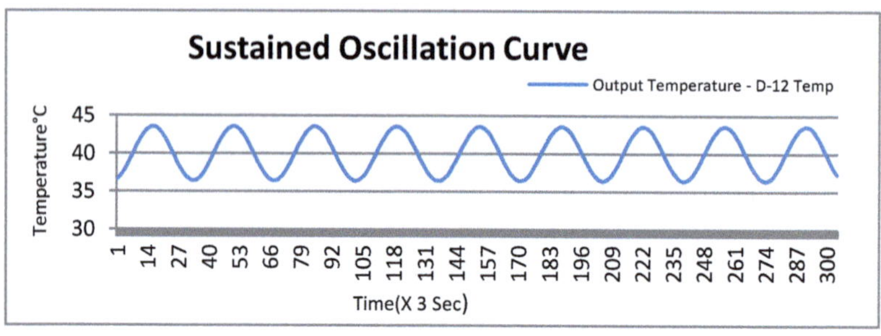

Fig. 2: Output Temperature of the system with Proportional gain $K_c = 11, T_i = 0$ and $T_d = 0$

The period of oscillation calculated from the response, $T_c = 1.65$ minutes. The recommended settings as per Z-M for PID are as follows.

$K_p = 0.60K_c \Rightarrow K_p = 0.6 \text{ X } 11 = 6.6$

$\tau_i = T_c/2.0 \Rightarrow 1.65/2 = 0.83$ Minutes

$\tau_d = T_c/8.0 \Rightarrow 1.65/8 = 0.20$ Minutes

The above obtained values were substituted in the PID controller given by the equation [2]

$$U = K_p e(t) + \frac{1}{T_i}\int_0^t e(t) + T_d \frac{d}{dt}e(t)$$

where, e (t) = Set point - Process variable.

Same tuning algorithm was implemented on Cooling Chamber with inlet temperature 60˚C and outlet temperature 50˚C. The following values were derived and implemented on parameters of the PID Controller.

$K_p = 0.60K_c \Rightarrow K_p = 0.6 \text{ X } 11 = 6.0$

$\tau_i = T_c/2.0 \Rightarrow 1.65/2 = 0.83$ Minutes

$\tau_d = T_c/8.0 \Rightarrow 1.65/8 = 0.20$ Minutes

4 Development and Implementation Of Sliding Mode control to Thermal Transfer Module Of SFHTS

4.1 Thermal Energy Exchange Process_1: Heating

The SFHTS's Thermal Transfer Module is designed with dedicated pipe line for thermal source and Therminol-D12. In the Thermal Transfer Module, the heat transfer takes place and the desired temperature is transferred to the Therminol-D12.

The real time values of the above parameters as obtained from the industry are

$U = 250 kg/m^3/°C; A = 2m^2; C_{pc} = 0.56$ Kcal/ kg;ρ_c=733Kg/m^3;V_h=200Ltrs;v_h=0-200lph

The dynamic model of the system was developed from equation (1) and written as.

$y = \delta$

$\dot{y} = \dot{\delta} = f + \Delta c,$ (3)

where

$$f = \frac{UA}{C_{ph}\rho_h V_h}*(\beta - \delta)$$

$$\Delta = \frac{1}{V_h} * (\gamma - \delta)$$

c is the control input.

Let α_s, W_s be the positive scalars and δ be the desired output of the system.
Define the sliding surface ε_s to be [3]

$$\varepsilon_s = \alpha(\gamma - \delta) \quad (4)$$

Scheme 1:

The Sliding mode controller was derived as,

$$C = \left\{ \frac{-1}{\frac{1}{V_h}*(\gamma-\delta)} \left(\frac{\frac{UA}{C_{ph}\rho_h V_h}*(\beta-\delta)+W_s sgn(\varepsilon_s)}{\alpha} \right) \right\} \quad (5)$$

Proof:

The sign function is defined as,

$$Sgn(x) = \begin{cases} +1 & if \ x > 0 \\ 0 & if \ x = 0 \\ -1 & if \ x < 0 \end{cases} \quad (6)$$

Considering the first order dynamics of the surface

$$\dot{\varepsilon}_s = \alpha \frac{d}{dt}(\gamma - \beta) \quad (7)$$

So,

$$\dot{\varepsilon}_s = \alpha(\dot{\delta}) \quad (8)$$

By substituting the value of C in the above equation yielded

$$\dot{\varepsilon}_s = -W_s sgn(\varepsilon_s) \quad (9)$$

The reachability condition of the controller was confirmed by the equation
given below [4].

$$\dot{s}s < \eta|s| \quad (10)$$

where η is small positive value.[4]

The controller designed above satisfied the reachbility and stability condition
[5].

Thermal Energy Transfer Process 2: Cooling

The dynamic model of the system was developed from Equation (2) and is
written as

$$\dot{\beta} = \frac{v_c}{V_c}(\alpha - \beta) + \frac{UA}{C_{pc}\rho_c V_c}(\delta - \beta) \quad (11)$$

The real time values of the above parameters are

$U = 220 kg/m^3/°C; A = 2m^2; C_{pc} = 0.51$ Kcal/kg$;\rho_c$=774Kg/m^3;V_c=200Ltrs;v_c=
0-200 *lph*

Simplifying the above model,

$$y = \beta$$
$$\dot{y} = \dot{\beta} = p + \psi q$$

where

$$p = \frac{UA}{C_{pc}\rho_c V_c}(\delta - \beta)$$

$$\psi = \frac{1}{V_c}(\alpha - \beta)$$

q is the control input.

The sliding mode controller for cooling purpose was derived and validated as was done for the previous case(Heating).[4]

$$C = \left\{ \frac{-1}{\frac{1}{V_c}*(\alpha-\beta)} \left(\frac{\frac{UA}{C_{pc}\rho_c V_c}*(\delta-\beta)+W_s sgn(\varepsilon_s)}{\alpha} \right) \right\} \qquad (12)$$

The above controller confirmed the definite reachbility and stability [6].

5 Development and implementation of Feedback linearization controller to thermal transfer module of SFHTS

For Heating: Recalling the dynamic model of the heating system, equation (3),

$$\dot{y} = \dot{\delta} = f + \Delta\, c,$$

where

$$f = \frac{UA}{C_{ph}\rho_h V_h}*(\beta - \delta)$$

$$\Delta = \frac{1}{V_h}*(\gamma - \delta)$$

c is the control input.

The feedback linearization controller for heating was designed for finite time reachbility and is given as

$$C = \left\{ \frac{-1}{(\frac{1}{V_h}*(\gamma-\delta))} \left(\frac{UA}{C_{ph}\rho_h V_h}*(\beta - \delta) + \mu\,(\beta - \delta) \right) \right\}$$

where μ is a positive scalar quantity.

For Cooling:

Recalling the dynamics of the cooling system,

$$y = \beta$$

$$\dot{y} = \dot{\beta} = p + \psi\, q$$

where

$$p = \frac{UA}{C_{pc}\rho_c V_c}(\delta - \beta)$$

$$\psi = \frac{1}{V_c}(\alpha - \beta)$$

q is the control input.

The feedback linearization controller for cooling was designed for finite time reachbility and is given as

$$C = \left\{ \frac{-1}{(\frac{1}{V_c}*(\gamma-\delta))} \left(\frac{UA}{C_{pc}\rho_c V_c}*(\delta - \beta) + \Pi\,(\alpha - \beta) \right) \right\}$$

where η is a positive scalar quantity.

6 Virtual Instrument

The LabVIEW software from National Instruments was used. The entire project
was developed on LabVIEW 12, Professional Development System. The PID,
Control System Design and Simulation toolkits were extensively used for this
project. The front panel of SFHTS with Concentric Jacketed Tank and Expansion
Tank with different controllers implemented is shown in the figure (3). The Virtual Instruments used Tab control to toggle the display between heat transfer
and level control module. Indicators were provided to display Flowrate, Sliding
surface, Temperature, Level in Concentric Jacket Tank and Expansion Tank. The
toggle operation was confined only to display whereas the control action was
simultaneous applied to both temperature and level controllers. The front panel
of the system is common to all the controllers .However, the block diagrams are
different.

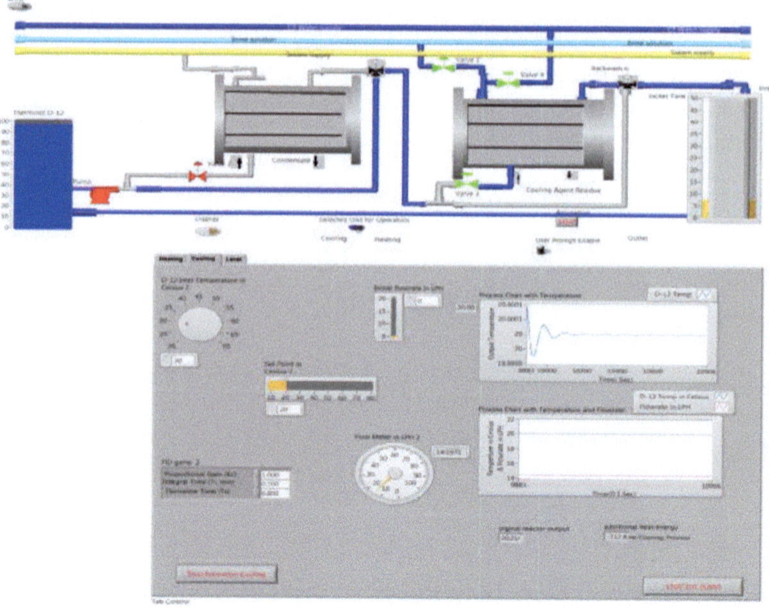

Fig. 3: Front Panel of SFHT system developed using LabVIEW

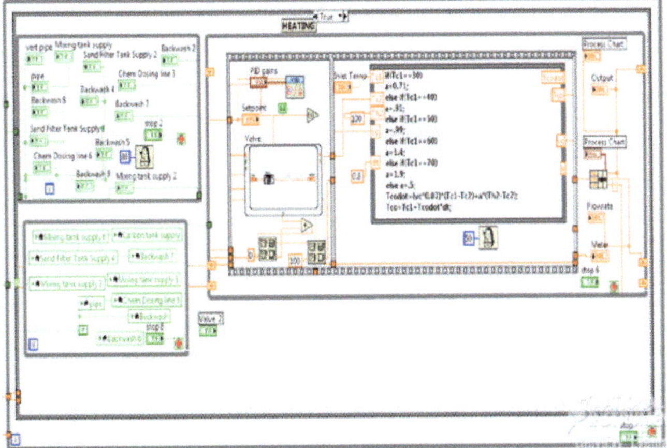

Fig.4: Block diagram of SFHT system developed using LabVIEW

7 Results

In the present investigation performance evaluation of three different controllers were carried out and their responses have been presented in figures 14 to 17.

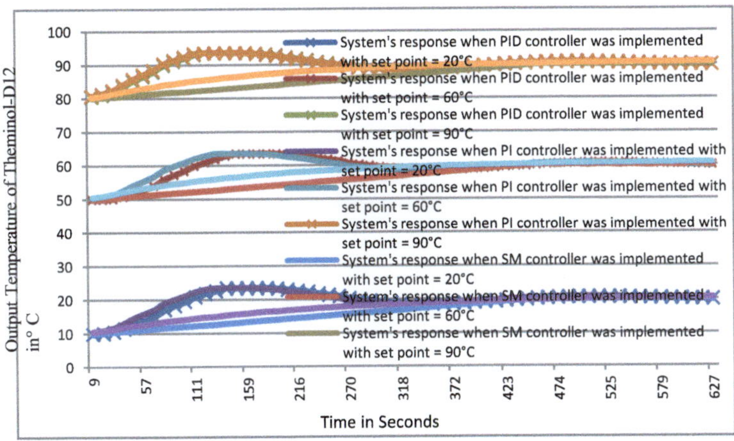

Fig.5: Output Temperature of Therminol-D12 when PID,PI and SM Controller were implemented for heating operation of SFHTS

Table 1. Numerical values inferred from the Fig. 5

S.No	Inlet Temperature of Process Variable (°C)	et Point °C)	Overshoot (%)				Approximate Settling time (Seconds)				Rise time (Seconds)			
			PID	PI	SMC	FBLC	PID	PI	SMC	FBLC	PID	PI	SMC	FBLC
1	10	20	30	30	Nil	Nil	250	400	420	270	110	115	390	230
2	50	60	30	30	Nil	Nil	250	400	420	270	110	115	390	230
3	80	90	30	30	Nil	Nil	250	400	420	270	75	80	390	230

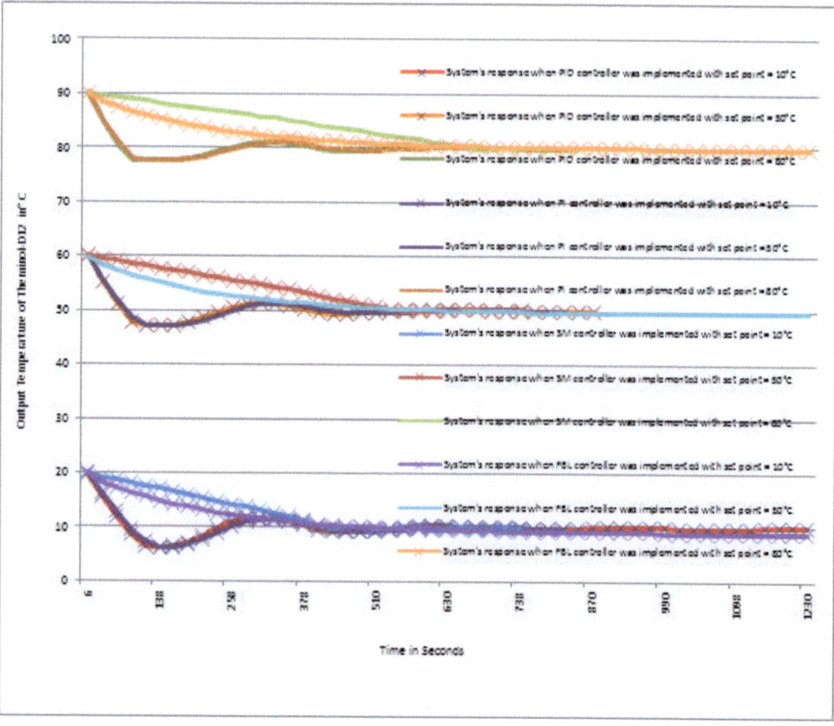

Fig. 6: Output Temperature of Therminol-D12 when PID,PI and SM Controller were implemented for cooling operation of SFHTS

Table 2. Numerical values inferred from Figure.6

S. No	Inlet Temperature of Process Variable (°C)	Set Point (°C)	Overshoot (%)				Approximate Settling time (Seconds)				Rise time (Seconds)			
			PID	PI	SMC	FBLC	PID	PI	SMC	FBLC	PID	PI	SMC	FBLC
1	20	10	20	20	Nil	Nil	250	250	360	370	110	110	350	330
2	60	50	30	30	Nil	Nil	250	250	460	400	100	100	430	370
3	90	80	40	40	Nil	Nil	250	250	560	440	75	75	520	400

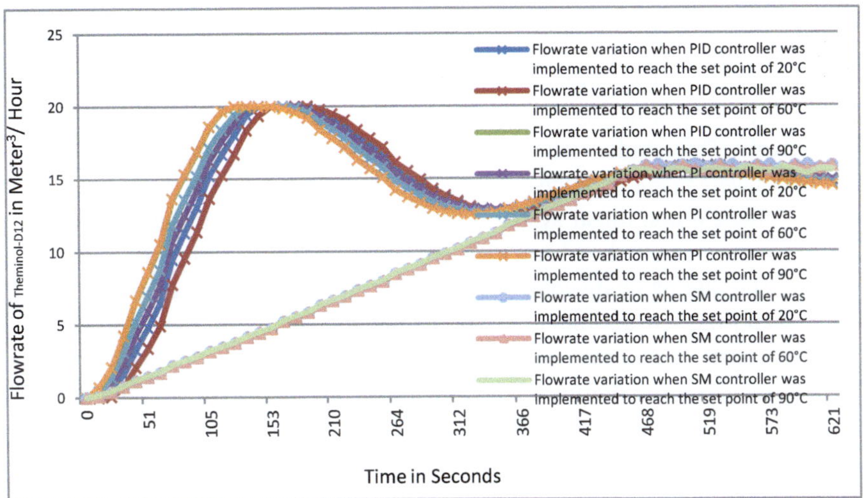

Fig.7: Flowrate variation when different controllers were used to control the heating operation

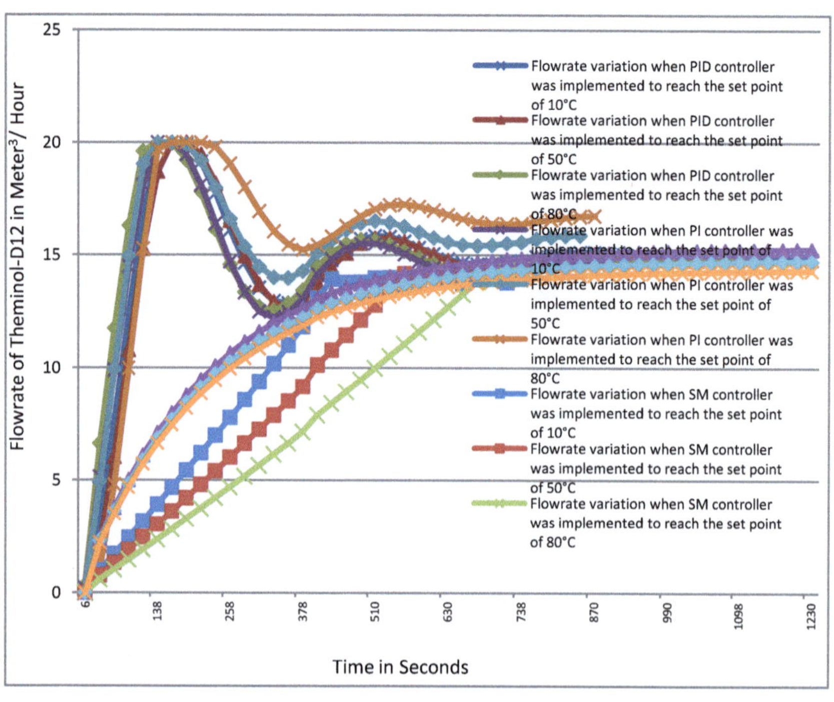

Fig.8: Flowrate variation when different controllers were used to control the cooling operation

To have a holistic view of the process three desired points 20°C, 60°C and 90°C, and 10°C, 50°C and 80°C were set for heating and cooling operation respectively. The tabular values were derived from the graphs. The following inferences were made from the data above

- Being a non linear system , the SFHTS's thermal transfer module response to both the PI and PID controllers were almost similar
- The Sliding Mode controller could completely eliminate the overshoot and the settling time was approximately equal to those of the PI controller. If the gain of the Sliding mode is increased it will lead to noticeable chattering effect. Hence a trade off exist between the settling time and chattering effect and the former was preferred than the latter.
- When the feedback linearization controller (FBLC) was implemented the overshoot was completely eliminated and the set point was reached in shorter time compared to the controllers considered here.
- The flowrate variation was also captured and presented

Acknowledgment

I would like to thank the Management, Principal and Head of Department of EIE, VNR Vignana Jyothi Institute of Engineering and Technology, Hyderabad and Department of EEE, JNTUH Hyderabad for helping me to carry out this research project.

References

1   K.M.Hangos, J.Bokor, G.Szederkenyi, "Analysis and control of Non linear Process systems, Springer-Verlag, London, 2004.
2   Curtis.D.Johnson,"Process Control Instrumentation",Prentice Hall of India Pvt Ltd., New Delhi,8th Edition, 2006.
3   Jean-Jacques E.Slotine ,Weiping Li, "Applied non linear control",Prentice Hall International, New Jersy,1991.
4   Hai Wang, Zhihong Man, Do Manh Tuan, Zhenwei Cao, Weixiang Shen, "Sliding Mode Control for Steer-by-Wire Systems With AC Motors in Road Vehicles", IEEE transactions on Industrial Electronics, vol. 61, no. 3, March 2014, pp 1596-1611.
5   Reham Haroun, Angel Cid-Pastor, Abdelali El Aroudi, Luis Martínez-Salamero , "Synthesis of Canonical Elements for Power Processing in DC Distribution Systems Using Cascaded Converters and Sliding-Mode Control" , IEEE transactions on Power Electronics, Vol. 29, No. 3, March 2014, pp 1366-1380.
6   Jian-Xin Xu, Zhao-Qin Guo, Tong Heng Lee, "Design and Implementation of Integral Sliding-Mode Control on an Underactuated Two-Wheeled Mobile Robot" , IEEE Transactions on industrial electronics, Vol. 61, No. 7, July 2014, pp 3671-3681.
7   Jeffery Travis and Jim Kring, "LabVIEW for Every one:Graphical programming made easy and fun", 3rd Edition,Prentice Hall International, New Jersy, 2006.
8   Ricardo de Castro, Rui Esteves Araújo, Diamantino Freitas, "Wheel Slip Control of EVs Based on Sliding Mode Technique With Conditional Integrators", IEEE Transactions on Industrial Electronics, Vol. 60, No. 8, August 2013, pp 3256-3271.
9   T. D. Eastop and A. Mc Conkey, Applied Thermodynamics for engineering technologist , 3rd Edition, Pearson Education in South Asia, 2012.

Sridevi.H.R [1], Aruna Prabha B S [2], H.M Ravikumar [3] and
P.Meena [4]

Implementation of SVPWM Controlled Three Phase Inverter for use as a Dynamic Voltage Restorer

Abstract: This paper presents the implementation of a two level inverter using Space Vector Pulse Width Modulation (SVPWM) technique by the method of sampling the phase reference amplitudes of the control signal. The objective of this work is to use the experimental set up to conduct a detailed study on a three phase inverter driven by the above method and explore the possibility of its use as a dynamic voltage restorer (DVR) in power supply systems to improve the power quality. The inverter operation when used as a DVR is to be initiated by a trigger generated in response to the detection of short duration voltage disturbances such as voltage sags in three phase power supply systems .The algorithm used for detection of sags is based on the concept of Space Vectors. The three phase inverter is simulated using MATLAB /SimPower Systems tool-box. The experimental set up has the gating signals generated using the TMS320F2812 Digital Signal Processor, which are amplified using a low power operational amplifier to drive the prototype of a three phase inverter built using Insulated Gate Bipolar Transistors (IGBTS).

Keywords: Dynamic voltage restorer (DVR), Space vector PWM technique (SVPWM), voltage source inverter (VSI), voltage sag detection

1 Dept of Electrical & Electronics, NMIT, Bangalore, India
www.nmit.ac.in
sridevi_hr@yahoo.co.in
2 Dept of Electrical & Electronics, NMIT, Bangalore, India
www.nmit.ac.in
bs.arunaprabha@gmail.com
3 Dept of Electrical & Electronics, NMIT, Bangalore, India
www.nmit.ac.in
hmrgama@gmail.com
4 Dept of Electrical & Electronics, B M S College of Engg, Bangalore, India
www.bmsce.ac.in
meenabms@gmail.com

1 Introduction

Power quality is one of the major concerns in electricity industry today. Main Power quality problems are voltage sags, swells, transients, noise and harmonic distortion[1]. Voltage sag is considered as the most serious problem of power quality. Voltage sag is a momentary decrease in rms voltage magnitude in the range of 0.1to 0.9 p.u. It is considered as the most serious problem of power quality. Voltage sag analysis is a complex stochastic issue since it involves a large variety of random factors, type of short circuits in the power system, location of faults, protective system performance and atmospheric discharges. If the voltage sag persists for more than 2 to 3 cycles ,the manufacturer's system making use of sensitive electronic equipments are likely to be affected leading to major problems. It ultimately leads to wastage of resources as well as financial losses.

It is possible to have good power quality only by ensuring that uninterrupted flow of power is maintained at proper voltage levels. Dynamic voltage restorer (DVR) [2] and Distribution static compensator (D- STATCOM) are recently being used as the active solution for voltage sag mitigation and has proven to be a competitive solution in medium and high power applications .Three phase inverters are used as DVRs in three phase supply systems. For low power applications, cost reductions are required to make technology even more competitive. Sinusoidal PWM and Space vector PWM control techniques are used for controlling the DVR.

Space vector Pulse Width Modulation (SVPWM) technique [3, 4] makes efficient utilization of DC bus voltage and generates fewer harmonics in the inverter output voltage. It refers to a special technique of determining the switching sequence of the semiconductor devices of a three phase voltage source inverter. The whole objective of SVPWM is to obtain the switching pattern for the top and bottom switches so that the average variation of the space vector is sinusoidal[5]. The tip of the space vector traverses a circular trajectory of different radii when the DC link voltage magnitude is varied.

The algorithm adopted for generation of gating pulses in this work makes use of the sampled values of the reference phase voltage amplitudes for obtaining the switching time instants [6]. It is a simple sorting algorithm which does not require sector identification and look up table which is otherwise required in SVPWM technique. Care has also been taken in the algorithm to provide a suitable dead time to prevent short circuit of dc bus voltage of a voltage source inverter. The switching on of upper switches of VSI and the switching off of lower switches of VSI is delayed by dead time value which is configurable [7,8].

2 Simulation in Matlab Simulink for Generation of Gating Pulses using SVPWM

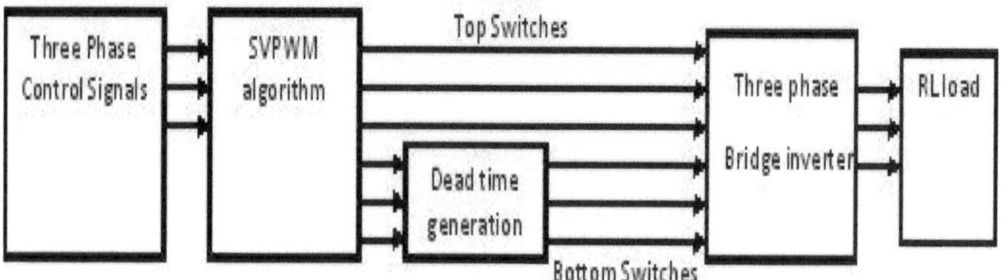

Fig. 1: Block diagram indicating the simulation process

The block diagram for the process of simulation is as shown in Fig.1.The reference voltages are three sine waves which are out of phase with each other by 120°.

Steps involved in the algorithm are as follows: 1) The reference voltages Va, Vb, Vc are sampled to obtain the switching time instants Tas, Tbs, Tcs.

$$Tas = Va * (Ts/Vdc)$$
$$Tbs = Vb*(Ts/Vdc)$$
$$Tcs = Vc* (Ts/Vdc)$$

Where Ts is the sampling period (In this work Ts is taken as 256 μs)

2)The effective time is calculated which is nothing but the maximum(maximum of Tas, Tbs and Tcs) of the sampled value and the minimum (minimum of Tas, Tbs and Tcs) of the sampled value.The minimum value is always negative where as the maximum value is always positive.

3)Then the Zero time 'T0' where T0 = Ts – Teff is calculated.The time period should be positive therefore an offset is added to shift it to positive side. The offset time Toffset = T0/2 – Tmin. This gives the corresponding gating signal to the top swiches in a sampling period. For the rest of the period bottom switch should turn on.

$$Tga = Tas + Toffset$$
$$Tgb = Tbs +Toffset$$
$$Tgc = Tcs + Toffset$$

4)At the end of the algorithm a modulating signal is obtained which is compared with the triangular wave to obtain the gating pulses. The simulation model is shown in figure 2.

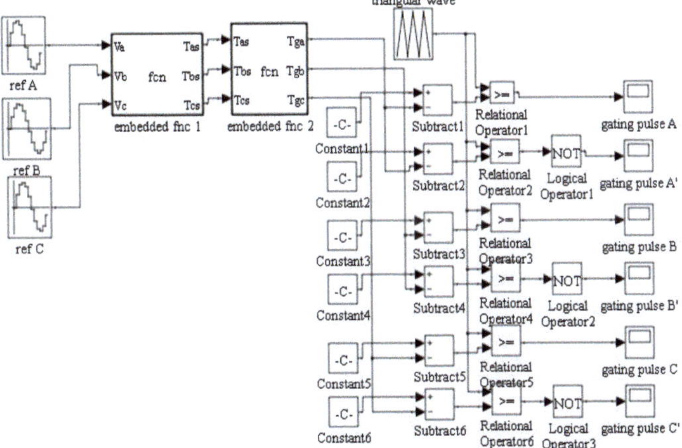

Fig. 2: Simulink model for generation of gating pulses with dead time using sampled values of reference phase voltages

The gating pulses thus generated are given to a three phase inverter circuit connected to an RL load and simulation is carried out using MATLAB Simulink and Sim Power Systems toolbox. The model is shown in fig. 3.

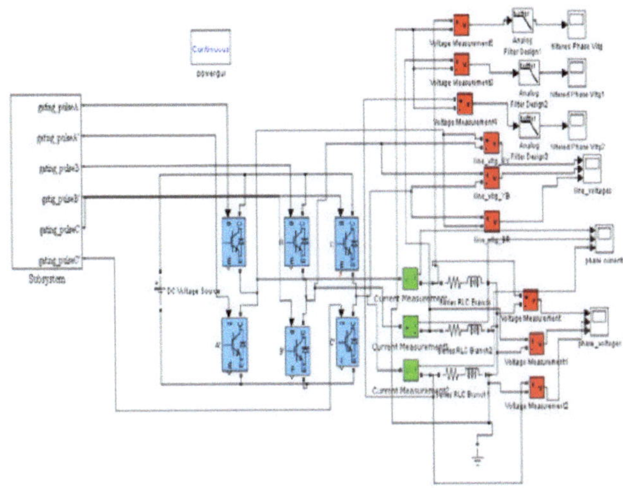

Fig.3: Three phase voltage source inverter with RL load

3 Study of Inverter Operation as a DVR

The block diagram of the operation of the inverter in response to short duration voltage disturbance such as voltage sag is indicated by an LED on the DSP kit. The disturbance is detected by an algorithm [9] which involves the transformation of the space vector of the three phase reference signals along the stationary reference frame α-β as two alternating components orthogonal to each other. The resultant of these components, the resultant space vector is sensed at every sampling instant. The magnitude of the resultant space vector has a fixed value under normal conditions. Deviations from this value during disturbances are measured. This is further processed using hysteresis comparators set to appropriate threshold values for detecting the disturbance (example 0.01-0.09 for sag) which results in the generation of a trigger signal.

The disturbance detection algorithm is implemented on the DSP kit and the LED glows up in response to the trigger generated during the disturbance indicating the occurrence of the disturbance. The block diagram in figure 4 shows the detection procedure involved.

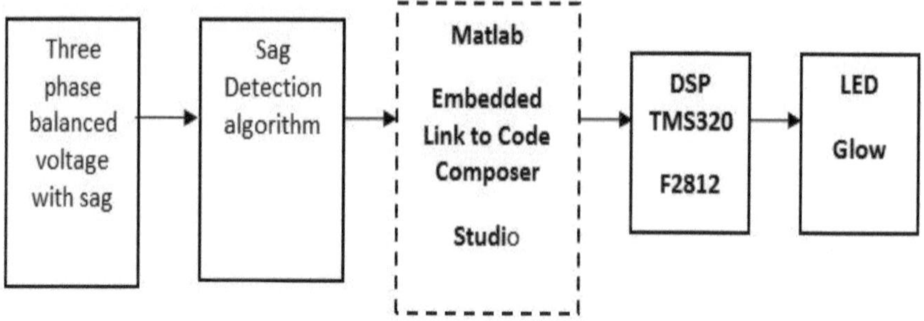

Fig.4: The block diagram of the disturbance detection set up using the LED on the DSP kit

The trigger thus generated initiates the inverter operation as a DVR. This is represented by the block diagram shown in Fig.5.

The control signals to the DVR are the ripple generated which is nothing but the deviations of the control signals from the reference value during disturbance. The voltage developed across the load connected to the inverter in presence of the ripple when added to the voltage across the load under sag conditions results in the voltage across the load to be restored to its normal value.

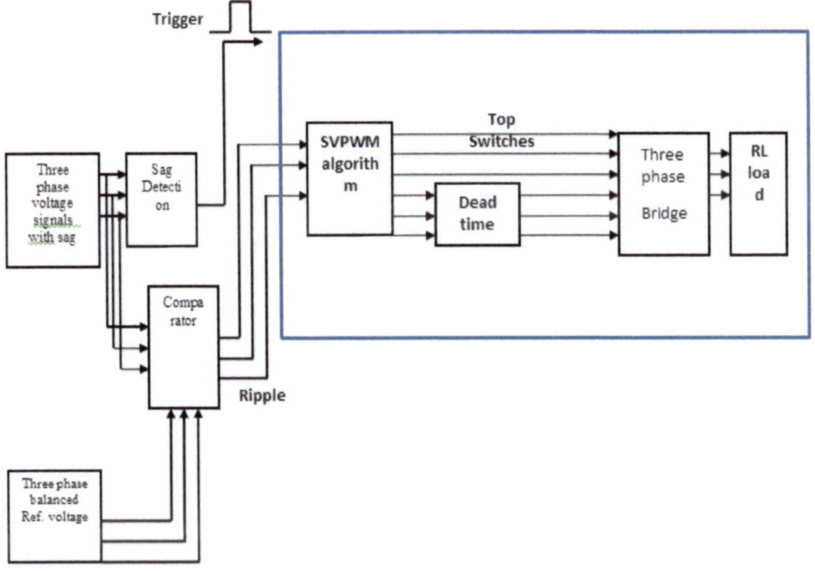

Fig.5: The block diagram of the simulation process of the DVR

4 Experimental Set – Up

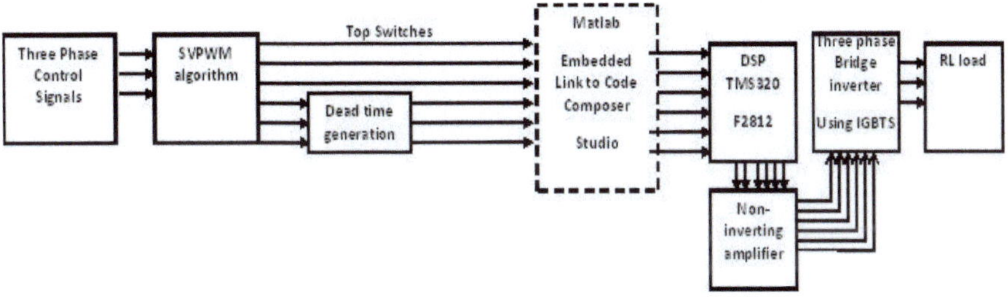

Fig.6: Block diagram indicating the hardware set-up of the three phase inverter

The block diagram of the experimental setup is shown in fig 6. It includes eZdsp2812 kit [10], six IGBT's (IRGBC20S, Ic = 10A, Vce = 600V, Vge = 15V)) connected to form a three phase bridge inverter and a balanced RL load. The PWM pulses generated by the DSP is around 4V therefore the pulses are amplified to 12V using a non-inverting amplifier to drive the IGBT's. The supply to the three phase bridge inverter is given from a rectifier through a capacitor.

5 Simulation Results

The simulation results of operation of the three phase inverter under normal conditions of the control signal and under sag conditions (sag is created from 0.06s to 0.03s) is shown in fig 7 and fig 8.

The results of the operation of the inverter in response to the ripple are as shown in fig 9. The waveforms of the injected voltage and the load voltage during sag are presented. It is seen that the ripple exists as long as the disturbance exists. The injected load voltage when added to the load voltage of the inverter during sag results in normal voltage restored across the load during disturbance, thus indicating successful operation as a dynamic voltage restorer.

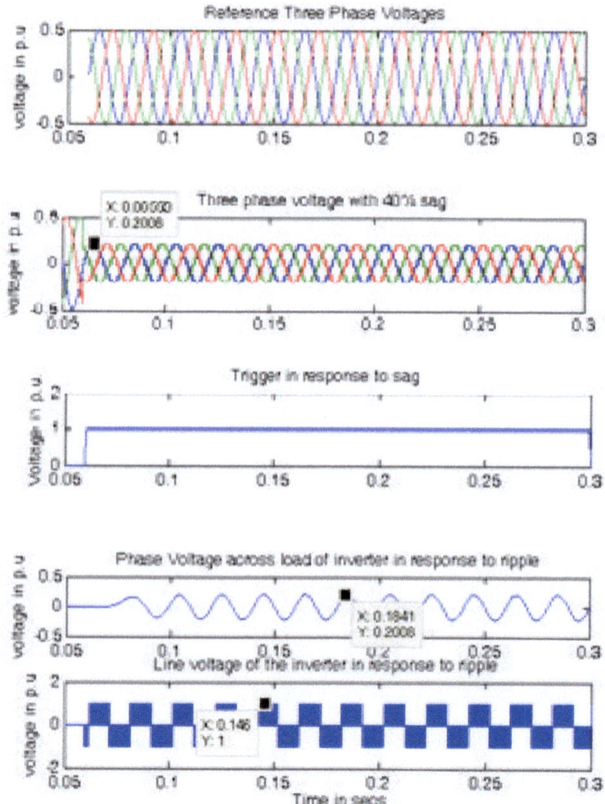

Fig.7: Operation of the inverter under normal conditions

6 Observations from the Simulation Results

The observations made from the simulation results are as follows

(i) As the peak value of the reference voltage increases, the peak value of the fundamental component of load phase voltage also increases which clearly verifies the operation of the inverter in the linear modulation region.

(ii) The peak value of load phase voltage during sag is proportional to the level of sag, indicating the operation of the inverter in the linear modulation region even during dynamic conditions of voltage variations.

(iii) The fundamental component of load phase voltage increases with increase in ripple. Also in absence of ripple, the inverter voltage is observed to be zero. This justifies quick response of the inverter to ripple voltage hence successful operation as a DVR.

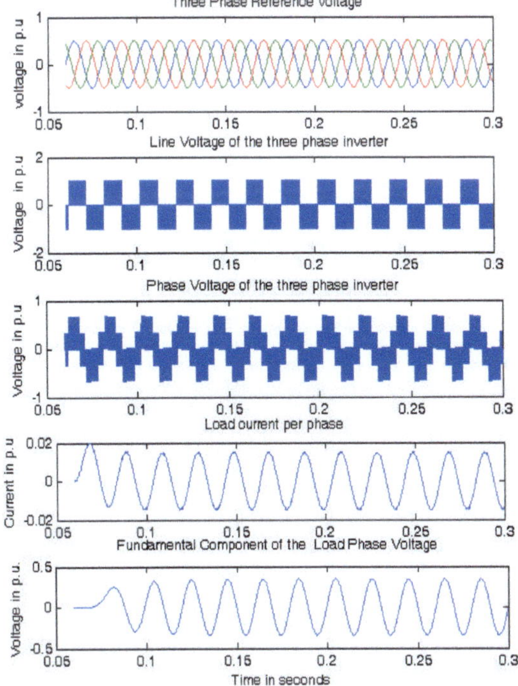

Fig. 8: Operation of the inverter in response to sag

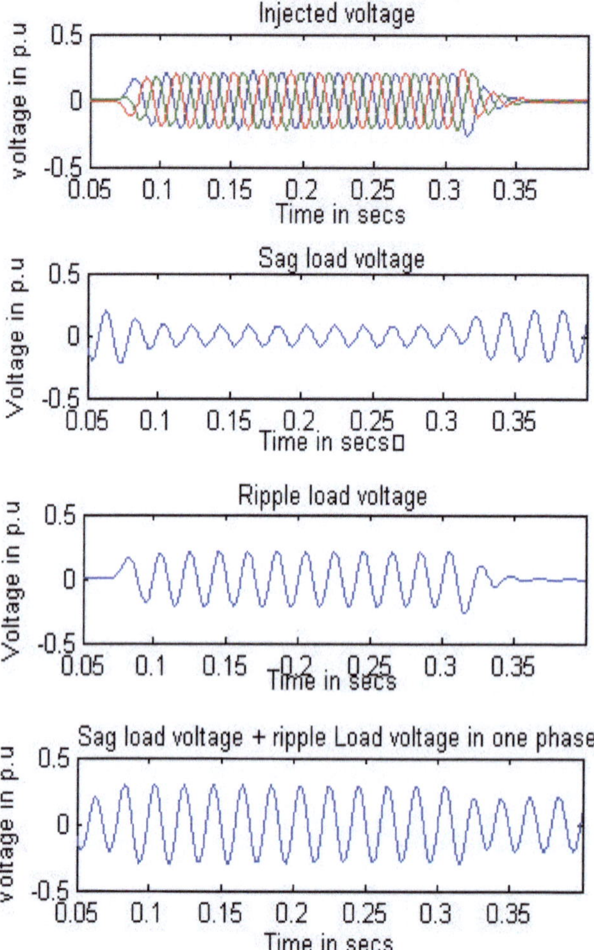

Fig. 9: Operation of the inverter as DVR in response to ripple

7 Experimental Results

The experimental results of dead time generation, gating pulses of the inverter and also the amplified gating pulses required for the IGBT's are shown in fig 10(a), 10(b) and 10(c) respectively.

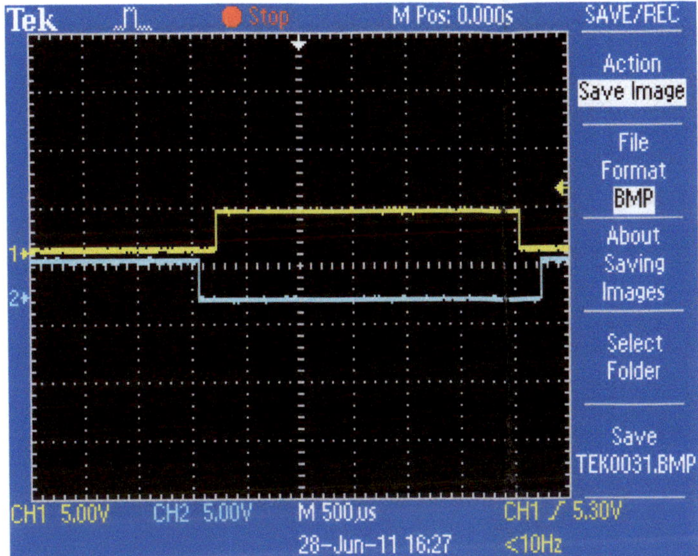

Fig. 10(a): Gating pulses with dead time

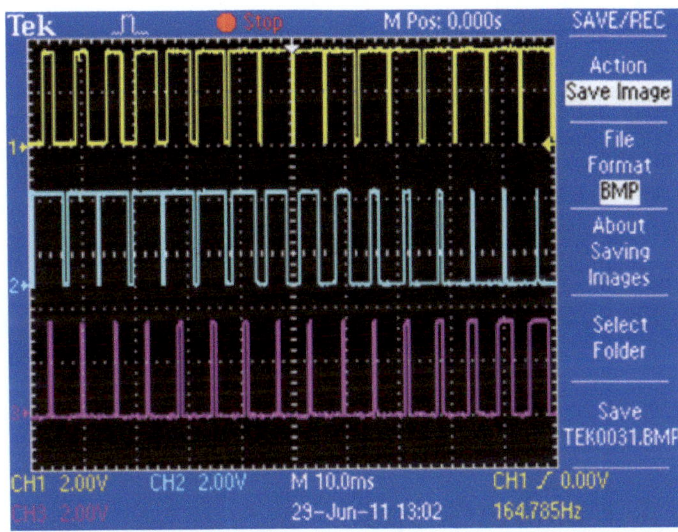

Fig. 10(b): Gating pulses for top switches

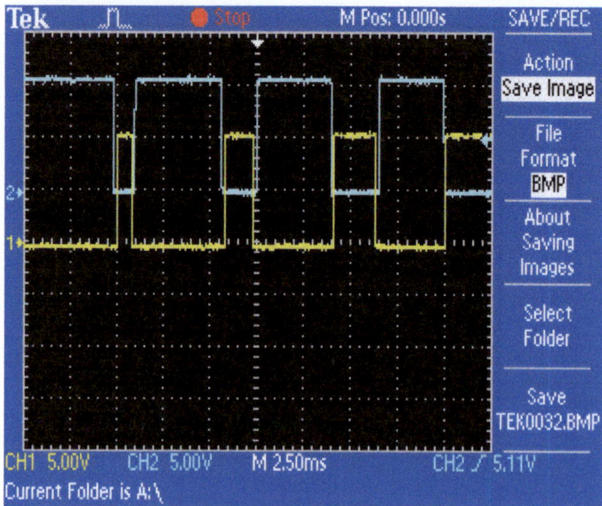

Fig. 10(c): Gating pulses for one leg amplified

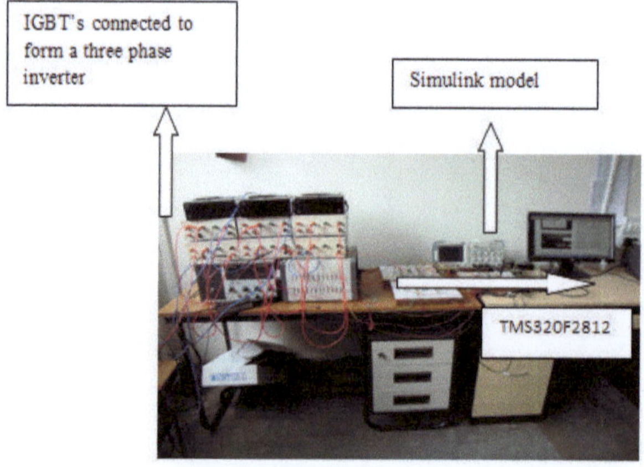

Fig. 11: Experimental set-up

Conclusion And Future Scope

The three phase inverter controlled by SVPWM technique was successfully implemented both in simulation as well as in hardware.The response of the inverter to dynamic variations in three phase voltages were observed and were appreciable. The response of the inverter to ripples or deviatons of the control signal from the reference values were proportional . This is suggestive of the use of the above inverter successfully as a dynamic voltage restorer operating during short duration voltage sags.However an extensive study on its operation is required to decide on its application as a DVR during the occurrence of other types of distrubances.

References

1   Math H.J.Bollen, Irene Y.H.Gu, Surya Santoso, Mark F McGranaghan,Pete A..Crossley, Moisės V.Ribeiro,and Paulo F.Ribeiro "Bridging the gap between signal and Power" IEEE Signal Processing Magazine[12] July 2009.

2   Atif Iqbal, S.K.Moin Ahmed, Mohammed Arif Khan, Haitham Abu – Rub, "Generalised simulation and experimental implentation of Space vector pwm technique for a three phase voltage source inverter", Internation Journal of Engineering, Science and technolgy Vol2 No.1,2010, pp. 1-12

3   Keliang Zhou and Danwei Wang, "Relationship between Space Vector – Modualtion and three phase Carrier based Pulse width mosulation a Comprehensive analysis, IEEE transactions on industrial electronics vol 49, No.1, Feb 2002.

4   k.Vinoth kumar, Prawin Angel Michael, Joseph.P.John and Dr.S.Suresh Kumar, "Simulation and comparison of SPWM and SVPWM control of three phase inverter, Asian research publishing network, vol5, No.7, july 2010.

5   Uppunoori Venkata Reddy, Paduchuri.Chandra Babu, S.S.Dash "Space Vector Pulse Width Modulation Based DVR to Mitigate Voltage Sag and Swell" 2013 International Conference on Computer Communication and Informatics (*ICCCI* -2013), Jan. 04 – 06, 2013, Coimbatore, INDIA

6   R.S.Kanchan, K.Gopakumar and R.Kennel "Synchronised carrier based SVPWM signal generation scheme for the entire modulation range extending upto 6 step mode using sampled amplitude of reference phase voltages", IET Electr. Power Appl.,vol 1, No.3, May2007.

7   DN Sonwane, M.S.Sutaone, B.N.Choudhari and ABhijeet Badurkar "FPGA implementatation of simplified SVPWM algorithm for three phase voltage source inverter", International journal of Computer and Electrical Engineering, Vol2, No.6, Dec2010 1793-8163

8   Joffie Jacob, Reshmi V, Nisha Prakash "Comparison of DVR Performance with Sinusoidal and Space Vector PWM Techniques" IEEE International Conference on Magnetics, Machines & Drives (AICERA-2014 iCMMD), 24 Jul - 26 Jul 2014, Kerala India

9   P.Meena, K.Uma Rao, Ravishankar Deekshit, "A simple al gorithm for fast detection and quantification of voltage deviations using Space Vectors" IEEE IPEC 2010 Conference, Singapore, 27-29 Oct. 2010.

10 Agileswari Ramaswamy, Vigna Kumaran Ramachandaramurthy, Rengan Krishna Iyer, Liew Zhan Liu " Control of Dynamic Volage Restorer using TMS320F2812" 9th international conference Electric Power quality and utilization , Barcelona, 9-11, Oct 2007.

Jithendra.K.B [1] and Shahana.T.K [2]

Enhancing the Uncertainty of Hardware Efficient Substitution Box based on Differential Cryptanalysis

Abstract: The researches in secure communication fields are in search of better security with lesser resources'. It is almost true that the security of a crypto system is in proportion with the hardware complexity. It is a real challenge to design a crypto system with high security with lesser hardware complexity. In this paper differential cryptanalysis of a part of hardware efficient Substitution box is carried out. The analysis proves that the security of the S Box is better than that of conventional one. Moreover, this paper gives some clues to enhance the uncertainty further, based on differential cryptanalysis.

Keywords: Encryption; Block Cipher; Security; Substitution Box; Differential Cryptanalysis.

1 Introduction

The Substitution box introduced by C.Shannon[1] is the most important element for the block ciphers since it contributes greatly to security through its nonlinear input-output characteristics. Normally the strength of the S box is decided by the complexity of the algorithm which determines the hardware complexity. Cryptanalysis is the method used to measure the strength of the S box. Mainly there are two types of cryptanalysis for block ciphers: Linear cryptanalysis and differential cryptanalysis explained by Heys.H[2], Musa et al. [3] and Sinkov.A[4]. The paper by Jithendra.K.B and Shahana.T.K [5](2014) introduces a new S box which claims to have better security with reduced hardware complexity. Here the security is measured through differential cryptanalysis

1 Department of Electronics and Communication, College of Engineering, Vadakara, Calicut (Dt), Kerala, INDIA
jithendrakb@gmail.com
2 Division of Electronics Engineering, School of engineering, Cochin University of Science and Technology, Kerala, INDIA
shahanatk@cusat.ac.in

and compared with that of the conventional systems. The analysis leads to new methods which can enhance the uncertainty to greater levels.

2 Differential Cryptanalysis of a Conventional S Box

The differential cryptanalysis[6] focuses on finding the relationships between the differences of input vectors and that of output vectors. Here the attacks on a cryptosystems are mounted on differentials.

Let x' and x'' are two input vectors to an S Box and y' and y'' be the corresponding output vectors. The Figure 1 shows a well defined 4 bit S box with Boolean variable set $\{ x_3, x_2, x_1, x_0 \}$ as input vector and Boolean variable set $\{ y_3, y_2, y_1, y_0 \}$ as output vector.

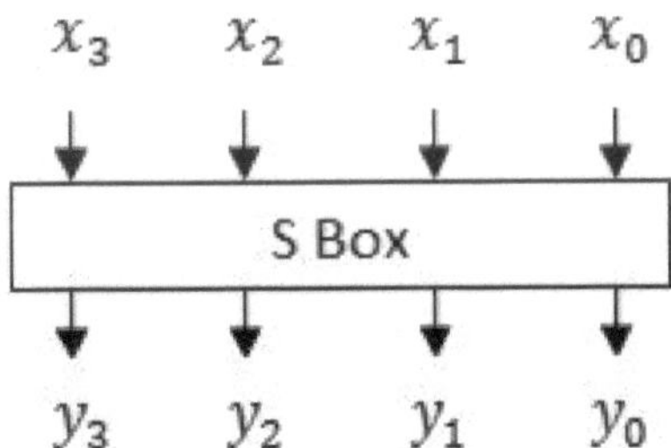

Figure. 1: S Box block diagram

Further let

$$x' \oplus x'' = \Delta x \quad (1)$$
$$y' \oplus y'' = \Delta y \quad (2)$$

Where Δx and Δy are the respective differentials. When an input differential results in the occurrence of a specific output differential with high probability it is said to be a weakness. As seen earlier, let

$$x' = \{ x_3, x_2, x_1, c \} \quad (3)$$
$$y' = \{ y_3, y_2, y_1, y_0 \} \quad (4)$$

If Δy is completely independent of Δx, each bit of Δy will have a probability of exact ½. Suppose $\Delta x = 1001$ and $\Delta y = 0011$. The weakness is measured by calculating how many times same value is occurred for Δy for the same value of Δx. More number of occurrences indicates grater weakness.

Suppose the S box-1 is defined as in Table 1.

Table1. S box – 1

x_3x_2 / x_1x_0	00	01	10	11
00	C	D	5	2
01	B	A	6	3
10	0	8	E	9
11	7	1	F	4

A sample differential data analysis is given in Table 2 for the S Box given in Table 1

Table 2. Sample differential data analysis for the S Box given in Table 1

		1				2		3	4	5
x_3	x_2	x_1	x_0	y_3	y_2	y_1	y_0	$x' \oplus \Delta x$	$y' \oplus \Delta y$	Δy
0	0	0	0	1	1	0	0	1100	0010	1110
0	0	0	1	1	0	1	1	0010	0000	1011
0	0	1	0	0	0	0	0	1001	0110	0110
0	0	1	1	0	1	1	1	0100	1101	1010
0	1	0	0	1	1	0	1	1011	1111	0010
0	1	0	1	1	0	1	0	1110	1001	0011
0	1	1	0	1	0	0	0	0011	0111	1111
0	1	1	1	0	0	0	1	1010	1110	1111
1	0	0	0	0	1	0	1	1111	0100	0001
1	0	0	1	0	1	1	0	0110	1000	1110
1	0	1	0	1	1	1	0	0001	1011	0101
1	0	1	1	1	1	1	1	0111	0001	1110
1	1	0	0	0	0	1	0	0000	1100	1110
1	1	0	1	0	0	1	1	1000	0101	0110
1	1	1	0	1	0	0	1	1101	0011	1010
1	1	1	1	0	1	0	0	0101	1010	1110

1. Column 1 and 2 shows the 4 bit input and output vectors for the S Box given in Table 1.
2. Column 3 gives the some sample sequences of random change in input vector. For example 0000 changes to 1100, 0001 changes to 0010 and so on. The list is not complete. Total number of possible changes for a 4 bit vector is 256.
3. Column 4 represents the data substituted in place of the data shown in column The substitution is done here with the data in S Box – 1.
4. Column 5 shows the difference Δy between output vectors corresponding to difference between input vectors Δx.

It is possible to find out the frequency distribution of Δy for each possible values of Δx. The Table 3 shows the complete differential distribution of the S Box-1.

Table 4 shows the calculation of probability for the weak trails identified to be true.

Suppose x is the input and y is the output of a system when mixed with a key K.

$y = x \oplus K$

when x is changed by Δx to x'

$x' = x \oplus \Delta x$

Correspondingly,

$y' = x' \oplus K$

$= x \oplus \Delta x \oplus K$

$= x \oplus K \oplus \Delta x$

$= y \oplus \Delta x$

Table 3. Differential distribution of S Box-1

		Output Differential														
		0	1	2	3	4	5	6	7	8	9	A	B	C	D	E	F
Input Differential	0	16	0	0	0	0	0	0	0	0	0	0	0	0	0	0	0
	1	0	4	0	2	0	0	0	6	0	2	0	0	0	2	0	0
	2	0	0	0	0	0	2	0	2	0	2	0	6	4	0	0	0
	3	0	0	2	0	0	0	2	0	2	0	4	4	2	0	0	0
	4	0	4	0	0	0	2	2	4	2	0	0	2	0	0	0	0
	5	0	2	0	0	2	0	8	0	0	0	2	0	0	0	0	2
	6	0	0	2	0	2	0	0	0	0	0	2	0	6	4	0	0
	7	0	2	0	2	0	0	0	0	0	0	4	0	0	6	0	2

8	0	2	0	0	0	2	0	0	2	4	0	0	0	2	2	2
9	0	0	0	0	0	0	0	0	4	2	2	0	2	0	4	2
A	0	2	4	0	4	2	0	0	0	0	2	0	0	0	2	0
B	0	0	2	6	0	2	2	0	0	2	0	2	0	0	0	0
C	0	0	0	2	0	0	2	0	4	2	0	0	2	0	4	0
D	0	0	0	0	2	0	0	2	0	2	0	2	0	0	2	6
E	0	0	2	2	2	4	0	2	0	0	0	0	0	2	0	2
F	0	0	4	2	4	2	0	0	2	0	0	0	0	0	2	0

Table 4. Input output differential vectors and probabilities for SPN-1

Input differential $\Delta p = \Delta w_0 = 0101\ 0000\ 0000\ 0000$		
S_3 in Round 0	$\Delta x = 5 \quad \rightarrow \quad \Delta y = 6$	P = 8/16
S_2 in Round 1	$\Delta x = 8 \quad \rightarrow \quad \Delta y = 9$	P = 4/16
S_1 in Round 1	$\Delta x = 8 \quad \rightarrow \quad \Delta y = 9$	P = 4/16
S_3 in Round 2	$\Delta x = 6 \quad \rightarrow \quad \Delta y = C$	P = 6/16
S_0 in Round 2	$\Delta x = 6 \quad \rightarrow \quad \Delta y = C$	P = 6/16
Output differential $\Delta w_3 = 1001\ 1001\ 0000\ 0000$		

From the above analysis it can be understood that whatever is the value of the Key it has no effect on Δx. Since differentials are immune to the key values, it is possible to find out the weak trails of the Substitution Permutation Network with the use of the Differential Distribution Table shown in Table 3.

Assume a sample plane text pair is having differential $\Delta p = 0101\ 0000\ 0000\ 0000$ is applied to the to the S Box shown in Table 1

Figure 2 shows the weak trails identified which are drawn in bold lines (C.K Shyamala et al.[7]). If the relationships indicated are fully valid, the input differential will results in output differential Δw as
$\Delta w_3 = 1001\ 1001\ 0000\ 0000$

Total Probability $= \dfrac{8}{16} \times \dfrac{4}{16} \times \dfrac{4}{16} \times \dfrac{6}{16} \times \dfrac{6}{16} = \dfrac{9}{2048}$

Total Probability $= \dfrac{8}{16} \times \dfrac{4}{16} \times \dfrac{4}{16} \times \dfrac{6}{16} \times \dfrac{6}{16} = \dfrac{9}{2048}$

The above results show that the dynamic S Box mentioned in the paper by Jithendra.K.B and Shahana.T.K (2014) enhances the security, with reduced hardware complexity. As shown in Table 4 and 7, the inputs with differential 0101 0000 0000 0000 results in output differential 1001 1001 0000 0000 with a probability of $\frac{9}{2048}$ or 0000 0100 0100 0000 with a probability of $\frac{3}{2048}$ depends on the key. Now it can be concluded that the uncertainty has been improved.

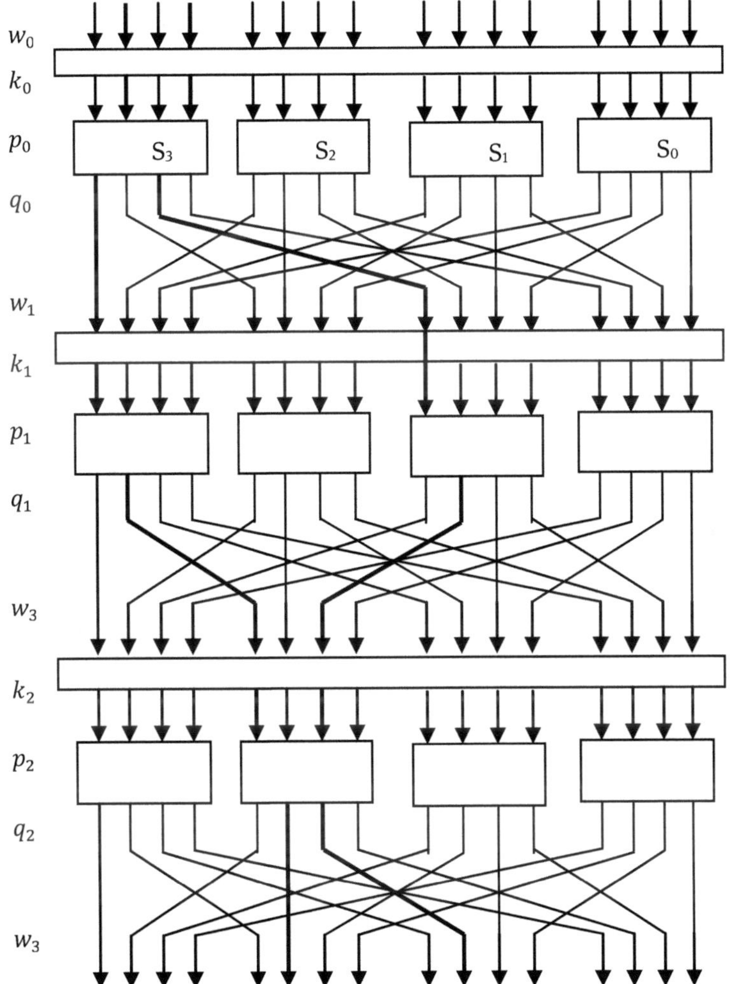

Figure. 3: SPN – 2. Weak trails identified with the S Box – 2

3 Security Enhancement of Design Mentioned in Previous Section

Based on the analysis done in the previous section, further uncertainty enhancement is possible. Let us assume all the S boxes in the Figure 4 are different

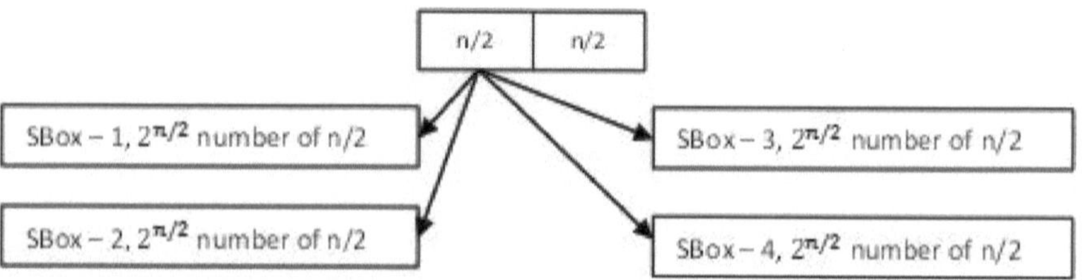

Figure. 4: Dynamic substitution with different S boxes

Let the S box − 3 be as shown in Table 5

Table 5. S box − 3

x_3x_2 / x_1x_0	00	01	10	11
00	6	7	5	2
01	B	A	D	C
10	4	E	8	9
11	3	1	F	0

The differential distribution table for S Box-3 is given in Table 6

Table 6. Differential Distribution Table for S Box-3

		Output Differential														
		0	1	2	3	4	5	6	7	8	9	A	B	C	D	E	F
Input	0	16	0	0	0	0	0	0	0	0	0	0	0	0	0	0	0
Differential	1	0	0	0	0	0	0	0	4	2	2	0	0	0	4	2	2
	2	0	0	4	0	0	0	0	0	2	2	0	4	2	2	0	0

3	0	0	2	0	2	6	2	0	0	0	2	0	0	0	0	2
4	0	8	2	0	0	0	0	2	0	0	2	0	0	0	0	2
5	0	0	0	0	0	2	2	0	2	2	0	0	4	2	0	2
6	0	0	0	4	0	0	0	0	2	2	4	0	2	2	0	0
7	0	0	0	0	6	4	0	2	0	0	0	0	0	2	2	0
8	0	2	0	2	0	2	4	2	0	0	0	0	4	0	0	0
9	0	0	0	0	0	0	0	0	4	0	0	8	0	0	4	0
A	0	2	0	0	2	0	0	0	0	0	2	0	2	2	6	0
B	0	0	2	6	0	0	2	2	0	4	0	0	0	0	0	0
C	0	0	2	2	2	0	2	4	0	0	0	0	0	2	2	0
D	0	2	0	0	2	0	0	0	0	4	6	0	0	0	0	2
E	0	0	0	0	0	2	2	0	0	0	0	4	2	0	0	6
F	0	2	4	2	2	0	2	0	4	0	0	0	0	0	0	0

Assume the same plane text pair which was applied in the previous case having differential

Δp = 0101 0000 0000 0000 is applied to the SPN in which S Box -2 is used

Figure 3 shows the weak trails identified which are drawn in thick lines. If the relationships indicated are fully valid, the input differential will results in output differential Δw as

Δw_3 = 0000 0110 0110 0110

Table 7 shows the calculation of probability for the weak trails identified to be true.

Table 7. Input output differential vectors and probabilities for SPN-3

Input differential $\Delta p = \Delta w_0$ = 0101 0000 0000 0000		
S_3 in Round 0	$\Delta x = 5 \;\rightarrow\; \Delta y = C$	$P = 4/16$
S_3 in Round 1	$\Delta x = 8 \;\rightarrow\; \Delta y = 6$	$P = 4/16$
S_2 in Round 1	$\Delta x = 8 \;\rightarrow\; \Delta y = 6$	$P = 4/16$
S_2 in Round 2	$\Delta x = C \;\rightarrow\; \Delta y = 7$	$P = 4/16$
S_1 in Round 2	$\Delta x = C \;\rightarrow\; \Delta y = 7$	$P = 4/16$
Output differential Δw_3 = 0000 0110 0110 0110		

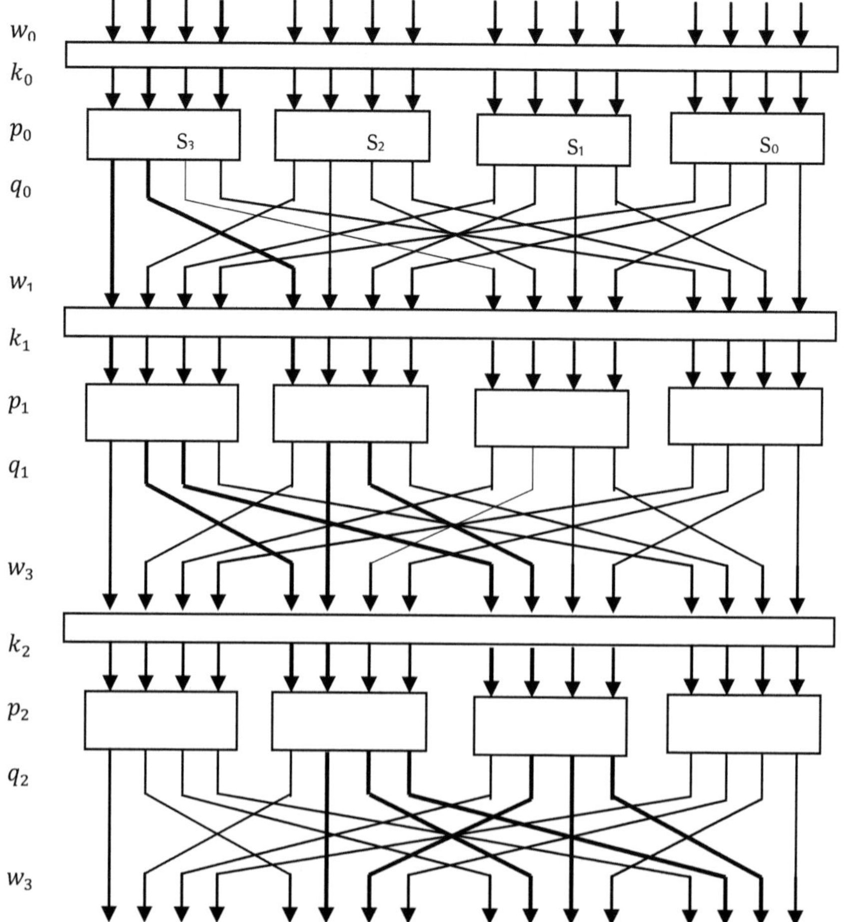

Figure. 5: SPN – 3. Weak trails identified with the S Box – 3

Total Probability $= \dfrac{4}{16} \times \dfrac{4}{16} \times \dfrac{4}{16} \times \dfrac{4}{16} \times \dfrac{4}{16} = \dfrac{1}{1024}$

Let the S Box – 4 be as shown in Table 8

Table 8. S box – 4

$x_3 x_2$ / $x_1 x_0$	00	01	10	11
00	9	4	8	3

01	2	A	C	1
10	6	E	5	0
11	D	F	B	7

The differential distribution table for S Box-3 is given in Table 9

Assume the same plane text pair which was applied in the previous case having differential

Δp = 0101 0000 0000 0000 is applied to the SPN in which S Box -2 is used

Figure 6 shows the weak trails identified which are drawn in thick lines. If the relationships indicated are fully valid, the input differential will results in output differential Δw as

Δw_3 = 0000 1110 0000 0000

Table 9 - Differential Distribution Table for S Box-4

		Output Differential															
		0	1	2	3	4	5	6	7	8	9	A	B	C	D	E	F
	0	16	0	0	0	0	0	0	0	0	0	0	0	0	0	0	0
	1	0	2	2	0	2	0	0	2	0	0	0	4	0	0	4	0
	2	0	0	0	2	0	2	2	2	0	0	2	0	0	2	0	4
	3	0	2	0	2	8	0	0	0	0	2	0	2	0	0	0	0
	4	0	0	2	0	0	2	0	0	4	0	0	2	2	4	0	0
Input Differential	5	0	0	2	4	0	0	2	0	0	4	0	2	0	0	0	2
	6	0	0	2	0	0	0	2	4	2	0	2	2	0	2	0	0
	7	0	0	0	0	2	0	2	0	2	2	0	0	6	0	0	2
	8	0	2	0	2	0	0	2	2	2	0	0	2	0	0	4	0
	9	0	0	0	0	0	4	0	0	2	4	2	0	0	2	0	2
	A	0	2	0	0	2	0	0	0	0	2	0	0	2	4	4	0
	B	0	0	2	2	0	2	0	2	0	0	4	0	2	0	0	2
	C	0	0	0	2	2	0	4	0	0	0	4	2	2	0	0	0
	D	0	4	2	0	0	2	0	0	4	0	2	0	0	2	0	0
	E	0	4	0	2	0	4	2	0	0	2	0	0	2	0	0	0
	F	0	0	4	0	0	0	0	4	0	0	0	0	0	0	4	4

Table 10. shows the calculation of probability for the weak trails identified to be true.

$$\text{Total Probability} = \frac{4}{16} \times \frac{4}{16} \times \frac{4}{16} \times \frac{4}{16} \times \frac{8}{16} \times \frac{8}{16} = \frac{1}{1024}$$

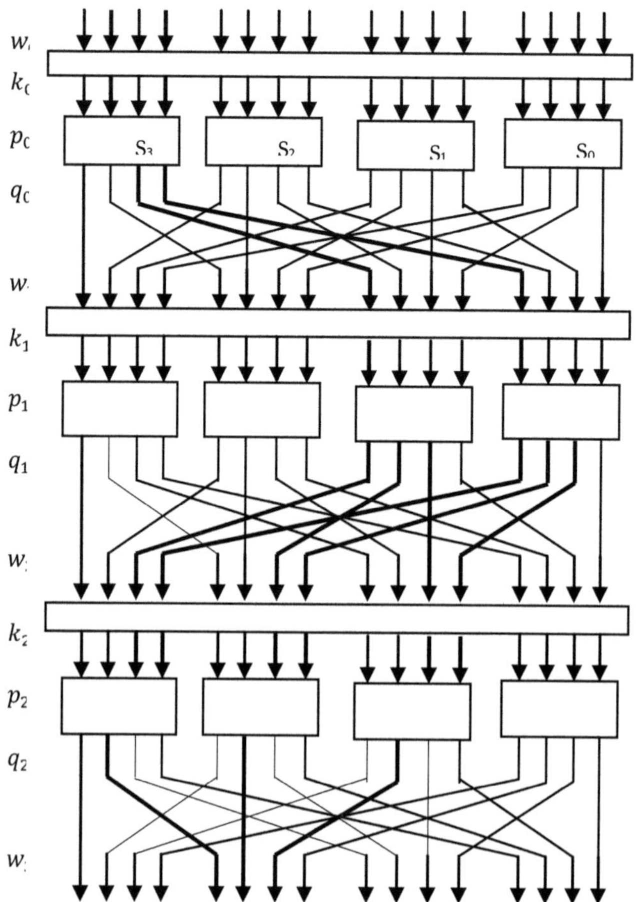

Figure. 6: SPN – 4. Weak trails identified with the S Box – 4

Table 10. Input output differential vectors and probabilities for SPN-3

Input differential $\Delta p = \Delta w_0 = 0101\ 0000\ 0000\ 0000$		
S_3 in Round 0	$\Delta x = 5 \rightarrow \Delta y = 3$	P = 4/16
S_1 in Round 1	$\Delta x = 8 \rightarrow \Delta y = E$	P = 4/16
S_0 in Round 1	$\Delta x = 8 \rightarrow \Delta y = E$	P = 4/16
S_3 in Round 2	$\Delta x = 3 \rightarrow \Delta y = E$	P = 4/16
S_2 in Round 2	$\Delta x = 3 \rightarrow \Delta y = 4$	P = 8/16
S_1 in Round 2	$\Delta x = 3 \rightarrow \Delta y = 4$	P = 8/16
Output differential $\Delta w_3 = 0000\ 1110\ 0000\ 0000$		

The differential cryptanalysis done on the S Box shown in Figure 4 clearly indicates that the uncertainty can further be enhanced with more LUTs. Tables 4, 7, 10 and 13 show that the inputs with differential 0101 0000 0000 0000 results in output differential

1001 1001 0000 0000 with a probability of $\frac{9}{2048}$ or

0000 0100 0100 0000 with a probability of $\frac{3}{2048}$ or

0000 0110 0110 0110 with a probability of $\frac{1}{1024}$ or

0000 1110 0000 0000 with a probability of $\frac{1}{1024}$ or

depends on the key. Now it can be concluded that the uncertainty has been improved considerably.

Conclusion and Futurescope

Differential cryptanalysis is carried out for a hardware efficient Substitution box and compared it with that of a Conventional S box. It is proved that the security of the hardware efficient S box with static as well as dynamic properties is better than that of the conventional system. The technique to enhance uncertainty further is also elaborated. It can be easily seen from the analysis that increasing the number of key bits increases the security.

References

1   Shannon.C, Communication theory of secrecy systems, Bell Systems Technical Journal, vol.28, 1949

2   Heys, H. "A Tutorial on Linear and Differential Cryptanalysis", Cryptologia, July 2002.

3   Musa, M.;Schaefer.;and Wedig, S. "A simplified AES Algorithm and Its Linear and Differential Cryptanalysises." Cryptologia, April 2003.Phoon, K. K. & Kulhawy, F. H. (1999). "Characterization of geotechnical variability", Canadian Geotechnical Journal, 36(4), 612-624.

4   Sinkov,A. "Elementary cryptanalysis: A Mathematical Approach". Washington,D.C.: The Mathematical Association of America, 1966

5   Jithendra.K.B, Shahana. T.K, "Hardware Efficient Parallel Substitution Box for Block Ciphers with Static and Dynamic Properties", Procedia Computer Science,Elsevier,Vol.46,pp.540-547, 2015

6   Heys, H.M and Tavares, S.E. Substitution Permutation Networks resistant to differential and linear cryptanalysis, Journal of Cryptology, pp 1-19, 1996

7   Shyamala C K, N Harini, T R Padmanabhan, Cryptography and Security, Wiley India Pvt ltd, New Delhi, 2011

Piyush Shegokar [1], Manoj V. Thomas [2] and K. Chandrasekaran [3]

Load Balancing of MongoDB with Tag Aware Sharding

Abstract: In the era of Big Data, traditional databases face many problems to process data effectively. To solve problems we have NoSQL databases which can solve the problems and are becoming very popular. MongoDB is a document oriented NoSQL database which can process large data sets with the help of sharding mechanism. MongoDB supports range-based sharding, hash-based sharding and tag aware sharding. Problem with range-based method is that one of the shard will have more load. In case of hash-based method, we can distribute the load among shards but related data aren't together. Tag aware sharding method is better than the other two methods. But this method need to be enhanced as tagged data can belong to more shards and balancer have to migrate the data to most appropriate shard, hence we are enhancing the tag aware sharding in MongoDB. Improper distribution of data can't allow us to use all the benefits of sharding. Tag aware sharding is administrator based method in which tags are mentioned by the administrator. According to tags, data is migrated but we need to balance the load. To solve this problem, Weighted Round Robin (WRR) load balancing algorithm is used and it has improved the writing and querying performance.

Keywords: Template; Load Balancing; MongoDB; NoSQL; Sharding; WRR

1 Introduction

Big Data is characterized by 4 V's-volume, variety, velocity and value. Huge volume of data is generated by Social networking organizations, E-commerce

1 Department of Computer Science and Engineering, National Institute of Technology Karnataka, Surathkal, Karnataka, India-575025
piyushshegokar13@gmail.com
2 Department of Computer Science and Engineering, National Institute of Technology Karnataka, Surathkal, Karnataka, India-575025
manojkurissinkal@gmail.com
3 Department of Computer Science and Engineering, National Institute of Technology Karnataka, Surathkal, Karnataka, India-575025
kch@nitk.ac.in

giants, Financial companies, etc. These organizations and industries generate data of size terabytes or gigabytes or exabytes which is very large amount of data [4]. Data generated by organizations is of different varieties, it can be structured, semi-structured and unstructured. The velocity of generating data is very high. As per the report given by International Data Corporation (IDC) in 2011, the total existing and copied data is 1.8 ZB and this figure will be nine times in the next 5 years [6]. In the case of Big data, value means the cost of technology and the value derived from the use of Big data. This is important factor as value refers to benefits derived from Big data and these benefits can be capital cost reduction, operational efficiency and business process enhancements [3].

Cloud computing has emerged to solve the problems of large data processing, storage and extracting it. Cloud computing has the capability to provide on demand resources and services [2]. Now cloud providers deliver Infrastructure as a Service(IaaS), Platform as a service(PaaS) and Software as a service (SaaS). This technology has reduced the capital cost, people can access their work worldwide using internet, anytime accessibility is possible [13]. From long time relational databases were used by different organizations. Relational databases are widely being used by organizations to store customer and employee records. Using relational database, we can store structured data in the form of tables and in a more organized way. The drawbacks of a relational database are that it needs maintenance; has a limited number of records and fields; images, multimedia, numbers and design cannot be categorized. Day by day size of data is increasing tremendously, relational databases are not able to process large data in less time and takes more cost to process it [9].

NoSQL or 'Not Only SQL' are the new representation of database class which has the ability to solve increasing volume, variety, velocity and value of Big Data [8]. NoSQL databases can be of type Key-Value, Document oriented, Column based and Graph based. NoSQL supports large volume of structured, semi-structured and unstructured data. These databases are more efficient, flexible and easy to use [11]. Before using relational databases, we need to design schema and we are restricted to same features while NoSQL supports dynamic schema. MongoDB is the document oriented database released by 10gen as an open source project [5]. MongoDB is highly used because it provides high availability in unreliable environment and supports for growing data. MongoDB supports Sharding method which has the ability to store large data on multiple machines. A database is partitioned into parts and these parts are stored on multiple machines [1]. Large data sets and high throughput applications can challenge the ability of a single server. High query rates can exhaust the CPU capacity of the server. Larger data sets exceed the storage capacity of a single machine. To solve such issues MongoDB uses sharding method. In this paper, tag-aware sharding with Weighted Round Robin load balancing algorithm is implemented. This paper discusses different sharding methods used in MongoDB, focuses on tag-aware sharding. Weighted Round Robin load balancing algorithm is used to balance the load among servers. This balancing strategy improves the clusters reading and writing performance.

The rest of the paper organized as, we compare the related work of MongoDB sharding in section II. Section III discusses the problem statement. Different types MongoDB sharding are explained in section IV. Section V consists of the proposed method, this contains framework and the load balancing algorithm used for the implementation. Results obtained for improved writing and querying performance are discussed in section VI and section VII concludes the paper.

2 Related Work

10gen has developed MongoDB as an open source project, previously relational databases were used for managing databases. Authors have inserted records into MongoDB and MYSQL database for comparing its querying time [17]. And they found that MongoDB has more advantages in inserting and querying data. Cloud providers provide different open source storage platforms for storing data. Hadoop, abiCloud, Cassandra and MongoDB has been compared on the basis of deployment, scalability, storage mechanism, cloud character and cloud form [7]. MongoDB is a document oriented database which has great flexibility to store large heterogeneous data. This database can be adopted in fast saving, searching temporal streams and real time playback [12].

MongoDB supports Auto-sharding feature by which data is partitioned and stored into multiple machines. The authors had proposed an algorithm based on the frequency of data operations in order to solve uneven distribution of data in auto-sharding [10]. They have considered operations insert, find, update, delete. But only first three operations insert, find and delete affect the performance of the system. Also, they have proposed balancing strategy to balance the load between the servers.

Overview of distributed databases and their features are discussed by the authors [11]. When requests of large data are sent to a relational database, they are not able to process it efficiently. To solve this problem authors has suggested to use NoSQL over relational databases. They have compared searching and indexing time of MongoDB with that of MySQL. Their analysis concludes that sharding and configuration in MongoDB have higher importance to get higher throughput. The authors had proposed heat based dynamic load balancing technique [16] to balance the load among servers. They had considered utilization of CPU, memory and bandwidth as parameters to find out overloaded and underloaded node. They had discussed VM overload, VM underload, Physical overload and Physical underload balancing methods.

The round robin scheme has been used for assigning jobs sequentially. There were other schemes which were used for fair queuing but has complexity of order O(log(n)). Authors had developed Deficit round robin scheme of order O(1) which fairly queue the jobs [14]. Weighted Round Robin strategy is extended version of the Round Robin scheme, which is used for cloud web services for balancing the load [15].

Specialists have concentrated on the information relocation issue in a re-balancing situation. They attempt to distinguish a better technique as indicated by the expense or the pace of information movement. Currently MongoDB using two sharding techniques, these are range-based sharding and hash-based sharding [1]. Issue with range based method is that one of the servers will have more load as most of the queries are serviced by it and others not. The hash-based technique can divide the load, but related will not be together as sharding is done on the basis of hash function.

3 Problem Statement

Auto sharding can reduce the load as it automatically partitions the data and migrate the data to other nodes. Currently MongoDB is using two sharding techniques. These are Range-based sharding and Hash-based sharding. The issue with the range-based method is that one of the servers will have more load as most of the queries are serviced by it and others not. Hash-based technique can divide the load, but related data will not be together as sharding is done on the basis of hash function. The Tag-based approach is better than these two, as partitions depend on tag defined by administrator. MongoDB runs two background process: splitting and balancer, which ensures balanced cluster. Balancer migrates data to appropriate shard. If the tag is updated then balancer has to migrate large data between shards, it will consist of many read, write, searching operations. This directly affects the performance of the system. To solve this problem, we need a balancing strategy to balance the load between servers.

4 Sharding in MongoDB

Sharding is the method of dividing databases into smaller parts which can be easily manageable. It can be used to reduce the number of operations and data stored on a partition. The MongoDB sharding is shown in the Figure 1.
Sharded cluster consists of following parts:

    (1) Shard: It stores the actual data. In case of availability and consistency each shard has a replica set.

    (2) Query Routers: Query routers or Mongos acts as intermediate between client and shards. It processes the client request, performs the operation on shards and return results to client.

    (3) Config Servers: These are used to store metadata which is mapping between cluster's data and shard.

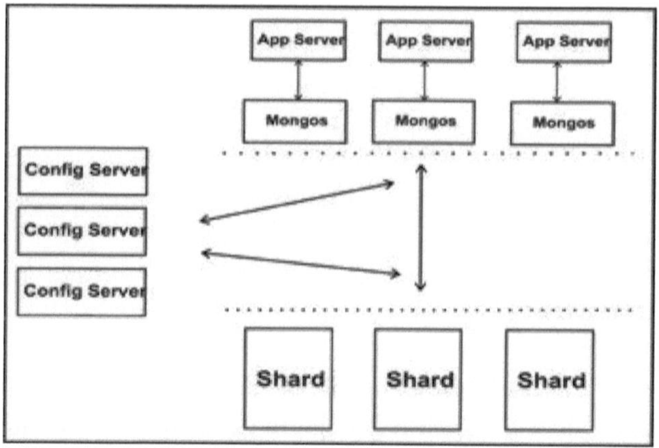

Fig. 1: MongoDB Sharding

The MongoDB supports different sharding methods. They are:
(1) Range Based Sharding: MongoDB partitions the data sets into ranges determined by the shard key values to provide range based portioning. Here we are considering numeric shard key. MongoDB partitions into small and non-overlapping ranges called chunks where a chunk is in between values from some minimum value to some maximum value. Close documents will be in same chunks, and hence on the same shard. Range based sharding is shown in the Figure 2.

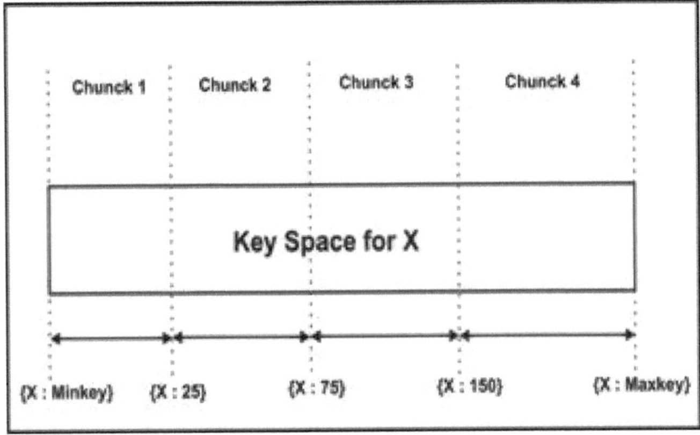

Fig. 2: Range Based Sharding

(2) Hash Based Sharding : In this method we have defined hash functions and then using hashing to create chunks. Two documents with close shard key values are unlikely to be in the same chunk. This ensures a more random distribution of a collection in the cluster. Random distribution can be seen in the Figure 3. Suppose there are 3 documents with keys 25,26 and 27. These will be stored in different chunks depending on the hash function used.

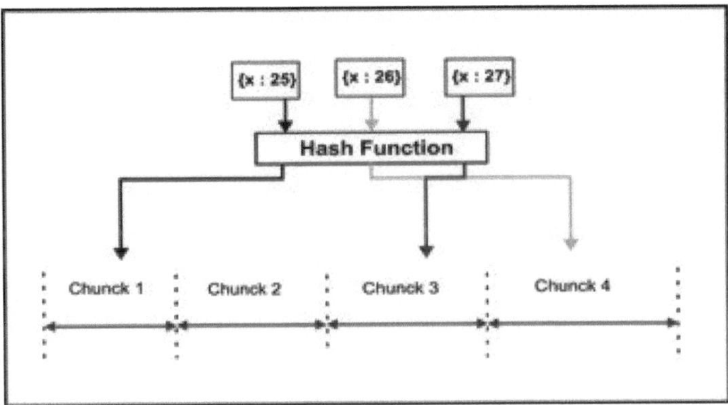

Fig. 3: Hash Based Sharding

(3)Tag aware Sharding: Balancing can be done using tag based partitioning. Administrators can create and assign tags with the ranges of the shard key, and then tags are assigned to the shards. Then, the balancer migrates data to the appropriate shards depending on tag and it ensures that the cluster always enforces the distribution of data that the tags describe. Using this method balancer can control the distribution of data.

Sharding divides the large data sets into multiple servers or shards. To shard the large data set we need to have shard key. Shard key is an indexed field or indexed compound field present in every document in the collection.

Using range based partition we are able to divide the large data, but in uneven distribution manner. Related data will be together, but most the request will be answered by some shards. This may deny benefits of sharding. In hash based method, we will have a random distribution of data depending on the hash function. Tagged data can belong to more than one shard. Balancer need to take care of migrating data to the most appropriate tag. Updating shard can create a mess in tag based method.

5 Proposed Method

The MongoDB framework consists of routing group, replica management group and decision making group. We have reformed this framework specially decision making group. Our load balancing algorithm is able to balance the load among the servers and it has improved the writing and querying perfor- mance. The proposed framework is shown in the Figure 4.

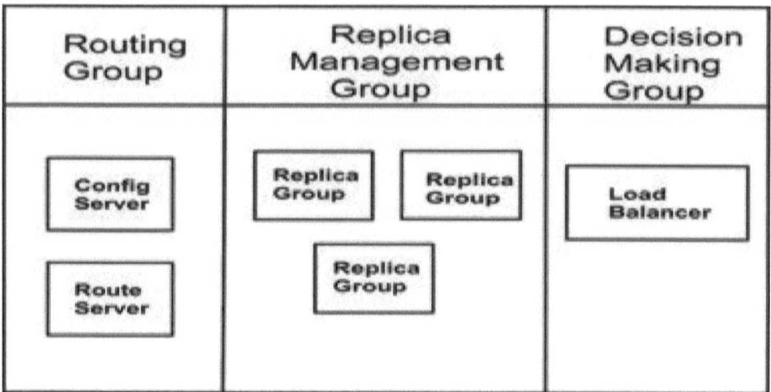

Fig. 4: Proposed Framework

5.1 Proposed Framework

Our proposed framework consists of routing group, replica management group and decision making group. Route decisions are made by routing group which consists of config server and route server. The Replica Management Group deals with managing replica sets. The Decision Making Group consists of load balancer which has improved the writing and querying performance. The proposed framework is discussed below.

(1) Routing Group: In MongoDB, the mongos instances acts as routing group. The application is not directly connected to shards, mongos instances are used for querying and writing to the shards. Routing group acts as intermediate between application and shards. Config server and Route server are the two building blocks of routing group. The Config Server stores the metadata between key range and servers.

(2) Replica Management Group: MongoDB is a distributed database, data is stored on multiple shards. Every shard acts as a replica group with primary and secondary nodes. The role of primary node is to write, update and read request while secondary node can serve read request only.

(3) Decision Making Group: The decision making group consists of load balancer which is used to balance the load between servers. In MongoDB, to balance the load between servers of a cluster, migrates the chunk between shards. But our load balancer avoids the chunk migration and balances the load between servers.

The server which stores the metadata, and the route server which has the routing information, are part of the routing group. The Replica Management Group consists of 3 replicas of data sets. The Decision Making Group consists of a load balancing algorithm. To achieve the objective, we need to implement tag-based sharding. This can be done by creating 3 config servers, then start these config server instances. After this, we can add shards to cluster and perform sharding on the database.

The next section describes the design strategy mentioned for load balancing. The capability of a server is challenged when all the requests are serviced by it, if jobs arrived to single server then its queue size increases. Hence server becomes overloaded, there is the need to balance the load among several servers. Balancing load among servers can increase the availability, improve resource utilization and can improve the overall performance of the cluster.

First, we need to find the capacity of servers, this can be identified from the factors CPU, memory, bandwidth of the server. Integrating all these factors of server, we can find the capacity of the server. The parameters used for finding the capacity are shown in the Table 1.

Table 1. Capacity Parameters for the Server

Parameter	Description
C_c	Capacity of CPU
C_m	Capacity of Memory
C_b	Capacity of Bandwidth
W_i	Weight of Server i

Integrated capacity of server can be found by the following equation:

$$W_i = C_c + C_m + C_b \tag{1}$$

The working of the Weighted Round Robin (WRR) load balancing is shown in the Figure 5. WRR load balancer acts as intermediate between clients request and servers. Server having high capacity serves the more request than less capacity servers.

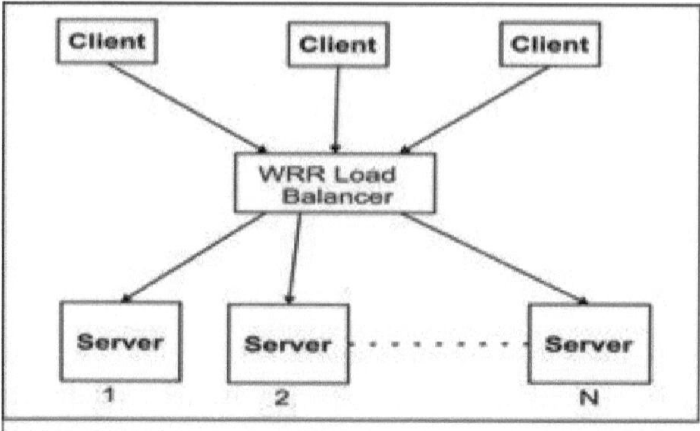

Fig. 5: Weighted Round Robin

5.2 Weighted Round Robin Load Balancing Algorithm

We are using Weighted Round Robin (WRR) load balancing algorithm to balance the load among Cluster. Input to this algorithm is the weight of the server and depending on the weight, number of requests are served by the server. WRR is shown in the Algorithm 1 below.

Algorithm 1 Weighted Round Robin

  Let S_i be the *Server i*
  Let W_i be the Weight of server i
  Let Q be the Request Queue
  Let $Total$ be the sum of Server Weights
  Initialize $Total$=0
  for $i \leftarrow 0$ to $N-1$ **do**
    Calculate the W_i for all S_i
    $Total \leftarrow Total + W_i$
  end for
  while TRUE **do**
    if Nonempty Request queue exist **then**
      for $i \leftarrow O$ to $N-1$ **do**
        Assign Requests W_i to S_i
      end for
    end if
    Dequeue Q
    $Q=Q\text{-}Total$
  end while

Here, W indicates the weight of the server, i is the server selected last time, and i is initialized to zero.

WRR is the extension of the Round Robin scheduling algorithm with weights. The Round Robin strategy is to distribute the load sequentially among the servers in a cluster. To implement this method, it is considered that all servers have equal resources and host identical applications. If the servers are not identical, then we can't use round robin methods to balance the load, and hence the WRR comes into the picture. The drawback of round robin is that weaker server gets a lesser chance to service the request, hence it becomes overloaded. WRR method has ability to balance out the weakness of the simple round robin. The weight of the server determines the number of requests served by the server. Server having more weight servers more request sequentially than server with less weight.

6 Experimental Results

6.1 Experimental Setup

In our experimental setup, we are using MongoDB of version is 2.4.11. To implement the proposed framework, we need to create shards. To do this, we have setup Auto-sharding cluster and Tag aware sharding cluster with 3 shards. And each of these shard has replica set. We have implemented Tag aware sharding with WRR and compared it with Auto-sharding of MongoDB. The database used for results is the Zip data, consisting 5 fields id, city, location, population and zip code.

6.2 Results and Analysis

To test the writing and querying performance, we have inserted Zip data to both Auto-sharding cluster and Tag aware sharding with the WRR cluster with the same number of concurrent number.

Writing performance of the cluster is evaluated by inserting the records into MongoDB database. Here we insert different number of records of the Zip data using tag aware sharding and as the concurrent number varies the writing performance also varies. For a less concurrent number and large number of record insertions, we get high writing performance. And as the concurrent number increases, the writing performance of auto-sharding of MongoDB decreases. But our tag aware sharding with WRR performs better than auto-sharding of MongoDB. Results of writing performance are shown the Figure 6, where the x-axis indicates the number of insertions and the y-axis indicates the concurrent number.

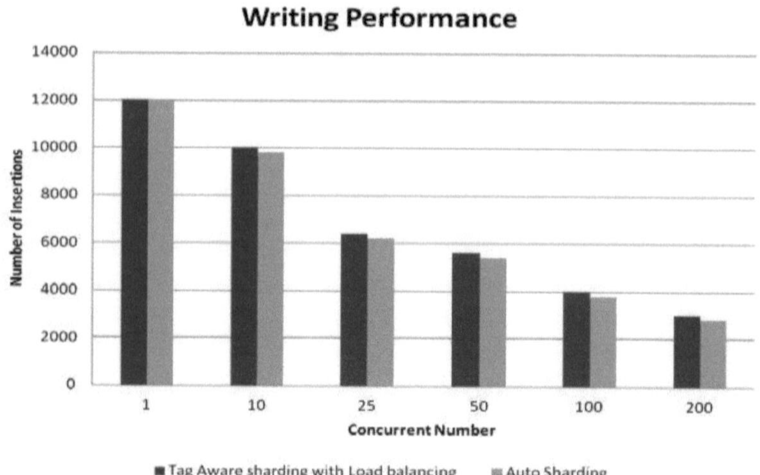

Fig. 6: Writing Performance

Querying performance of the cluster is tested on the inserted records of MongoDB database. Here we inserted a different number of records of the Zip data using tag aware sharding and as the concurrent number varies the querying performance also varies. For a large concurrent number and record insertions, we get high querying performance. The concurrent number and records increases, the querying performance also increases, but our tag aware sharding with WRR performs well than auto-sharding of MongoDB. The results of writing performance are shown the Figure 7, where the x-axis indicates the number of data queries and the y-axis indicates the concurrent number.

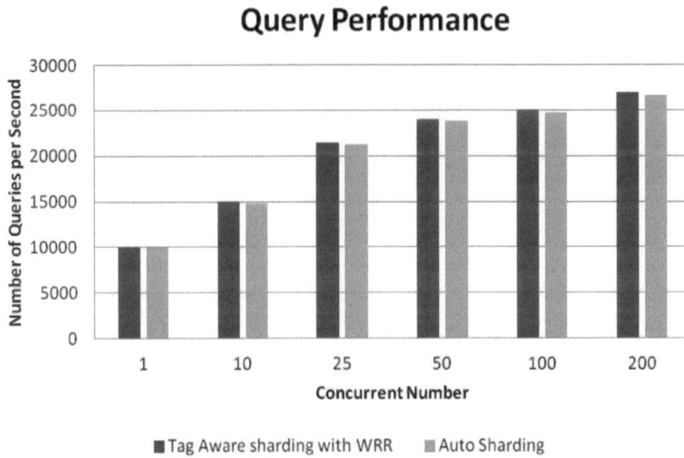

Fig. 7: Querying Performance

Conclusion and Future Work

The MongoDB makes use of sharding methods so that it can process large data sets and can perform high throughput operations. Currently, MongoDB is using the range-based sharding and hash-based sharding. Range-based method results in uneven distribution of data, while hash based method results in the random distribution of data. Due to improper distribution of data, we can't use all the benefits of sharding. Tag based sharding is administrator based method in which tags are mentioned by the administrator. According to tags, data is migrated but we need to balance the load.

Weighted round robin load balancing algorithm is used to balance the load between shards. In the experiment, we have inserted Zip data into the Auto-sharding cluster and Tag aware sharding cluster, then their writing and querying performance are compared. Tag aware sharding cluster with WRR has shown improvements in writing and querying performance. In the future, we are implementing this mechanism for real-time applications.

References

1   10gen. MongoDB. https://www.mongodb.org/, 2015.

2   Michael Armbrust, Armando Fox, Rean Griffith, Anthony D Joseph, Randy Katz, Andy Konwinski, Gunho Lee, David Patterson, Ariel Rabkin, Ion Stoica, et al., A view of cloud computing, Communications of the ACM, 53(4): 50–58, 2010.

3   Dan Vesset Carl W. Olofson, Big data: Trends, strategies, and sap technology, IDC, 2012.

4   Min Chen, Shiwen Mao, and Yunhao Liu, Big data: A survey, Mobile Networks and Applications, 19 (2): 171–209, 2014.

5   Kristina Chodorow, MongoDB: the definitive guide, O'Reilly Media, Inc, 2013.

6   John Gantz and David Reinsel, Extracting value from chaos, IDC review, (1142): 9–10, 2011.

7   Genqiang Gu, Qingchun Li, Xiaolong Wen, Yun Gao, and Xuejie Zhang, An overview of newly open-source cloud storage platforms, in Proc. IEEE International Conference on Granular Computing (GrC), pp. 142–147, 2012.

8   Jing Han, E Haihong, Guan Le, and Jian Du, Survey on nosql database, in Proc. 6th IEEE international conference on Pervasive computing and applications (ICPCA), pp. 363–366, 2011.

9   Yishan Li and Sathiamoorthy Manoharan, A performance comparison of sql and nosql database, in Proc. IEEE Pacific Rim Conference on Communications, Computers and Signal Processing (PACRIM), pp. 15–19, 2013.

10 Yimeng Liu, Yizhi Wang, and Yi Jin, Research on the improvement of mongodb auto-sharding in cloud environment, in Proc. 7th IEEE International Conference on Computer Science & Education (ICCSE), pp. 851–854, 2012.

11 Suyog S Nyati, Sanjay Pawar, and Rajesh Ingle, Performance evaluation of unstructured nosql data over distributed framework, in Proc. IEEE International Conference on Advances in Computing, Communications and Informatics (ICACCI), pp. 1623–1627, 2013.

12 Haller Piroska, Farkas Gyula, and Szanto Ioan-Cosmin, Data storage for smart environment using non-sql databases, in Proc. IEEE International Conference on Intelligent Computer Communication and Processing (ICCP), pp. 305–308, 2012.

13 Bhaskar Prasad Rimal, Eunmi Choi, and Ian Lumb, A taxonomy and survey of cloud computing systems, in Proc. Fifth IEEE International Joint Conference on INC, IMS and IDC, NCM'09, pp. 44–51, 2009.

14 Madhavapeddi Shreedhar and George Varghese, Efficient fair queuing using deficit round-robin, IEEE/ACM Transactions on Networking, 4(3), pp. 375–385, 1996.

15 Weikun Wang and Giuliano Casale, Evaluating weighted round robin load balancing for cloud web services, in Proc. 16th IEEE International Symposium on Symbolic and Numeric Algorithms for Scientific Computing (SYNASC), pp. 393–400, 2014.

16 Xiaolin Wang, Haopeng Chen, and Zhenhua Wang, Research on improvement of dynamic load balancing in mongodb, in Proc. 11th IEEE International Conference on Dependable, Autonomic and Secure Computing (DASC), pp. 124–130, 2013.

17 Zhu Wei-ping, Li Ming-Xin, and Chen Huan, Using mongodb to implement textbook management system instead of mysql, in Proc. 3rd IEEE International Conference on Communication Software and Networks (ICCSN), pp. 303–305, 2011

Dr. R. Arulmozhiyal[1], P. Thirumalini[2] and M. Murali[3]

Implementation of Multilevel Z-Source Inverter using Solar Photovoltaic Energy system for Hybrid Electric Vehicles

Abstract: This paper extends a topology for cascaded multilevel Z source inverter employs of z source inverter unit block with bidirectional switches. The conventional z source inverter consist network impedance with multiple indictors and capacitors arranged Z type topology performs either buck or boost the input voltage for non linear load depending upon duty ratio and modulation index in a multiple stage conversion with the help of impedance source passive network (L and C Z source inverter reduces harmonic distortion and overcomes limitations of conventional inverters. The proposed multilevel Z source inverter reduces harmonics further and provides a novel topology for power conversion. Capability of proposed configuration in producing all odd and even output voltage level source with input to the power source and increase the power efficiency is proved by simulation result for 3-level inverter. The output of the proposed system is simulated using MATLAB-SIMULINK and implemented in real simulation hardware for electric vehicles using dSPIC controller.

Keywords: ZSI, total harmonic distortion, MATLAB, multilevel z-source inverter,dSPIC

1 Introduction

Drastically change in environment and increase in oil pricing leads to carry effective research work in hybrid electric vehicles. Renewable energy sources are naturally available huge amount in India; such has wind, photo-voltaic (PV)

1 Professor, Department of Electrical & Electronics, Sona College of Technology, Salem, India
arulmozhiyal@gmail.com

2 PG Scholar, Department of Electrical & Electronics, Sona College of Technology, Salem, India
thirumalinip@gmail.com

3 Research Scholar, Department of Electrical & Electronics, Sona College of Technology, Salem, India
muralimunraj@gmail.com

and fuel cells and being cheaper. Nowadays, renewable energy sources are widely used in industrial and residential applications to reduce power scarcity and economic welfare. In particular, photo-voltaic systems attracts due to it's very often maintenance, free from pollution and zero fuel cost. In present scenario the energy generation through PV cell has grown rapidly increased from 40% to 45% per annum over the past 20 years. This is mostly because of development in nano-technology field which reduces the PV panels initial cost and build-up [1-2].The power electronic converter and inverter are other prominent areas which bring the significant improvement in overall generation system. Power electronic converter and inverter improve efficiency with switching sequence. Voltage and Current source inverter increases the harmonic contents with third and seventh harmonic distortion. Due to the apparent inherent advantages in Z source inverter, it is used in both voltage of buck and boost, it has been taken up for various applications, Z source type of inverter variable Z type performs efficient power transfer between dc to ac. In a conventional Z source inverter performs bidirectional ZSI has been proposed with better steady state operating character tics compare with conventional converters. In the centralized configuration, Z source inverter with inductance and capacitance values [4-5]. The Z-source increase efficiency connected individual to source maximizing the possible energy harvesting. Therefore, reducing the output of a single z source inverter has a minimal impact on the overall system performance. With the use of the topology as that of a conventional VSI, a voltage-type Z-source inverter can presume all active (finite output voltage) and null (0V output voltage) switching states of VSI. Voltage-type Z-source inverter has the unique feature of performs power switches of a phase-leg to be turned ON simultaneously (shoot-through state) without damaging the inverter [1] In addition the reliability of the inverter for effective solar power generation due to shoot through state caused in Z source parameters by Electro Magnetic Interference (EMI) noise can reduce the circuit power delivery for longer periods[7]. The appropriate switching permits and addition of capacitor voltages or inductor in the output port reduce the harmonic of THD. Z source inverter has advantages in multi-levels has the following merit: (i) sinusoidal output waveform, (ii) a efficient filter size and (iii) a minimized EMI noises. The main advantage of using Z source inverter is to reduce third and seventh order harmonics in the output voltage. Multi generation PWM strategy is adopted in modulating strategy for inverter. Multicarrier PWM is several cycle of triangular carrier signals are compared with a cycle of sinusoidal modulating signal. The number of carriers is necessary to produce m-level output is (m-1). All carriers have the similar peak to peak amplitude Ac and frequency fc [10].The main objective of this paper is to implement multilevel z-source inverter for electric vehicles interfacing renewable power generation system as shown in Figure.1.

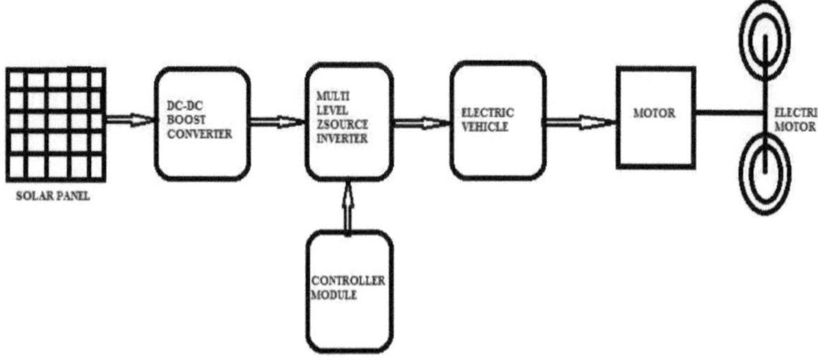

Figur e 1. Block diagram of proposed system

The function of multilevel z-source inverter is to synthesis a decided output ac voltage from several levels of dc input voltage with minimum harmonic distortion. The magnitude of output ac voltage of conventional inverter is limited to dc source voltage and with more harmonic distortion and this limitation can be resolved by cascaded multilevel z-source inverter. The proposed multilevel z-source inverter shoots through to boost dc link voltage. A comparison analysis has been made with conventional z-source and multilevel z-source network in MATLAB/SIMULINK environment and its performance has been analysed using real time controller.

2 Photovoltaic Cell

The equivalent model of a photovoltaic cell can be represented by figure 2 as shown below. To obtain maximum power from solar are connected in series or in parallel, which forms a module. Again, these types of modules are connected in parallel or in series to get required voltage and current. The characteristics of PV cell can be derived using the equation given below

$$I = I_{pv} - I_o \left[e^{\frac{V+R_sI}{V_t a}} - 1 \right] - \left(\frac{V+R_sI}{R_p} \right) \tag{1}$$

Where I_{pv} = photo voltaic current, I_0 = Nil or saturation current, $V_t = N_s kT/q$, array thermal voltage, N_s= pv cell connected in series, T = PV panel diode temperature, k = constant of Boltzmann, q = electron charge, R_s= equivalent resistance of the series connected array, R_p= equivalent 1 resistance of parallel connected array, a = ideality constant of diode. The resistance connected in series (R_s) can be regulated from the value of shunt resistance R_p either high or low when compared to the value of series resistance R_s. The saturation current generated by the light is linearly depends upon the solar irradiation from sun and temperature which, influence the generation of photovoltaic current in the solar cell which is derived by the following equation.

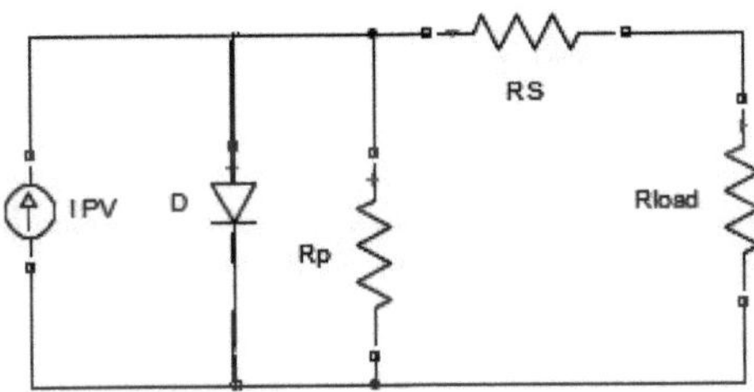

Figure 2. Photo Voltaic cell equivalent circuit

$$Ipv = [\,Ipv,n + K_I\Delta T\,]\,\frac{G}{Gn} \tag{2}$$

Where I_{pv}= current generation of photo voltaic cell at the nominal condition($25°$C and W/m^2) , $\Delta T = T - Tn$, T = temperature in actual condition [K] , T_n = nominal temperature[K] , K_I = current coefficients of PV cell, G = irradiation of the device surface[W/m^2] ,G_n = nominal irradiation.

2.1 DC – DC Boost Converter

The power generated by the PV panel fed in to converter for domestic user applications. The power obtained from PV source is unregulated and required to maintain optimum outputs from the converter. The energy will fed to the load with the dual stage power electronic system comprising as converting medium to boost power as type of DC-DC converter and inverter. In order to maintain a constant voltage to the load, a DC-DC step up converter is introduced in process of power generation. The Photovoltaic array and the inverter needs maintain constant power output to domestic consumers. The conduction mode of the DC-DC converter can be given by the following equation.

$$\frac{Vo}{Vin} = \frac{1}{1-D}$$

(3)

Where D = duty cycle, the value of the duty cycle D will always between 0 and 1 and hence it is important to maintain the output voltage higher than the input voltage in magnitude.

3 Z-Source Inverter

An Impedance inverter is termed as Z-source inverter and through its control method to implement operation of dc-to-ac power conversion. The Z-source inverter is a unique resistance capacitance and inductance network to deliver maximum power form solar modules and an inverter circuit, the conventional Z source inverter has drawbacks of more harmonic distortion in power outputs voltage sag and swell. Thus, to overcome drawbacks in conventional Z source inverter a modified multilevel converter is proposed in this topic. The Z-source inverter overcomes the theoretical and hypothetical barriers and restrictions of the traditional VSI and CSI inverters as it provides efficient power conversion techniques. The Z-source concept is also a bidirectional capability inverting circuit applied to all dc-to-ac, ac-to-dc, ac-to-ac, and dc-to-dc power conversion with less quality issues. Z-source inverter for dc-ac power conversion is efficient applications in fuel cell power generation. The Z source-resistance-capacitance-inductance network is to couple the inverter circuit topology in multiple cascaded configurations to the link with power source and thus providing uni-que concept cascaded multilevel feature as shown in figure.3. The control stra-tegies with the insertion of shoot-through states in Z-source inverter with LC filter is analyzed. There were exists two conventional inverters, voltage-source inverter(VSI) and current-source converters(CSI), to perform either rectifying or inverting operation depending on power flow directions. There are some limita-tions in the quality of power output with higher Total Harmonic Distortion (THD).

Total harmonics distortion affects overall output of system. In specific third distortion is standard frequency of harmonics which is generates three times of fundamental frequency harmonics in the system it more severe than to create EMI problems in switching patterns; five times the fundamental is fifth harmonic leads over heating with sag and swell issues;. The harmonics in a system can be defined generally using from the eq.4.

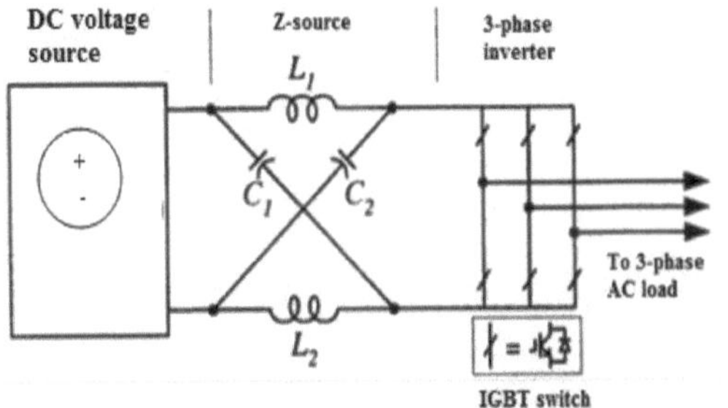

Figure 3. Z source Inverter Circuit configuration

$$f_h = hf_{ac} \tag{4}$$

Where fh is the number of harmonics and fac is the frequency of fundamental system harmonics. In domestic interface power lines higher order harmonics are not given much importance which does create harm to system. The important and harm creating harmonics in system are 3rd, 5th, 7th, 9th, 11th and 13th. The harmonics waveforms in general is given in eq.5

$$v_n = v_{rn} \sin(n\omega t) \tag{5}$$

Where, Vrn is the voltage in rms of particular harmonic frequency (harmonic or power line).

3.1 Analysis of Z source Inverter

The Z-source network operates in "shoot-through zero state" which is possible to provide the unique boost-buck operation in the aspect of inverter [6]. The Z-source inverter is operated in two modes. i) Shoot through mode: In the shoot-through state, switch S7 is triggered and diode Da is off state. The equivalent circuit of shows the shoot-through state with corresponding ON and OFF switches is shown in Figure 4. The operational analysis is expressed as:

$$V_L = V_C \tag{6}$$

$$V_{dc} = V_L - V_C \tag{7}$$

ii)Non shoot through mode: In non-shoot-through mode switch S7 is triggered and as with capacitance and inductance in parallel as shown in Figure 5. Output of LC filter network and inductor voltage can be calculated as:

$$Vin = Vc\text{-}V_L \qquad (8)$$

The boost coefficient factor B is determined by the Modulation Index (MI). The shoot-through zero state is affects PWM control of the inverter. As it generates equivalently the zero voltage to the load terminal and the creates a shoot-through period varying by modulation index. In Z-source inverter is cascaded in to two levels to design a multilevel-Z-source inverter. Advantages of multi level Z Source Inverter are: (i) either increase or decrease in the voltage for PV energy process to the load, (ii) as it eliminates order harmonics leads to reduces the mismatch of switching patterns and EMI distortions, (iii) provides ride-through during power quality issues like voltage sags and swell without need for additional circuits, (iv) improved domestic user power factor (PFC), (v) reduced harmonic current and distortion, (vi) reduced common-mode voltage, (vii) cheaper in implementation, (viii) improves reliability and (ix) highly efficient for cascaded structure.

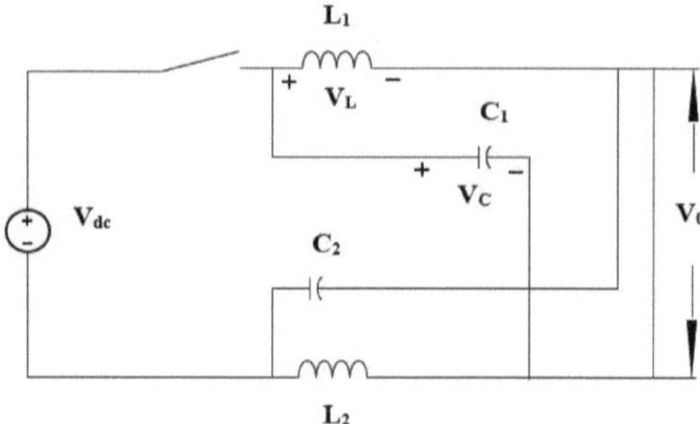

Figure 4. circuit of shoot through mode

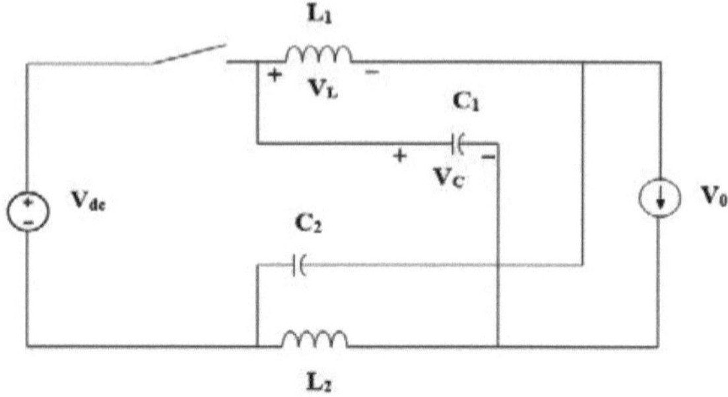

Figure 5. circuit of shoot through mode

3.2 Cascade multilevel Z-source inverter

The topology consists of a series single phase Hbridge inverter units, Z impedances and DC voltage sources. It is supplied from several DC sources, which may be obtained from batteries, fuel cells, solar cells or ultra - capacitors. Each H-bridge inverter can generate three different voltage outputs +Vin, 0, -Vin. The number of output voltage levels in the suggested topology is 2n+1 where n is the number of Z impedances or DC voltage sources. As compared to cascaded H-bridge inverters, the proposed topology has an extra switching state: shoot through state. During the shoot-through state, the output voltages of Z networks are zero. In this paper, each Hbridge is turned in to shoot-through state when the output voltage level is traditional zero therefore some or all of the zero states are changed to shoot-through state. Table 1 indicates the values of V0 for states of switches.

Several modulation strategies have been proposed for multilevel converters. Among these methods, the most common used is the sinusoidal pulse width modulation (SPWM). The principle of the SPWM method is based on a comparison of a sinusoidal reference waveform, with shifted carrier triangular waveforms. Proposed Modulation waveforms to switching of proposed inverter are shown in Fig. 4. For each H-bridge two triangular waveform is assumed, method uses one straight line (VP) to control the shoot-through state time, when one of the triangle waveforms is greater than VP, the related H-bridge turns in to shoot-through state and otherwise it operates just as traditional PWM. For example when C1 or C3 is greater than Vp, then upper staircase in Fig. 3 is turned in to shoot-through state, and when C2 or C4 is greater than VP, then down staircase is changed to shoot-through state. Therefore boost factor of proposed inverter can be controlled by the value of VP. Fig. 6 leads to the relation (9).

$$\frac{T_1}{T_{ac}} = \frac{V_{ac} - V_p}{V_{ac}} \tag{9}$$

Where Vca is the peak value of triangle waveform and T1 is shoot-through time during half of triangles period time, Tca. It is clear that:

$$\frac{T_{sh}}{T} = \frac{T_1}{T_{ac}} \tag{10}$$

boost factor, B, is obtained as:

$$B = \frac{1}{1 - 2\left(\frac{V_{ac} - V_p}{V_{ac}}\right)} \tag{11}$$

It is considered that by decreasing VP, boost factor B increases, as a result the Z network output voltage, i.e. Vinj increases and the load voltage is controlled.

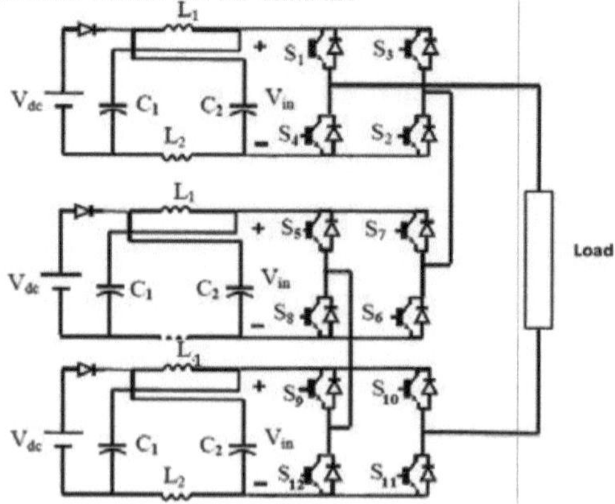

Figure 6. Circuit diagram of a single phase multi level based Z-Source inverter

4 Simulation results

The cascaded 3 level Z source inverter is simulated using MATLAB/SIMULINK tool with Z configuration. The PV cell is modeled from library and simulated with its equivalent circuit. Temperature and irradiation considered as the changing parameters to change the output of the solar cell. Two PV cells of 12V and 24V were taken as input for the multilevel Z source inverter configuration as per Figure 2. The switching circuit consists of reverse blocking switches using MOSFETs and diodes which are taken from SIMULINK library. A two winding 12V/230V is used to connect the multilevel Z source inverter to the load side. Resistive load was considered for the simulation and the switching sequence were provided according to the operation described above. A 50Hz, 20V peak voltage waveform was obtained at inverter output. The simulation diagram of multilevel Z source inverter is shown in figure 6. The output of 48volts from Dc to Dc converter is fed in multilevel Z source inverter to perform multiple boost operation with modulating PWM switching sequence as shown in figure.7. The output voltage and current waveforms are shown in figure 8 & 9 respectively. The output waveform boosts the converter voltage with reduced harmonics and utilized for electric vehicles to drive sensitive load (actuator). The harmonic analysis is carried and shown in figure.10 which enlightens the less distortion and performance analysis is compared with Z source inverter and multilevel Z sourceinverter shows the effective THD values.

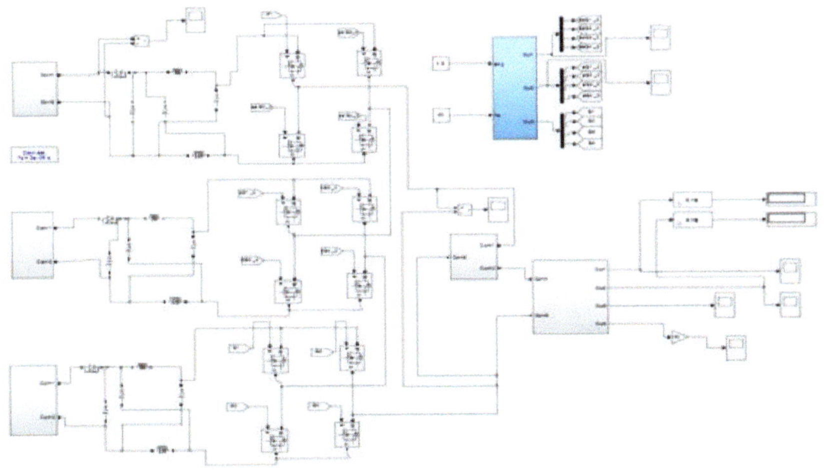

Figure 8. MATLAB circuit for multilevel Z source inverter

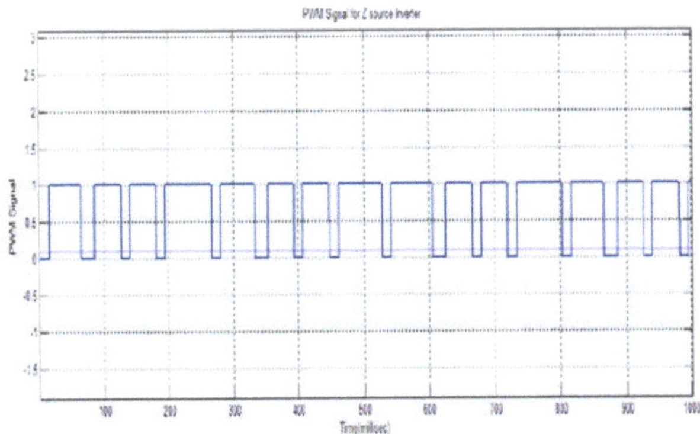

Figure 7. PWM switching sequence for multilevel inverter

5 Real Time implementation of proposed multilevel Inverter

The proposed multilevel inverter is verified with real time dSPIC controller used to interface the software simulation with the hardware MATLAB simulink and

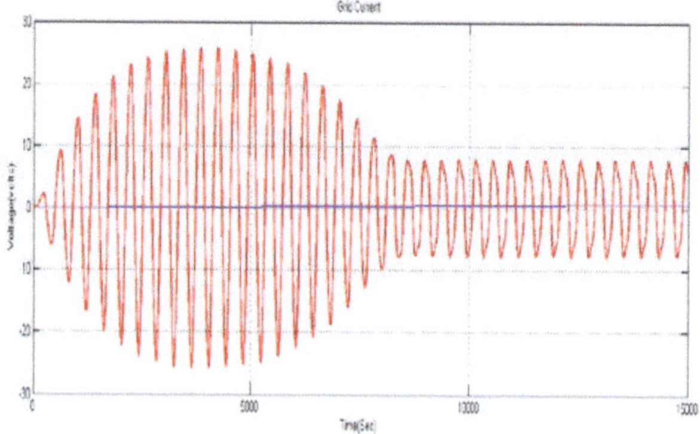

Figure 8. Output current of multilevel z-source inverter

various input/output ports as shown in figure.10. Three photo voltaic panels of different ratings are connected in the input of multilevel inverter.

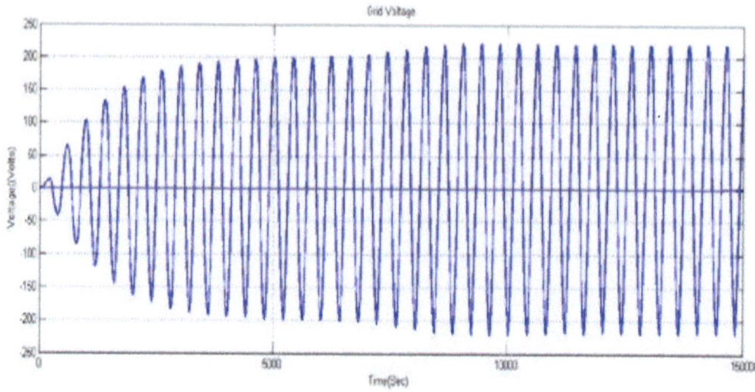

Figure 9. Output voltage of multilevel z-source inverter

The graphically modeled simulation is done in MATLAB/SIMULINK domain for simulating digital pulses and interprets the analog inputs from the inverter.

The dSPIC controller switching gate pulses through opto-coupler circuits to isolate the gate signals from power signals to multi level ZSI as shown in figure.11. The dSPIC controller provides six digital output control signals for six reverse blocking switches. The PC is interfaced with the dSPIC controller for alteration of control signals according to the required output. A single phase 50Hz 12V/230V transformer is connected at the output of proposed dSPIC controller to provide single phase 50Hz, 230V AC output. The output of the inverter is captured back to the dSPIC controller to calculate the Total Harmonic Distortion (THD). The output voltage and current waveforms are shown in figure 12 & 13 respectively.

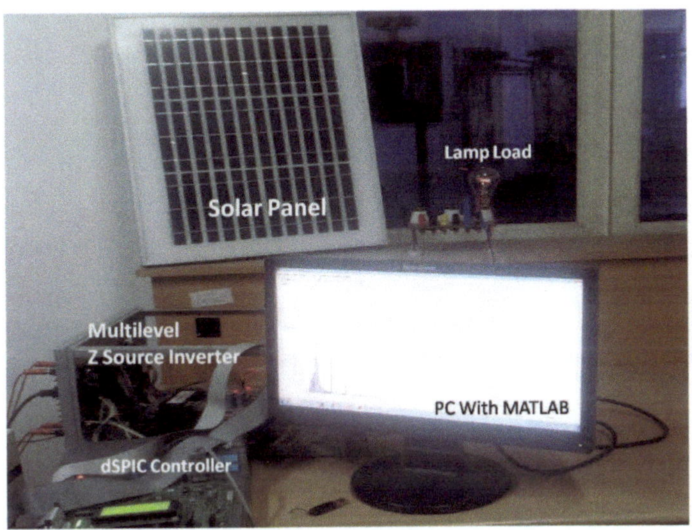

Figure 10. multilevel Z source inverter with dSPIC controller

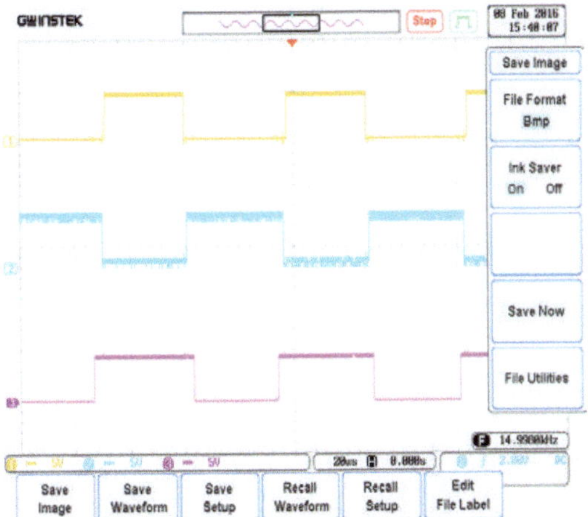

Figure 11. switching sequence of multilevel Z source inverter

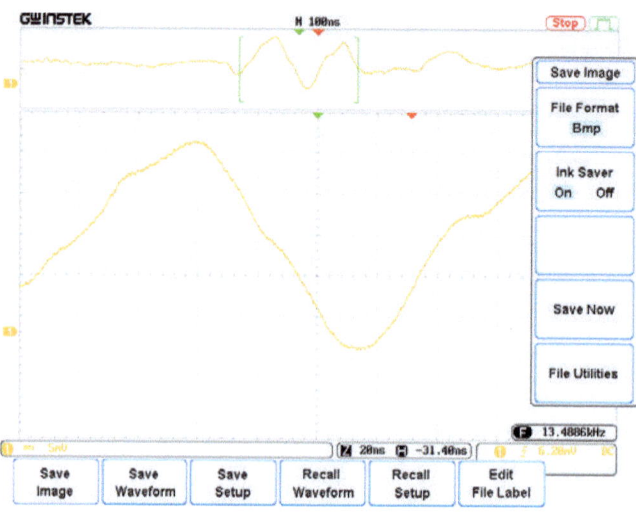

Figure 12. Output current of multilevel z-source inverter

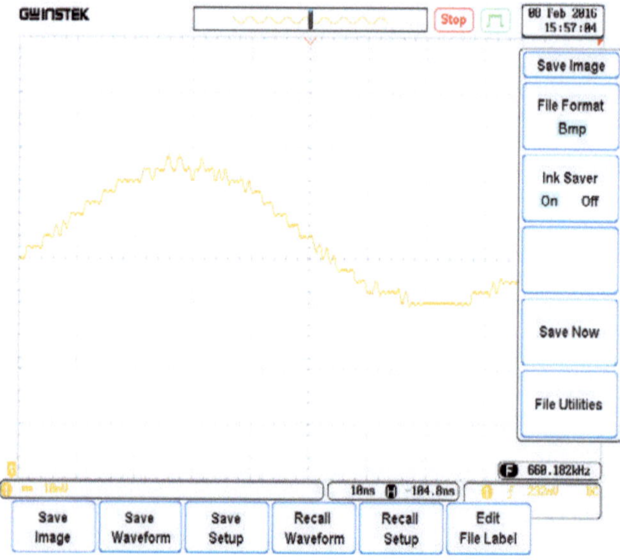

Figure 13. Output voltage of multilevel z-source inverter

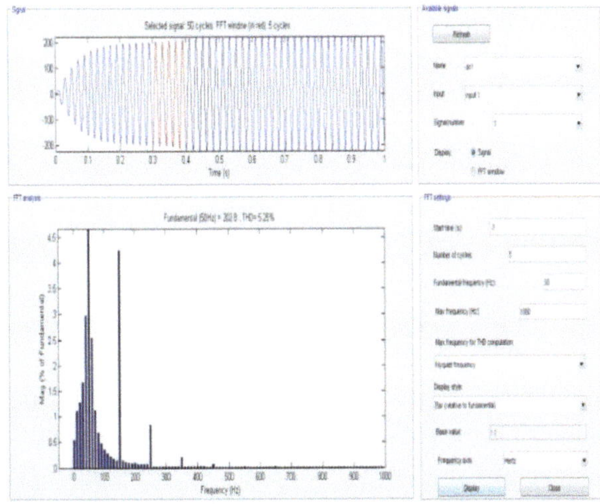

Figure 14. FFT analysis of multilevel z-source inverter

Table 1. THD analysis of Z-source inverters

S.NO	INVERTER	THD% analysis of VOLTAGE in MATLAB simulink	THD% analysis of CURRENT MATLAB simulink	THD% analysis of VOLTAGE in Real time dSPIC controller	THD% analysis of CURRENT In Real time dSPIC controller
1.	Z-Source inverter	18.59%	18.80%	36.58%	38.12%
2.	Multilevel(3 level) z –source inverter	5.25%	4.88%	24.58%	21.95%

The output waveform from multilevel inverter with reduced harmonics and utilized for electric vehicles to drive sensitive load (actuator). Z source inverter and multilevel Z source inverter shows the effective THD values shown in TABLE.1 The voltage and current harmonics range in three phase inverter is 18.59% and 18.80% respectively. Similarly the voltage and current harmonics range in z-source inverter is 10.89% and 10.85% respectively. The voltage and current harmonics range in three phase inverter is 18.59% and 18.80% respectively.

Similarly the voltage and current harmonics range in z-source inverter is 10.89%
and 10.85% respectively. By this, the multi z-source inverter finds to be more
efficient when compared with single stage z source inverter.

Conclusion

A multilevel Z source inverter for PV systems was simulated using
MATLAB/SIMULINK tool. This multi-rated source inverter can have an arbitrary
number of PV sources and is able to obtain the maximum possible power of
each delivery independently. The PV sources can be of different electrical parameters and working conditions. The multilevel Z source inverter is responsible
for boosting the voltage to required output level. Z network capacitors and
inductor on primary side in inverter realize soft-switching operation. The
control scheme and the switching algorithm of the proposed inverter were
described. The output of inverter is validated with THD using MATLAB FFT
analysis. THD of the output waveform was reduced to 10.85% using filter
circuits.

References

1   Fang Zheng Peng, *Senior Member, IEEE*. Z-Source Inverter. IEEE Transactions On Industry
    Applications, Vol.39, No.2, March/April 2003.
2   The Renianssance of Solar Power Panels Energy By Olatunji.Adetunji.
3   SudhirRanjan, Sushma Gupta, Ganga Agnihotri. Department of Electrical Engg, IPEC GZB
    UP, India, Department of Electrical Engineering, MANIT Bhopal, India. International
    Journal of Chem Tech Research CODEN (USA): IJCRGG ISSN: 0974-4290 Vol.5, No.2, pp 993-
    1002, April-June 2013.
4   Suresh.L, G.R.S.NagaKumar, M.V.Sudarsan & K.Rajesh. Department of Electrical &
    Electronics Engineering. Vignan's Lara Institute of Technology & Science, Vadlamudi-
    52213, India.
5   Muhammad H.Rashid, 1993. Power Electronics Circuits Devices and Applications. 2nd Edn.
    Englewood Cliffs, N.J., Prentice Hall.
6   Umesh Sinha, 1994. Network Analysis and Synthesis. Satya prakasan, Incorporating Tech
    India Publications. 5th Edn., New Delhi.
7   Robert,L. and Boylestad Louis Nashelsky,2000. Electronics Devices and Circuit Theory.
    Prentice Hall of India, New Delhi.
8   Sajith Shaik, I.Raghavendar / International Journal of Engineering Research and
    Applications (IJERA) ISSN: 2248-9622 . Vol. 2, Issue 6, November- December 2012. Power

Quality Improvement of Three Phase Four Wire Distribution System Using VSC With a Zig-Zag Transformer.

9   Implementation Of Impedance Source Inverter System For Photovoltaic Applications. Nisha K.C.R & T.N.Basavaraj. Electrical & Electronics Department, Sathyabama University, Chennai, India. Electronics & Communication Department, New Horizon College Of Engineering, Bangalore, India.

10   Power Factor Correction of Non-Linear Loads Employing a Single Phase Active Power Filter: Control Strategy, Design Methodology and Experimentation. Fabiana Pottker and Ivo Barbi Federal University of Santa Catarina. Department of Electrical Engineering. Power Electronics Institute

11   Florianopolis - SC - Brazil.

12   Active Power Filter Techniques for Harmonics Suppression in Non Linear Loads. Kanchan Chaturvedi, Dr. Amita Mahor, Anurag Dhar Dwivedi, Department of Electrical & Electronics Engineering. N.R.I. Bhopal (M.P.) India . Senior System Engineer Infosys.

Sarwade A N [1], Katti P K [2] and Ghodekar J G [3]

Use of Rogowski Coil to overcome X/R ratio effect on Distance Relay Reach

Abstract: This paper presents use of Rogowski Coil (RC) to prevent mal-operation of the distance relay caused by increased X/R ratio of primary fault path. Increased X/R ratio saturates the current transformer (CT) at earlier stage and distorts the current signals fed to distance relay.

Conventional CT can be replaced by RC due to its linear and wide operating range. Distance protection scheme along-with ideal CT, conventional CT and RC is modeled and simulated by using Power System Computer Aided Design (PSCAD) software. In this paper, the X/R ratio of the system is increased by increasing value of source inductance. The results given by use ideal CT, conventional CT and Rogowski coil in distance protection scheme are compared and discussed.

Keywords: X/R ratio, CTsaturation, Rogowski coil, Under reach, Mho relay, PSCAD

1 Introduction

Distance relays are used for protection of high voltage transmission line as they provide more secure, faster and reliable protection. Fault rate is usually much higher in transmission lines as compared to other components of the power system. The occurrence of the different types of the faults produces very large currents in the power system. Conventional CT utilize an iron core and winding ratio to step down these currents to a more manageable level for the secondary devices such as protective relays and meters. [1,2,3] The signal fed to relays and meters gets distorted due to saturation of iron core. In this saturation process most of the transformed current is absorbed by magnetizing branch. So the CT will not be capable for supplying and reproducing the primary component to

1 Research Scholar, Dr. Babasaheb Ambedkar Technological University, Lonere
asarwade@yahoo.com

2 Head of Electrical Engineering Dept, Dr. Babasaheb Ambedkar Technological University, Lonere
pk_katti2003@yahoo.com

3 Retd Principal, Government College of Engineering, Karad

secondary side until it comes out of saturation.[7] The bigger the difference between the primary and secondary current, the less information is given to a distance relay which causes it to under reach. This saturation effect strongly depends on system X/R ratio and the instant at which the fault takes place.[6] One possible solution of this problem is increasing the core size which would allow measurements of a higher current. This is yet unacceptable from commercial and economical point of view. [7]

RC operates on the same principle as conventional iron-core current transformer (CT). The main difference between RC and CT is that RC windings are wound over an (non-magnetic) air core, instead of over an iron core. As a result, RC's are linear since the air core cannot saturate. However, the mutual coupling between the primary conductor and the secondary winding in RC's is much smaller than in CTs. Therefore, RC output power is small, so it cannot drive current through low-resistance burden like CTs are able to drive. RC can provide input signals for microprocessor-based devices that have a high input resistance. As the distance relay require current as one of the input, the RC coil voltage output need to be integrated with the help of integrator. [4,5]

2 Theory of operation

2.1 Distance Protection Scheme

The distance relay collects the voltage and current signals from voltage transformer (VT) and CT respectively and calculates the apparent impedance which is proportional to distance of transmission line up to fault point (figure 1). The relay compare the setting impedance with the calculated apparent impedance to determine if the fault is inside or outside the protected zone and immediately release a trip signal or with certain time delay. [3]Thus the signals received from VT and CT plays an important role for correct operation of distance protection scheme.[1,2]

Current Transformer: The current transformer is used to replicate primary current I_p on secondary side with certain transformation ratio based on the ratings of burden (meters or relays) as shown in figure 2. When the fault occurs, the fault current invariably has DC offset or in other words the AC component is superimposed over DC offset. The DC in CT primary causes the CT core to saturate. In this saturation process most of the transformed current get diverted through magnetizing branch. [4] The saturation of iron core has tremendous

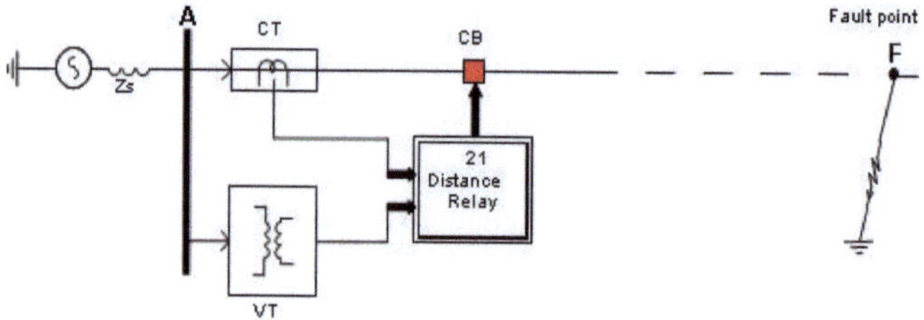

Figure 1. Basic Distance Protection Scheme

implications on operation of distance relay. This relay needs this ac component in first two or three cycles, however it may not be available due to saturation. [2]

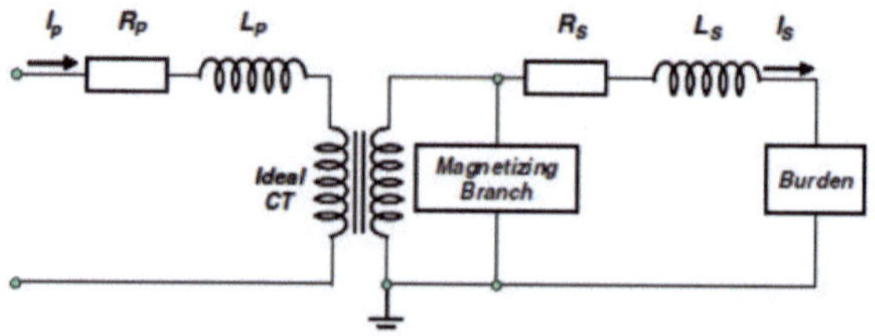

Figure 2. Equivalent circuit diagram of current transformer

Rogowski Coil: Generally RC are torroidal coils with air core (non-magnetic core). The absence of core makes them lighter in weight as compared to current transformer which is having iron core. The air core consideration in the RC construction makes the relative permeability as unity. Rogowski coil is placed around the current whose current is to be sensed. RC arrangement and equivalent circuit is shown in figure 3. [4,5]

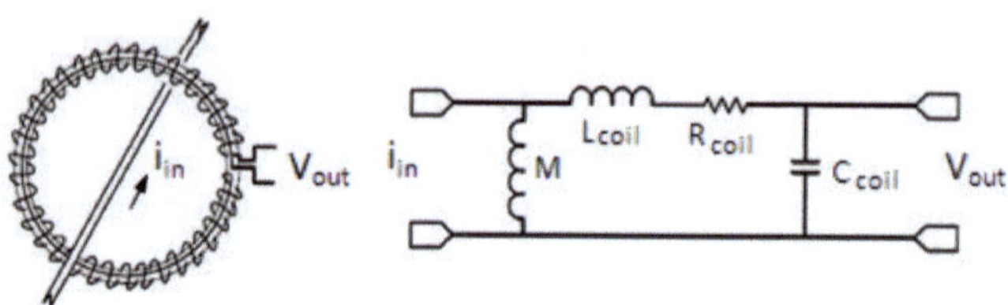

Figure 3. Rogowski Coil and its equivalent circuit

The Voltage induced at the coil output is equal to (1).

$$V(out) = -M \frac{di(t)}{dt} \tag{1}$$

Where, 'i' is the current which is to be sensed and M is the mutual coupling between primary conductor and secondary winding.

X/R ratio: X/R ratio is an important parameter when considering CT saturation because it is responsible for the decaying DC component in fault current. Since this DC component produces an almost constant magnetic flux (in comparison with magnetic flux produced by a 50 Hz sine wave) it significantly contributes to CT saturation. Table I shows values used for this simulation. Since it was desired to show only the influence of the X/R ratio the steady state fault current value had to be the same for every simulation. This can be achieved only when both (2) and (3) are satisfied at the same time. Table I shows values obtained after solving this set of equations.[6]

$$Z_T = \sqrt{(R_s + X_S)^2 + (R_L + X_L)^2} = Constant \tag{2}$$

$$\frac{X}{R} = \frac{(X_L + X_S)}{(R_L + R_S)} \tag{3}$$

Table 1. Line and Source Parameters

| Sr. No. | X/R ratio of the Primary Fault path (with CT burden 0.5Ω) | | | | | | |
	Line Resistance R_L (Ω)	Line Reactance X_L (Ω)	Source Resistance R_S (Ω)	Source Reactance X_S (Ω)	Source Inductance L_S (H)	Total Impedance Z_T (Ω)	X/R ratio
1	2.79	46.76	2.8	32.044	0.102	79	14.09
2	2.79	46.76	2.07	32.09	0.1022	79	16.22
3	2.79	46.76	1.18	32.14	0.1023	79	19.85
4	2.79	46.76	0.02	32.19	0.1025	79	28.09

3 Modeling and analysis of distance protection scheme

Figure 4, gives the stages considered while developing a distance protection scheme

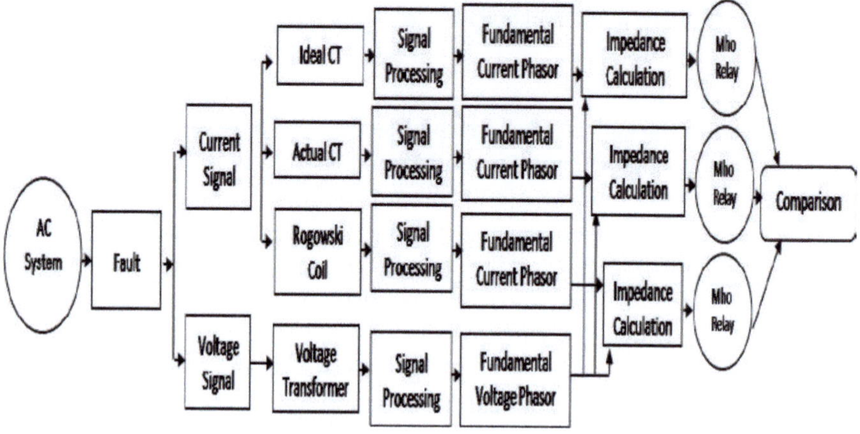

Figure 4. Distance protection scheme stages

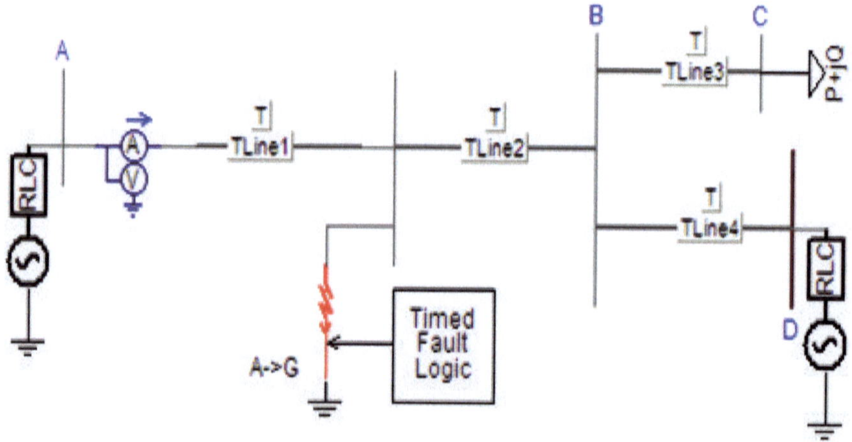

Figure 5. A two source AC power System model

3.1 Modeling of an AC system

An AC power system with two sources is used. The line to neutral fault is created on transmission line AB on which the distance relay is installed (figure 5). The details of the model are given in table II. [8]

Table 2. Details of AC System

Sr. No.	Parameters	Specifications
1	Source Voltage	220 kV, 50Hz
2	Source Impedance	$32.15\angle85^0\ \Omega$
3	System MVA	100 MVA
4	Length of AB	200 km
5	Positive sequence impedance(per km)	$0.2928\angle86.57^0\ \Omega$
6	Zero sequence Impedance (per km)	$1.11\angle74.09^0\ \Omega$
7	Load	(75+j25) MVA
8	Zone 1 setting (80% of line AB)	$4.63\angle79.21^0\ \Omega$
9	Ground compensation factor	1.8

3.2 Modeling of an Actual Current transformer

The actual CT with following specifications is used (figure 6 & table III).[8]

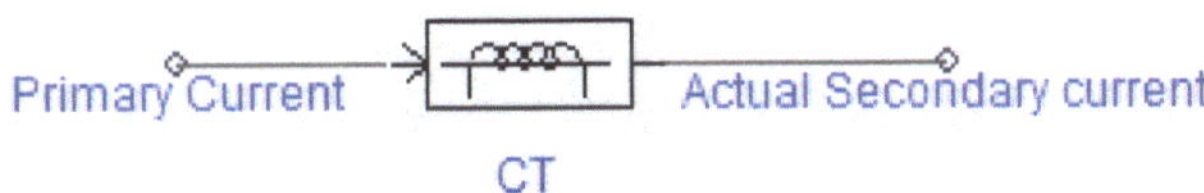

Figure 6. Actual CT Model

Table 3. CT Details

Sr. No	Parameter	Specifications
1	CT ratio	270/1
2	Secondary winding Resistance	$0.5\ \Omega$
3	Secondary winding Reactance	0.8mH
4	Magnetic Core Area	$2.6\times10^{-3}\ mm^2$
5	Magnetic Path Length	0.677mtr
6	CT Burden	**2.1** to $10\ \Omega$

3.3 Modeling of an Ideal Current transformer

The primary current is divided by number of turns of the actual current to get ideal value of secondary current (fig.7).[8]

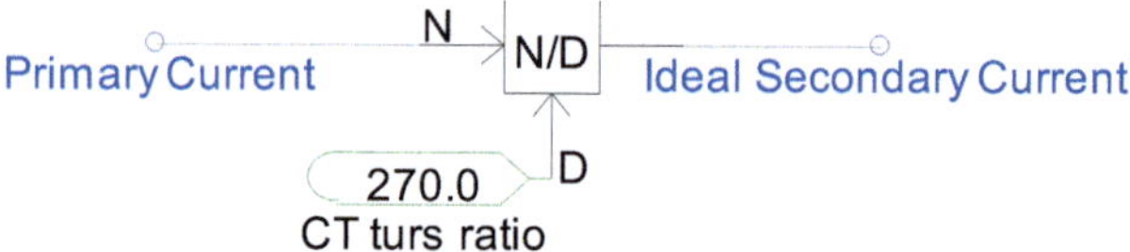

Figure 7. Ideal CT model

3.4 Modeling of Rogowski Coil with Integrator

Figure 8 shows model of RC equivalent circuit along with the integrator and table IV gives parameters used in RC model. [4]

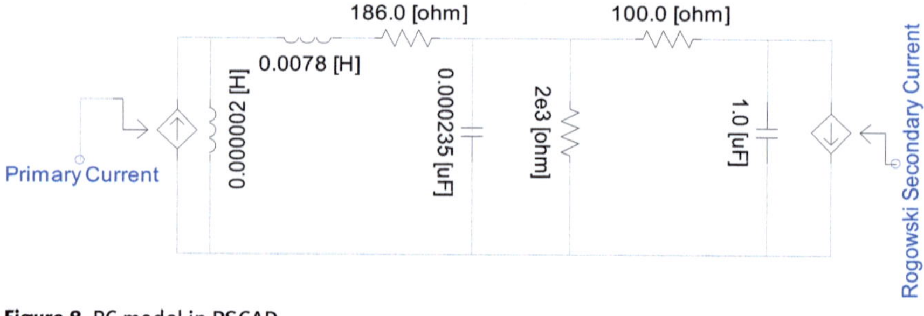

Figure 8. RC model in PSCAD

Table 4. Rogowski Coil Parameters

Sr. No.	Parameter	Specifications
1	Mutual Inductance	2 µH
2	L of Rogowski Coil	7.8 mH
3	R of Rogowski Coil	186 Ω
4	C of Rogowski Coil	235pF
5	Z of Rogowski Coil	2000 Ω
6	R of integrator	100 Ω
7	C of Integrator	0.1µF

4 Simulation Results

4.1 Impact of X/R ratio

Secondary currents waveforms: Figures 9(a) to 9(e) shows the secondary current waveforms generated by of Ideal CT (blue), Actual CT (red) and RC (green). In figure 9a, with X/R ratio as 14.097, no saturation is observed, as the fault is created when the primary voltage is maximum.

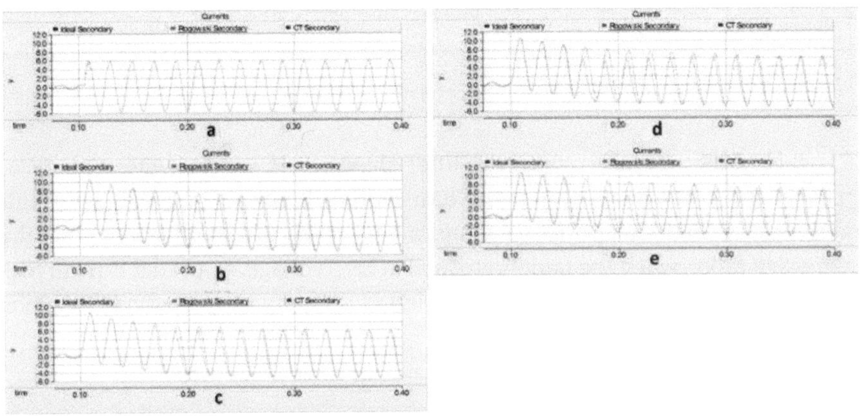

Figure 9. Secondary Current waveforms with

    a.   X/R=14.097 & fault at v=Vmax
    b.   X/R=14.097 & fault at v=0
    c.   X/R= 16.22 & fault at v=0
    d.   X/R= 19.85 & fault at v=0
    e.   X/R= 28.09 & fault at v=0

With the same ratio, when the fault is created at zero voltage, DC offset responsible for saturation of CT is produced (figure 9(b)). It is observed that RC can transform primary current faithfully on secondary side as it is overlaying on secondary current produced by ideal CT. But some distortions are observed on actual CT secondary current. As the X/R is increased (14.097-28.09) by increasing the reactive part of the source the waveform distortions were more visible (Figure 9(c) to 9(e)). Table V, gives the comparision of the secondary current with different X/R ratios. It is observed that the secondary current of actual CT goes on reducing with increase in system X/R. But ideal CT and RC secondary current remains equal & constant.

Table 5. Comparision of Secondary Current

	Secondary Currents				
Instant of Fault	v =Vmax	v = 0			
X/R Ratio	14.097	14.097	16.22	19.85	28.09
Without CT (Amp)	4.21	4.25	4.31	4.43	4.69
With CT (Amp)	4.21	3.72	3.56	3.36	3.06
With Rogowski Coil (Amp)	4.22	4.26	4.31	4.436	4.71

B-H Curves: Figures 10(a) to 10(e) shows, B-H curves generated by of magnetisation of actual CT. In figure 10(a), with X/R ratio as 14.097 and the fault is created at maximum primary voltage, a linear B-H curve is observed. With the same X/R ratio , when the fault is ctreated at zero voltage, saturated B-H curve is observed. Figures 10(b)-10(e) shows saturated B-H curves when the X/R ratio is incresed from 14.097-28.09.

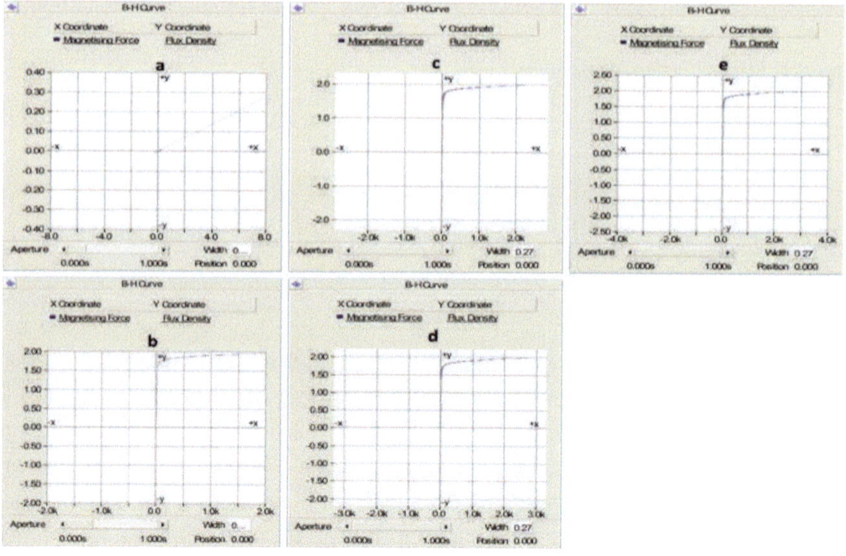

Figure 10. B-H curves with

a. X/R=14.097, v=Vmax
b. X/R=14.097, v=0
c. X/R=16.22, v=0
d. X/R=19.85, v=0
e. X/R=28.09, v=0

With increased X/R ratio, CT requires more amp-turns to produce same amount of flux density as shown in table VI. As the RC is air core transducer, always it gives linear characteristics.

Table 6. B-H Parameters at different X/R Ratio

	Secondary Burden				
Rb (Relay Burden)	0.5Ω				
Instant of Fault	v =Vmax	v = 0			
X/R ratio	14.097	14.097	16.22	19.85	28.09
B(Flux Density) Wb/m^2	0.27	2	2	2	2
H (Magnetizing Force) AT/m	7.75	1955	2597	3296	3840

Apparent Impedance: Figures 11(a) to 11(e) shows apparent impedance trajectories with ideal CT (green), actual CT (red) and RC (blue) on mho element when the fault is created inside Zone-1 of line AB, ie. at 150km. In fig. 11(a), it is observed that all the impedance trajectories (Ideal CT-Green, actual CT-red & RC-blue) are ovelaying on each other due to unsaturated CT. Figures 11(c), 11(d) & 11(e) shows that the distance element impedance trajectories are significantly impacted by the CT saturation with increased X/R ratio. When the CT saturates, secondary current decreases and the measured impedance increases. As the CT recovers from saturation, the measured impedance matches that of the unsaturated plot. Therefore, the distance element appears to have a tendency to under-reach.

a. X/R=14.097& fault at v=Vmax
b. X/R=14.097& fault at v=0
c. X/R=16.22& fault at v=
d. X/R=19.85 & fault at v=0
e. X/R=28.09 & fault at v=0

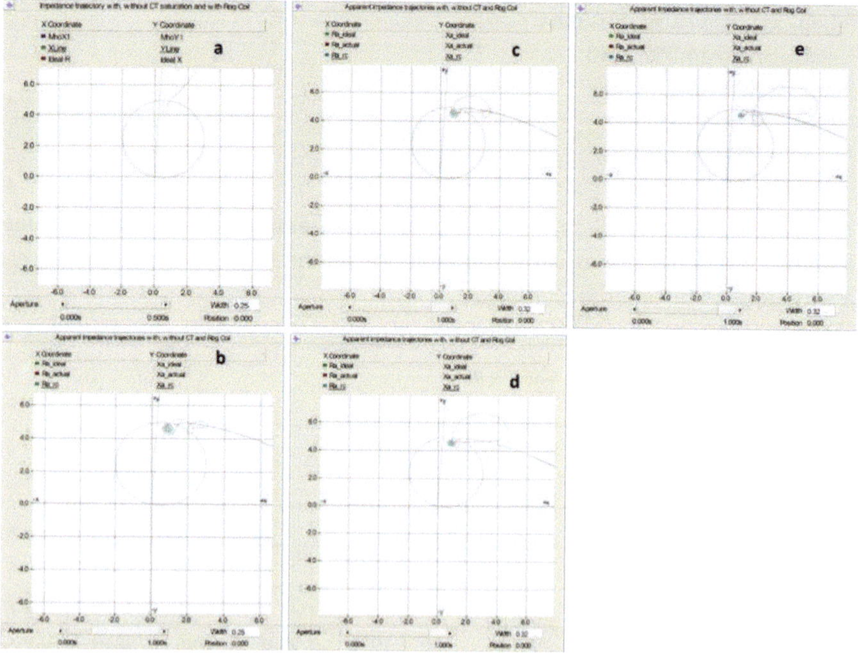

Figure 11. Impedence Trajectory with

Table VII gives the values of apparent impedances obtained with different fault instants and increased burdens.

Table 7. Apparent Impedance at Different X/R ratio

	Apparent Impedance				
Instant of Fault	v =Vmax	v = 0			
X/R Ratio	14.097	14.097	16.22	19.85	28.09
Without CT	4.63∠79.21⁰	4.69∠78.16⁰	4.69∠78.16⁰	4.69∠78.16⁰	4.69∠78.16⁰
With CT	4.64∠79.01⁰	5.612∠56.61⁰	5.87∠53.20⁰	6.25∠48.97⁰	6.92∠42.94⁰
With Rogowski Coil	4.62∠79.34⁰	4.68∠78.30⁰	4.68∠78.30⁰	4.68∠78.30⁰	4.68∠78.30⁰

Operating time: The operating time of a distance element can be critical to ensure high-speed tripping. When a CT saturates, it can delay operation of the distance element and result in slower than expected tripping times. The system X/R ratio of the system CT is increased from 14.097-28.09 until the operating time exceeds a specified value. (Figures12(a) to 12(e)).

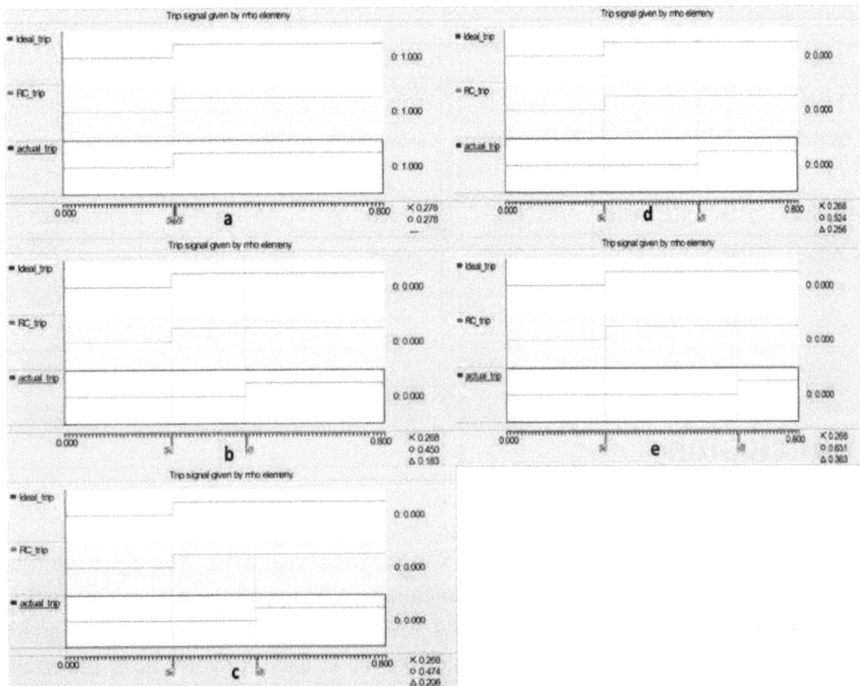

Figure 12. Tripping signals with

    a. X/R=14.097& fault at v=Vmax
    b. X/R=14.097 & fault at v=0
    c. X/R=16.22 & fault at v=0
    d. X/R=19.85& fault at v=0
    e. X/R=28.09 & fault at v=0

Table VIII gives the time delay the distance protection scheme will take when the X/R ration is increased from 14.097-28.09.

Table 8. Tripping Time of the Relay at Different X/R ratio

	Tripping Time				
Instant of Fault	v =Vmax	v = 0			
X/R Ratio	14.097	14.097	16.22	19.85	28.09
Without CT	Instantaneous	Instantaneous	Instantaneous	Instantaneous	Instantaneous
With CT	Instantaneous	After 0.183 S	After 0.206 S	After 0.256 S	After 0.363 S
With Rogowski Coil	Instantaneous	Instantaneous	Instantaneous	Instantaneous	Instantaneous

Conclusion

Conducted simulations show the importance of replacing conventional CT by Rogowski coil. Influence of system X/R ratio was investigated and it proved that not taking this factor into account can cause a CT to produce a highly distorted secondary current. After changing the X/R ratio from 14.097 to 16.22 a small indication of core saturation was observed for at least 4 cycles after the fault. After setting X/R ratio to 28.09 distortions were present during 12 cycles after the fault and they caused RMS current to be smaller than in fact it was. Rogowski coil with integrator produces exact replication of primary current without distorting it with any amount of load burden and able to overcome under reaching issue of the distance relay.

References

1   Stanley H. Horowitz, Arun G. Phadke, "Power System Relaying", John Wiley & Sons Ltd, ISBN: 978-0-470-05712-4, Third Edition, 2008.
2   Ashish S. Paramane, Avinash N. Sarwade, Pradeep K. Katti and Jayant G. Ghodekar, "Rogowski Coil - A Novel Transducer for Current Measurement", 6th International Conference on Power System Protection and Automation, CBIP, New Delhi, India, 27-28 February 2014, pp 80-88.

3   Avinash N. Sarwade, Pradeep K. Katti andJayant G. Ghodekar, "Optimum Setting of Distance Protection Scheme for HV Transmission Line", Journal of Power Electronics and Power Systems, STM, Volume 3, Issue 2, 2013, pp 23-30.

4   Veselin Skendzic and Bob Hughes, Schweitzer Engineering Laboratories, Inc. "Using Rogowski Coils Inside Protective Relays", 66th Annual Conference for Protective Relay Engineers College Station, Texas, April 8–11, 2013

5   IEEE PSRC report, "Practical Aspects of Rogowski Coil Applications to Relaying", Power System Relaying Committee of the IEEE Power Engineering Society, September 2010, pp 1-72

6   Piotr Sawko, "Impact of Secondary Burden and X/R Ratio on CT Saturation" Wroclaw University of Technology, Faculty of Electrical Engineering, 2008, pp 1-3.

7   Joe Mooney, " Distance Element Performance Under Conditions of CT Saturation" 11th Annual Georgia Tech Fault and Disturbance Analysis Conference as an alternate, Atlanta, Georgia, 19-20 May 2008, pp 1-7.

8   Power System simulation software, "PSCAD/EMTDC 4.2.1", Manitoba HVDC Research Centre Inc., Canada, 2008.

Mani Mishra [1], Aasheesh Shukla [2] and Vinay kumar Deolia [3]

Stability Analysis of Discrete Time MIMO Nonlinear System by Backstepping Technique

Abstract: Stability analysis of discrete time MIMO NL system is presented in this interpretive paper. Robustness is the major concern while designing the controller to stabilize the nonlinear system. Backstepping NL technique is used to construct the controller for the system which makes system output to pursue a desired trajectory. N-step ahead predictor is used to avoid the problem of causality contradiction which arises in discrete time backstepping technique. Neural network plays crucial part in order to approximate the unknown functions in the system. In this paper, radial basis function neural network (RBFNN) is used in order to approximate the unknown functions of the considered system. A basic notion about input nonlinearities which could arise in the system such as dead zone, saturation etc. is also presented along with the introductory idea of lyapunov stability issues to prove stability of NL systems. Here a descriptive stability analysis of Multiple Input Multiple Output nonlinear system is presented.

Keywords: Discrete-time MIMO system, lyapunov stability, backstepping technique controller, neural network, input nonlinearity

1 Introduction

Stability of NL control systems has been an important issue since last few decades and therefore the control techniques for NL systems has drawn increasing attention of research towards this domain. Constructing discrete time NL controller has become effective space of research due to the fundamental complexity that raises challenges in controller design. In order to approximate

1 M.tech Scholar, GLA University, Mathura
mani.mishra_mtec14@gla.ac.in
2 Assistant Professor, GLA University, Mathura
aasheesh.shukla@gla.ac.in
3 Associate Professor, GLA University, Mathura
vinaykumar.deolia@gla.ac.in

the nonlinear complex functions, neural networks have been used which possess the adaptation, approximation and learning abilities that are very much essential for approximating unknown functions [1]-[3]. Neural network has been successfully used along with various NL control techniques such as backstepping NL control technique [4], [5], sliding mode [6]. The three most commonly accepted neural networks are high-order NN [7], multilayer NN and radial basis function NN [8]. The practical systems possess enormous amount of nonlinear characteristics such as uncertainties, disturbances and different types of errors. Extensive amount of research have been done in single-input-single-output i.e., SISO NL systems [9], [10], and multiple-Input-multiple-output i.e., MIMO NL systems [7], [8], [11] and lot more developments are required in the field of MIMO NL control systems because there are many difficulties related to MIMO NL control system such as input coupling, complex nature etc. In [9], dead zone nonlinearity has been taken which is not known in SISO system and stability analysis has been done with the help of backstepping NL control technique. In [10], a stochastic NL pure feedback form SISO system has been considered along with the backstepping control technique.

Tremendous amount of work has been done in the field of continuous domain related to MIMO NL system [12], [13]. In [12], dead zone input constraints have been taken on discrete time system and adaptive neural network tracking has been achieved. In [13], NN control of MIMO NL system along with some input nonlinearities have been considered. The stability analysis of MIMO NL systems can be performed by taking several nonlinearities such as dead zone, saturation, backlash, etc. Control approach of backstepping has been done for discrete time NL control system with the involvement of nonlinearities at input. In [8], [12], the analysis of MIMO NL system is done with dead zone nonlinearity. In discrete time backstepping control design, there was a problem of causality contradiction when constructing controller for strict feedback form systems but the solution has been provided in [7] by the application of n-step ahead predictor.

In this paper, a discrete time MIMO NL system with input nonlinearity such as saturation or dead zone along with external disturbance is considered and stability analysis is being carried out with the help of backstepping NL control system. Since dead zone, saturation is important phenomenon which commonly interviewed during several control system actions. The unknown function is approximated by radial basis function neural network (RBFNN) which uses algorithm of adaptive computations that is very much required for the fast computation of several applications. The stability analysis is motivated by enormous work already done in the area of discrete time MIMO along with backstepping NL technique [7], [8], etc.

The remaining portion of is arranged in a manner as follows: Section II comprises of formulation of system with its prelims of discrete time MIMO NL system. Section III gives introduction to considered nonlinearity and neural network for approximation of the unknown functions. Section IV includes brief introduction to backstepping technique and controller design of MIMO NL system. Section V involves the introduction to lyapunov functions used and stability analysis. Section VI states the conclusion.

2 MIMO Dynamics and Prelims

The systems which comprises of more than one input and one output are known as MIMO systems.

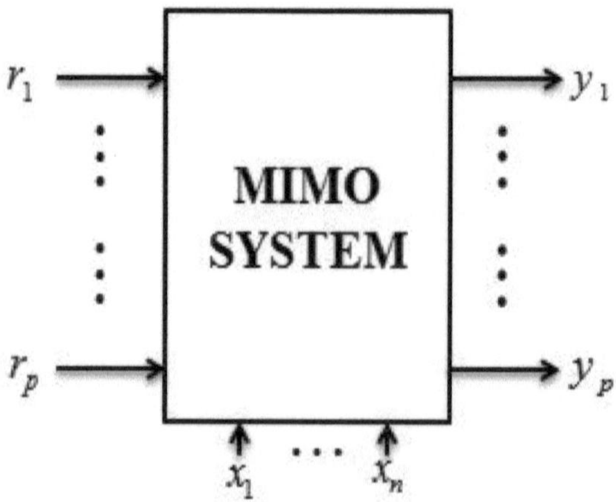

Figure 1. MIMO system [14]

Where, r_1......r_p are the input, y_1......y_p are the output and x_1......x_p are state variable of the system. The state space statement of discrete time Multiple-Input-Multiple-output system is:

$$\varphi_{1,i_1}(k+1) = F_{1,i_1}(\overline{\varphi}_{1,i_1}(k)) + G_{1,i_1}(\overline{\varphi}_{1,i_1}(k))\varphi_{1,i_1+1}(k)$$
$$\varphi_{1,n_1}(k+1) = F_{1,n_1}(\varphi(k)) + G_{1,n_1}(\varphi(k))U_1(k) + D_1(k)$$
$$1 \le i_1 \le n_1 - 1$$
$$\vdots$$
$$\varphi_{2,i_2}(k+1) = F_{2,i_2}(\overline{\varphi}_{2,i_2}(k)) + G_{2,i_2}(\overline{\varphi}_{2,i_2}(k))\varphi_{2,i_2+1}(k)$$
$$\varphi_{2,n_2}(k+1) = F_{2,n_2}(\varphi(k)) + G_{2,n_2}(\varphi(k))U_2(k) + D_2(k)$$
$$1 \le i_2 \le n_2 - 1$$
$$\vdots$$
$$\varphi_{j,i_j}(k+1) = F_{j,i_j}(\overline{\varphi}_{j,i_j}(k)) + G_{j,i_j}(\overline{\varphi}_{j,i_j}(k))\varphi_{j,i_j+1}(k)$$
$$\varphi_{j,n_j}(k+1) = F_{j,n_j}(\varphi(k)) + G_{j,n_j}(\varphi(k))U_j(k) + D_j(k)$$
$$1 \le i_j \le n_j - 1$$
$$\vdots$$
$$\varphi_{n,i_n}(k+1) = F_{n,i_n}(\overline{\varphi}_{n,i_n}(k)) + G_{n,i_n}(\overline{\varphi}_{n,i_n}(k))\varphi_{n,i_n+1}(k)$$
$$\varphi_{n,n_n}(k+1) = F_{n,n_n}(\varphi(k)) + G_{n,n_n}(\varphi(k))U_n(k) + D_n(k)$$
$$1 \le i_n \le n_n - 1 \tag{1}$$

The output equation is-

$$Y_j(k) = \varphi_{j,1}(k), \quad 1 \le j \le n \tag{2}$$

The above equation shows the MIMO system with n inputs and n outputs where state of the jth subsystem is presented as $\varphi_j(k) = [\varphi_{j,1}(k)...\varphi_{j,n_j}(k)]^T$, the full system state is $\varphi(k) = [\varphi_1^T(k)...\varphi_N^T(k)]^T$, i_j^{th} j^{th} subsystem stage is given by $\overline{\varphi}_{j,i_j}(k) = [\varphi_{j,1}(k)...\varphi_{j,i_j}(k)]^T \in R^{i_j}$, $Y_j(k) = \varphi_{j,1}(k)$ is jth system's output, the unknown NL functions are F_{j,i_j} and G_{j,i_j}, extraneous disturbance is given by $D_j(k)$, and i_j, j, n_j are integers which are positive in nature. The control law derived is denoted by $U_j(k)$ which is output of the nonlinearity considered and the input of the system.

3 Input nonlinearity and neural network

Any part of the control system may comprise of nonlinearity. Some non-linearity are intrinsically present, some are knowingly inserted into the system to change the characteristics of the system. In engineering application, there are two common nonlinearities which mostly come into existence and influence the performance of the control system, namely, saturation and dead zone non linearity [8], [13], etc.

3.1 Saturation nonlinearity

Saturation is a condition in a device whose output saturates at a particular value and it does not have any effect on further increase of input, this phenomenon commonly occurs in actuators.

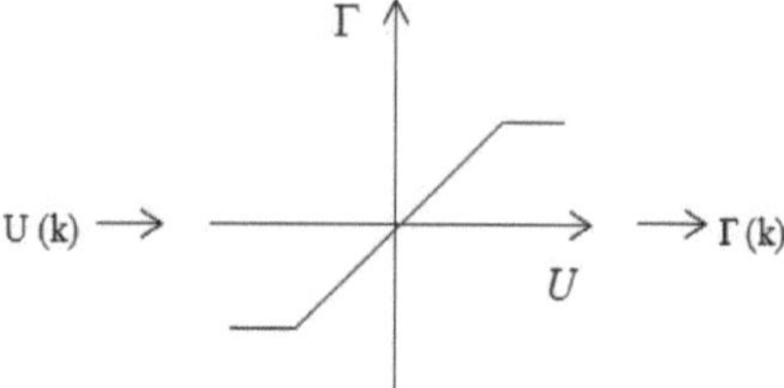

Figure 2. Saturation nonlinearity [15]

Mathematically, the output is given by-

$$\Gamma(k) = A * \tanh(U(k))$$

(3)

The control U (k) is deal with amplitude which has some limitation.

3.2 Dead zone nonlinearity

It is the condition in which device output becomes zero on exceeding the input by certain limit of a value.

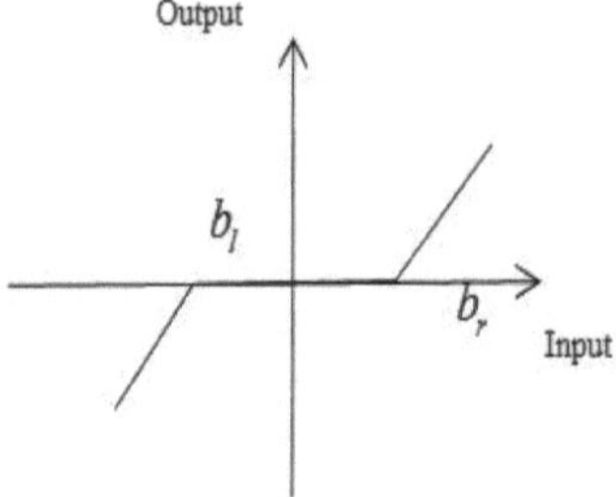

Figure 3. Dead zone nonlinearity [14]

The equation for dead zone is given by-

$$U_j(k) = D(V_j(k)) = \begin{cases} m_{rj}(V_j(k) - b_{rj}), & if & V_j(k) \geq b_{rj} \\ 0, & if & -b_{lj} < V_j(k) < b_{rj} \\ m_{lj}(V_j(k) + b_{lj}), & if & V_j(k) \leq -b_{lj} \end{cases} \qquad (4)$$

Where b_{rj} and b_{lj} denotes the breakpoints, m_{rj} and m_{lj} stands for right and left slopes, $U_j(k)$ is the output of the dead zone and input of the jth subsyetm and input of the dead zone is given by $V_j(k)$. It can be rewritten as-

5)
$$U_j(k) = m_j(k)V_j(k) + b_j(k) \qquad (5)$$

3.3 Neural network (RBFNN) [8]

It is a double layered artificial NN that comprises of one hidden layer whose nodes execute a set of Gaussian function which is actually the function of radial basis for the network. The function approximation by RBFNNs is given by-

$$F^{NN}(y) = W^T S(y) \qquad (6)$$

Where y represents input variable of the NN, W shows the weight vector, $S(y)$ is the vector of basis function. The generally used Gaussian function is given by-

$$S_i(y) = \exp\left[\frac{-(Y - \mu_i)^T(Y - \mu_i)}{v_i^2}\right], \quad i = 1,\ldots,l \qquad (7)$$

Where v_i is the width and μ_i is the center of the Gaussian function. According to the basis function vector chosen, the above equation can be written as-

$$F^{NN}(y) = W^{*T} S(y) + e^*(y) \qquad (8)$$

Where W^* is the constant weight which is ideal in nature, $e^*(y)$ is the approximation error.

4 Controller design by backstepping

In this section, backstepping NL control technique is used to construct the control law for the MIMO NL system. This technique came into existence in 1990 and main contributor was Peter V. Kokotovic. It is mainly pertinent to strict feedback form systems. The basic concept behind backstepping is that the stability analysis of upper order systems is very challenging, hence they are partitioned into subsystems and control law is constructed for each one of them. At last coordinate all the control laws to achieve the final controller.

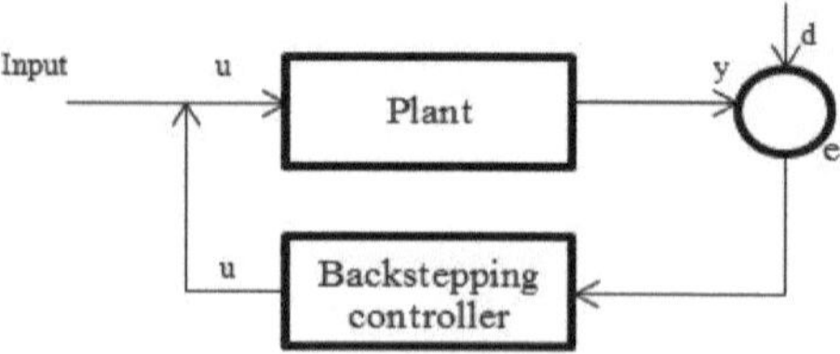

Figure 4. Backstepping control technique

The derived control law is the input of the system and output of the controller. The basic idea of controller design is given by-

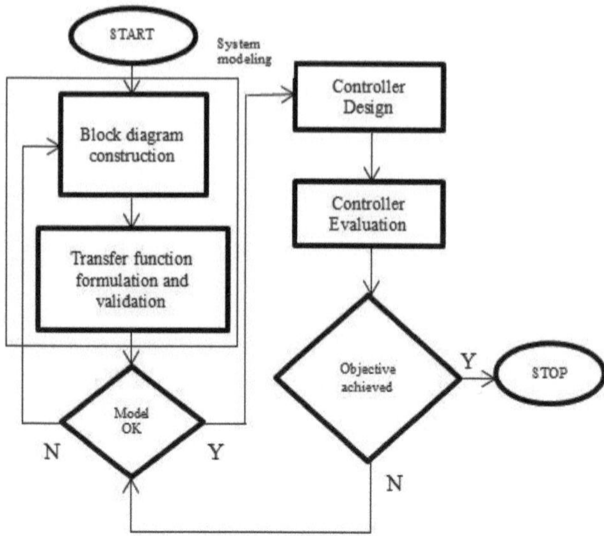

Figure 5. Controller Design

To design the control action for MIMO NL system, consider the first subsystem of equation (1),

$$\varphi_{1,i_1}(k+1) = F_{1,i_1}(\overline{\varphi}_{1,i_1}(k)) + G_{1,i_1}(\overline{\varphi}_{1,i_1}(k))\varphi_{1,i_1+1}(k)$$
$$\varphi_{1,n_1}(k+1) = F_{1,n_1}(\varphi(k)) + G_{1,n_1}(\varphi(k))U_1(k) + D_1(k) \tag{9}$$
$$1 \le i_1 \le n_1 - 1$$

In order to apply backstepping control technique, rewriting the equation (9) in strict feedback form-

$$\varphi_{1,1}(k+1) = F_{1,1}(\overline{\varphi}_{1,1}(k)) + G_{1,1}(\overline{\varphi}_{1,1}(k))\varphi_{1,2}(k),$$

$$\varphi_{1,2}(k+1) = F_{1,2}(\overline{\varphi}_{1,2}(k)) + G_{1,2}(\overline{\varphi}_{1,2}(k))\varphi_{1,3}(k)$$

$$\vdots$$

$$\vdots$$

$$\varphi_{1,n_1-1}(k+1) = F_{1,n_1-1}(\overline{\varphi}_{1,n_1-1}(k)) + G_{1,n_1-1}(\overline{\varphi}_{1,n_1-1}(k))\varphi_{1,n_1}(k) \tag{10}$$

$$\varphi_{1,n_1}(k+1) = F_{1,n_1}(\varphi(k)) + G_{1,n_1}(\varphi(k))U_1(k) + D_1(k)$$

Step 1-

Controller construction by backstepping is based on error dynamics, i.e., the actual output difference of the system and desired response.

$$e_{1,1}(k) = \varphi_{1,1}(k) - \varphi_{1,1d}(k)$$

$$e_{1,1}(k+1) = \varphi_{1,1}(k+1) - \varphi_{1,1d}(k+1) \tag{11}$$

Substituting the value of equation (10) in above equation,

$$e_{1,1}(k+1) = [F_{1,1}(\overline{\varphi}_{1,1}(k)) + G_{1,1}(\overline{\varphi}_{1,1}(k))\varphi_{1,2}(k)] - \varphi_{1,1d}(k+1) \tag{12}$$

The fictitious controller for subsystem 1 of equation (10) is -

$$\varphi_{1,2d}(k) = \frac{1}{G_{1,1}(\overline{\varphi}_{1,1}(k))}[-F_{1,1}(\overline{\varphi}_{1,1}(k)) + \varphi_{1,1d}(k+1) - k_{1,1}e_{1,1}(k)] \tag{13}$$

Equation (13) shows the fictitious controller of the subsystem 1, where $k_{1,1} > 1$ .To prove stability in the sense of Lyapunov i.e., $e_1(k+1) < 0$, substituting the above value in equation (12) results-

$$e_{1,1}(k) = -k_{1,1}e_{1,1}(k) \tag{14}$$

Where $k_{1,1} > 1$ which shows that subsystem 1 is stabilized with fictitious controller as shown in equation (13).

Step 2-

Considering the error dynamics of subsystem 2of equation (10),

$$e_{1,2}(k) = \varphi_{1,2}(k) - \varphi_{1,2d}(k)$$

$$e_{1,2}(k+1) = \varphi_{1,2}(k+1) - \varphi_{1,2d}(k+1) \tag{15}$$

Substituting the value of equation (10) in above equation,

$$e_{1,2}(k+1) = [F_{1,2}(\overline{\varphi}_{1,2}(k)) + G_{1,2}(\overline{\varphi}_{1,2}(k))\varphi_{1,3}(k)] - \varphi_{1,2d}(k+1) \tag{16}$$

The fictitious controller for subsystem 2 of equation (10) is –

$$\varphi_{1,3d}(k) = \frac{1}{G_{1,2}(\overline{\varphi}_{1,2}(k))}[-F_{1,2}(\overline{\varphi}_{1,2}(k)) + \varphi_{1,2d}(k+1) - k_{1,2}e_{1,2}(k)] \tag{17}$$

The term $\varphi_{1,2d}(k+1)$ in equation (17) is the future value of the fictitious controller $\varphi_{1,2d}(k)$ which is not known. So, the fictitious controller which is shown in equation (17) will be infeasible in practice. This above stated problem is known as *causality contradiction*. And the actual controller derived in this way will not be feasible in practice due unavailability of information related to future. The problem of causality contradiction can be averted by transforming the equations based on n-step ahead predictor [7]. The transformed equations [7] are given as-

$$\varphi_{j,1}(k+n_j) = F_{j,1}(\overline{\varphi}_{j,n_j}(k)) + G_{j,1}(\overline{\varphi}_{j,n_j}(k))\varphi_{j,2}(k+n_j-1),$$

$$\vdots$$

$$\varphi_{j,i_j}(k+n_j-i_j+1) = F_{j,i_j}(\overline{\varphi}_{j,i_j}(k)) + G_{j,i_j}(\overline{\varphi}_{j,i_j}(k))\varphi_{j,i_j+1}(k+n_j-i_j) \tag{18}$$

$$\vdots$$

$$\varphi_{j,n_j}(k+1) = F_{j,n_j}(\varphi(k)) + G_{j,n_j}(\varphi(k))U_j(k) + D_j(k)$$

$$Y_j(k) = \varphi_{j,1}(k)$$

Now again designing the control law by transformed equation (18) by avoiding the problem of causality contradiction.

Step 1-

Error dynamics of subsystem 1 of equation (18) is given by-

$$e_{j,1}(k) = \varphi_{j,1}(k) - \varphi_{j,1d}(k)$$

$$e_{j,1}(k+n_j) = \varphi_{j,1}(k+n_j) - \varphi_{j,1d}(k+n_j) \tag{19}$$

Substituting the value of equation (18) in above equation,

$$e_{j,1}(k+n_j) = F_{j,1}(\overline{\varphi}_{j,n_j}(k)) + G_{j,1}(\overline{\varphi}_{j,n_j}(k))\varphi_{j,2}(k+n_j-1) - \varphi_{j,1d}(k+n_j) \tag{20}$$

The fictitious controller for subsystem 1 of equation (18) is -

$$\varphi_{j,2d}(k+n_j-1) = -\frac{1}{G_{j,1}(\overline{\varphi}_{j,n_j}(k))}[F_{j,1}(\overline{\varphi}_{j,n_j}(k)) - \varphi_{j,1d}(k+n_j)] \tag{21}$$

Equation (21) shows the fictitious controller of the subsystem 1 of equation (18).

Step 2-

Error dynamics of subsystem 2 of equation (18) is given by-

$$e_{j,2}(k) = \varphi_{j,2}(k) - \varphi_{j,2d}(k)$$
$$e_{j,2}(k+n_j) = \varphi_{j,2}(k+n_j-1) - \varphi_{j,2d}(k+n_j-1) \tag{22}$$

Substituting the value of equation (18) in above equation

$$e_{j,2}(k+n_j-1) = F_{j,2}(\overline{\varphi}_{j,n_j}(k)) + G_{j,2}(\overline{\varphi}_{j,n_j}(k))\varphi_{j,3}(k+n_j-2) - \varphi_{j,2d}(k+n_j-1) \tag{23}$$

The fictitious controller for subsystem 2 of equation (18) is -

$$\varphi_{j,3d}(k+n_j-2) = -\frac{1}{G_{j,2}(\overline{\varphi}_{j,n_j}(k))}[F_{j,2}(\overline{\varphi}_{j,n_j}(k)) - \varphi_{j,2d}(k+n_j-1)] \tag{24}$$

In the similar manner control law for all the subsystem of equation (18) are as follows-

$$\beta_{j,2}^*(k) \overset{\Delta}{=} \varphi_{j,2d}(k+n_j-1) = -\frac{1}{G_{j,1}(\overline{\varphi}_{j,n_j}(k))}[F_{j,1}(\overline{\varphi}_{j,n_j}(k)) - \varphi_{j,1d}(k+n_j)]$$

$$\beta_{j,3}^*(k) \overset{\Delta}{=} \varphi_{j,3d}(k+n_j-2) = -\frac{1}{G_{j,2}(\overline{\varphi}_{j,n_j}(k))}[F_{j,2}(\overline{\varphi}_{j,n_j}(k)) - \varphi_{j,2d}(k+n_j-1)] \tag{25}$$

$$\vdots$$

$$\beta_{j,n_j}^*(k) \overset{\Delta}{=} \varphi_{j,n_jd}(k+1) = -\frac{1}{G_{j,n_j-1}(\overline{\varphi}_{j,n_j}(k))}[F_{j,n_j-1}(\overline{\varphi}_{j,n_j}(k)) - \varphi_{j,(n_j-1)d}(k+2)]$$

$$U_j^*(k) \overset{\Delta}{=} U_j(k) = -\frac{1}{G_{j,n_j}(\varphi(k))}[F_{j,n_j}(\varphi(k)) + D_j(k) - \varphi_{j,n_jd}(k+1)]$$

The above equation (25) shows all the virtual controller by $\beta_{j,2}^*(k), \beta_{j,3}^*(k)....\beta_{j,n_j}^*(k)$ and actual controller by $U_j^*(k)$ with the help of which the system can be stabilized in every step without the problem of causality contradiction. The basic aim of designing ideal or actual controller is to direct the output of the system to pursue a bounded and known trajectory.

5 Stability Analysis of MIMO NL system

The stability analysis of MIMO NL system is carried out by lyapunov stability theory along with the controller designed by the backstepping technique. Lyapunov theory shows the manner in which energy functions vary over time and resultant prove a system as stable or not by selecting appropriate lyapunov function. The ideal or virtual control actions derived above by backstepping control technique can track the desired output if known definite system model is there without any disturbances. But the MIMO system comprises of unknown functions as well as disturbances. Therefore, radial basis function is applied to approximate the unknown functions.

Step 1-

Using RBFNN, the fictitious controller $\varphi_{j,2d}(k+n_j-1)$ of equation (25) would be approximated as [8]-

$$\beta^*_{j,2}(k)=\varphi_{j,2d}(k+n_j-1)=W^{*T}_{j,1}S_{j,1}(Z_{j,1}(k))+\zeta^*_{j,1}(Z_{j,1}(k)) \tag{26}$$

Where $W^*_{J,1}$ the ideal is constant weight, $S_{j,1}(Z_{j,1}(k))$ is basis function vector, $\zeta^*_{j,1}(Z_{j,1}(k))$ is optimal approximation error and $Z_{j,1}(k)$ is the input vector of the neural network.

Let

$$\tilde{W}_{j,1}(k)=\hat{W}_{j,1}(k)-W^*_{j,1}(k) \tag{27}$$

Where $\tilde{W}_{j,1}(k)$ is the estimation of $W^*_{j,1}(k)$.

Virtual control is designed as-

$$\beta^*_{j,2}(k)=\varphi_{j,2d}(k+n_j-1)=\hat{W}^T_{j,1}(k)S_{j,1}(Z_{j,1}(k)) \tag{28}$$

The adaptation law for weight of NN is-

$$\hat{W}_{j,1}(k+1)=\hat{W}_{j,1}(k_1)-\tau_{j,1}[S_{j,1}(Z_{j,1}(k_1))e_{j,1}(k+1)+\Gamma_{j,1}\hat{W}_{j,1}(k_1)] \tag{29}$$

Where $\tau_{j,1}$ and $\Gamma_{j,1}$ are the parameters of design and $k_1=k-n_j+1$. Subtracting $W^*_{j,1}$ and using equation (27), equation (29) can be written as-

$$\tilde{W}_{j,1}(k+1)=\tilde{W}_{j,1}(k_1)-\tau_{j,1}[S_{j,1}(Z_{j,1}(k_1))e_{j,1}(k+1)+\Gamma_{j,1}\hat{W}_{j,1}(k_1)] \tag{30}$$

Now writing-

$$\varphi_{j,2}(k+n_j-1)=e_{j,2}(k+n_j-1)+\beta^*_{j,2}(k) \tag{31}$$

Substituting the value of equation (31) in equation (20) and on rearranging,

$$e_{j,1}(k+n_j)=G_{j,1}(\overline{\varphi}_{j,n_j}(k))[-\beta^*_{j,2}(k)]+e_{j,2}(k+n_j-1)+\beta^*_{j,2}(k) \tag{32}$$

Substituting the value of equation (26) and (28) in equation (32),

$$e_{j,1}(k+n_j)=G_{j,1}(\overline{\varphi}_{j,n_j}(k))[\tilde{W}^T_{j,1}(k)S_{j,1}(Z_{j,1}(k))-\zeta_{j,1}(Z_{j,1}(k))+e_{j,2}(k+n_j-1)]$$

or

$$\tilde{W}^T_{j,1}(k)S_{j,1}(Z_{j,1}(k))=\frac{e_{j,1}(k+n_j)}{G_{j,1}(\overline{\varphi}_{j,n_j}(k))}+\zeta_{j,1}(Z_{j,1}(k))-e_{j,2}(k+n_j-1) \tag{33}$$

Now defining lyapunov functions-

$$V_{j,1}(k) = \frac{1}{\overline{G}_{j,1}} e_{j,1}^2(k) + \sum_{p=0}^{n_j-1} \widetilde{W}_{j,1}^T(k_1 + p)\tau_{j,1}^{-1}\widetilde{W}_{j,1}(k_1 + p) \tag{34}$$

Where $k_1 = k - n_j + 1$.

By forward difference equation-

$$\Delta V_{j,1}(k) = V_{j,1}(k+1) - V_{j,1}(k) \tag{35}$$

From equation (34) and (35),

$$= \frac{1}{\overline{G}_{j,1}}[e_{j,1}^2(k+1) - e_{j,1}^2(k)] + \widetilde{W}_{j,1}^T(k+1)\tau_{j,1}^{-1}\widetilde{W}_{j,1}(k+1) - \widetilde{W}_{j,1}^T(k_1)\tau_{j,1}^{-1}\widetilde{W}_{j,1}(k_1) \tag{36}$$

Substituting the value of equation (30) in equation (36),

$$= \frac{1}{\overline{G}_{j,1}}[e_{j,1}^2(k+1) - e_{j,1}^2(k)] + e_{j,1}^T(k+1)S_{j,1}^T(Z_{j,1}(k_1))\tau_{j,1}S_{j,1}(Z_{j,1}(k_1))e_{j,1}(k+1) \tag{37}$$
$$+ e_{j,1}^T(k+1)S_{j,1}^T(Z_{j,1}(k_1))\tau_{j,1}\Gamma_{j,1}\hat{W}_{j,1}(k_1) + \hat{W}_{j,1}(k_1)\Gamma_{j,1}^T\tau_{j,1}S_{j,1}(Z_{j,1}(k_1))e_{j,1}(k+1)$$
$$+ \hat{W}_{j,1}^T(k_1)\Gamma_{j,1}^T\tau_{j,1}\Gamma_{j,1}\hat{W}_{j,1}(k_1) - e_{j,1}^T(k+1)S_{j,1}^T(Z_{j,1}(k_1))\widetilde{W}_{j,1}(k_1) - \widetilde{W}_{j,1}^T(k_1)S_{j,1}(Z_{j,1}(k_1))e_{j,1}(k+1)$$
$$- \hat{W}_{j,1}^T(k_1)\Gamma_{j,1}\hat{W}_{j,1}(k_1) - \hat{W}_{j,1}^T(k_1)\Gamma_{j,1}^T\widetilde{W}_{j,1}(k_1)$$

$$= \frac{1}{\overline{G}_{j,1}}[e_{j,1}^2(k+1) - e_{j,1}^2(k)] - 2\widetilde{W}_{j,1}^T(k_1)[S_{j,1}(Z_{j,1}(k_1))e_{j,1}(k+1) + \Gamma_{j,1}\hat{W}_{j,1}(k_1)] \tag{38}$$
$$+ [S_{j,1}(Z_{j,1}(k_1))e_{j,1}(k+1) + \Gamma_{j,1}\hat{W}_{j,1}(k_1)]^T \tau_{j,1}[S_{j,1}(Z_{j,1}(k_1))e_{j,1}(k+1) + \Gamma_{j,1}\hat{W}_{j,1}(k_1)]$$

$$= \frac{1}{\overline{G}_{j,1}}[e_{j,1}^2(k+1) - e_{j,1}^2(k)] - 2\widetilde{W}_{j,1}^T(k_1)S_{j,1}(Z_{j,1}(k_1))e_{j,1}(k+1)$$
$$- 2\widetilde{W}_{j,1}^T(k_1)\Gamma_{j,1}\hat{W}_{j,1}(k_1) + S_{j,1}^T(Z_{j,1}(k_1))\tau_{j,1}S_{j,1}(Z_{j,1}(k_1))e_{j,1}^2(k+1) \tag{39}$$
$$+ 2\Gamma_{j,1}^T\hat{W}_{j,1}^T(k_1)\tau_{j,1}S_{j,1}(Z_{j,1}(k_1))e_{j,1}(k+1) + \Gamma_{j,1}^2\hat{W}_{j,1}^T(k_1)\tau_{j,1}\hat{W}_{j,1}(k_1)$$

Since the value of i_j ranges from $1 \le i_j \le n_j - 1$ so considering from first value and substituting equation (33) in equation (39),

$$= \frac{1}{\overline{G}_{j,1}}[e_{j,1}^2(k+1) - e_{j,1}^2(k)] - 2\frac{e_{j,1}^2(k+1)}{G_{j,1}(\overline{x}_{j,1}(k))} - 2\zeta_{j,1}(Z_{j,1}(k))e_{j,1}(k+1)$$
$$+ 2e_{j,2}(k)e_{j,1}(k+1) - 2\widetilde{W}_{j,1}^T(k_1)\Gamma_{j,1}\hat{W}_{j,1}(k_1) + S_{j,1}^T(Z_{j,1}(k_1))\tau_{j,1}S_{j,1}(Z_{j,1}(k_1))e_{j,1}^2(k+1) \tag{40}$$
$$+ 2\Gamma_{j,1}\hat{W}_{j,1}^T(k_1)\tau_{j,1}S_{j,1}(Z_{j,1}(k_1))e_{j,1}(k+1) + \Gamma_{j,1}^2\hat{W}_{j,1}^T(k_1)\tau_{j,1}\hat{W}_{j,1}(k_1)$$

By the definition of $G_{j,1}(\overline{x}_{j,n_j}(k))$ and assumption [8],

$$\frac{-2e_{j,1}^2(k+1)}{G_{j,1}(\overline{\varphi}_{j,1}(k))} \le \frac{-2e_{j,1}^2(k+1)}{\overline{G}_{j,1}} \tag{41}$$

Now applying young's inequality,

$$S_{j,1}^{T}(Z_{j,1}(k_1))\tau_{j,1}S_{j,1}(Z_{j,1}(k_1))e_{j,1}^2(k+1) \leq \overline{\psi}_{j,1}l_{j,1}e_{j,1}^2(k+1)$$

$$-2\zeta_{j,1}e_{j,1}(k+1) \leq \frac{\overline{\psi}_{j,1}e_{j,1}^2(k+1)}{\overline{\overline{G}}_{j,1}} + \frac{\overline{g}_{j,1}\overline{\zeta}_{j,1}^2}{\overline{\psi}_{j,1}}$$

$$2\Gamma_{j,1}\hat{W}_{j,1}^T(k_1)\tau_{j,1}S_{j,1}(Z_{j,1}(k_1))e_{j,1}(k+1) \leq \frac{\overline{\psi}_{j,1}l_{j,1}e_{j,1}^2(k+1)}{\overline{\overline{G}}_{j,1}} + \overline{G}_{j,1}\Gamma_{j,1}^2\overline{\psi}_{j,1}\left\|\hat{W}_{j,1}(k_1)\right\|^2$$

$$2\tilde{W}_{j,1}^T(k_1)\hat{W}_{j,1}(k_1) = \left\|\tilde{W}_{j,1}(k_1)\right\|^2 + \left\|\hat{W}_{j,1}(k_1)\right\|^2 - \left\|W_{j,1}^*\right\|^2$$

$$2e_{j,2}(k)e_{j,1}(k+1) \leq \frac{\overline{\psi}_{j,1}e_{j,1}^2(k+1)}{\overline{\overline{G}}_{j,1}} + \frac{\overline{G}_{j,1}\overline{e}_{j,2}^2}{\overline{\psi}_{j,1}}$$

$$(42)$$

Where neural network node number is denoted by $l_{j,1}$ and Eigen value of matrix is given by $\overline{\psi}_{j,1}$. Now applying inequalities of equation (42) and assumption of equation (41) in equation (40),

$$\Delta V_{J,1}(k) \leq -\frac{1}{\overline{\overline{G}}_{j,1}}e_{j,1}^2(k+1) - \frac{1}{\overline{\overline{G}}_{j,1}}e_{j,1}^2(k) + 2\frac{\overline{\psi}_{j,1}e_{j,1}^2(k+1)}{\overline{\overline{G}}_{j,1}} + \frac{\overline{G}_{j,1}\overline{\zeta}_{j,1}^2}{\overline{\psi}_{j,1}} + \frac{\overline{G}_{j,1}\overline{e}_{j,2}^2}{\overline{\psi}_{j,1}} - \Gamma_{j,1}\left\|\hat{W}_{j,1}(k_1)\right\|^2$$

$$+\Gamma_{j,1}\left\|W_{j,1}^*\right\|^2 + \overline{\psi}_{j,1}l_{j,1}e_{j,1}^2(k+1) + \frac{\overline{\psi}_{j,1}l_{j,1}e_{j,1}^2(k+1)}{\overline{\overline{G}}_{j,1}} + \overline{G}_{j,1}\Gamma_{j,1}^2\overline{\psi}_{j,1}\left\|\hat{W}_{j,1}(k_1)\right\|^2$$

$$+\Gamma_{j,1}^2\tau_{j,1}\left\|\hat{W}_{j,1}(k_1)\right\|^2$$

$$(43)$$

It can also be written in the form-

$$\Delta V_{J,1}(k) \leq -A_{j,1}\frac{1}{\overline{\overline{G}}_{j,1}}e_{j,1}^2(k+1) - \frac{1}{\overline{\overline{G}}_{j,1}}e_{j,1}^2(k) + B_{j,1}$$

$$+\frac{\overline{G}_{j,1}\overline{\zeta}_{j,1}^2}{\overline{\psi}_{j,1}} - \Gamma_{j,1}C_{j,1}\left\|\hat{W}_{j,1}(k_1)\right\|^2$$

$$(44)$$

Where $A_{j,1} = 1 - 2\overline{\psi}_{j,1} - \overline{\psi}_{j,1}l_{j,1} - \overline{G}_{j,1}\overline{\psi}_{j,1}l_{j,1}$ $B_{j,1} = \dfrac{\overline{G}_{j,1}\overline{\zeta}_{j,1}^2}{\overline{\psi}_{j,1}} + \Gamma_{j,1}W_{j,1}^{*2}$ and

$$C_{j,1} = 1 - \Gamma_{j,1}\overline{\psi}_{j,1} - \overline{G}_{j,1}\Gamma_{j,1}\overline{\psi}_{j,1}$$

Step i_j -

Similar procedure is being carried out to prove i_j^{th} subsystem stable along with lyapunov function. And the final equation results-

$$\Delta V_{J,ij}(k) \leq -A_{j,ij}\frac{1}{\overline{\overline{G}}_{j,ij}}e_{j,ij}^2(k+1) - \frac{1}{\overline{\overline{G}}_{j,ij}}e_{j,ij}^2(k) + B_{j,ij}$$

$$+\frac{\overline{G}_{j,ij}\overline{\zeta}_{j,ij}^2}{\overline{\psi}_{j,ij}} - \Gamma_{j,ij}C_{j,ij}\left\|\hat{W}_{j,ij}(k_1)\right\|^2$$

$$(45)$$

Where $A_{j,ij} = 1 - 2\overline{\psi}_{j,ij} - \overline{\psi}_{j,ij}l_{j,ij} - \overline{G}_{j,ij}\overline{\psi}_{j,ij}l_{j,ij}$, $B_{j,ij} = \dfrac{\overline{G}_{j,ij}\overline{\zeta}_{j,ij}^2}{\overline{\psi}_{j,ij}} + \Gamma_{j,ij}W_{j,ij}^{*2}$ and

$$C_{j,ij} = 1 - \Gamma_{j,ij}\overline{\psi}_{j,ij} - \overline{G}_{j,ij}\Gamma_{j,ij}\overline{\psi}_{j,ij}$$

Step n_j -

The input nonlinearity will appear in the n_j^{th} subsystem. Nonlinearity may be of any type either dead zone, saturation, backlash etc. Considerable amount of work has been done by considering these types of nonlinearities [8], [9], [13], [15]. For example in [8], stability of MIMO NL system has been proved by taken dead zone nonlinearity into account. Here in this paper analyzing simple MIMO NL system without taking any dead zone, saturation into account. Hence in a similar manner the final equation will result -

$$\Delta V_{J,n_j}(k) \le -A_{j,n_j}\frac{1}{\overline{G}_{j,n_j}}e^2_{j,n_j}(k+1) - \frac{1}{\overline{G}_{j,n_j}}e^2_{j,n_j}(k) + B_{j,n_j}$$

$$+\frac{\overline{G}_{j,n_j}\overline{\zeta}^2_{j,n_j}}{\overline{\psi}_{j,n_j}} - \Gamma_{j,n_j}C_{j,n_j}\left\|\hat{W}_{j,n_j}(k_1)\right\|^2 \tag{46}$$

Where $A_{j,n_j} = 1 - 2\overline{\psi}_{j,n_j} - \overline{\psi}_{j,n_j}l_{j,n_j} - \overline{G}_{j,n_j}\overline{\psi}_{j,n_j}l_{j,n_j}$, $B_{j,n_j} = \frac{\overline{G}_{j,n_j}\overline{\zeta}^2_{j,n_j}}{\overline{\psi}_{j,n_j}} + \Gamma_{j,n_j}W^{*2}_{j,n_j}$

and $C_{j,n_j} = 1 - \Gamma_{j,n_j}\overline{\psi}_{j,n_j} - \overline{G}_{j,n_j}\Gamma_{j,n_j}\overline{\psi}_{j,n_j}$

Conclusion

This explanatory paper presents the more detailed analysis of MIMO NL system in strict feedback form and hence backstepping control technique is applied to construct the controller. Basic controller design procedure has been shown with the help of a flowchart. Causality contradiction problem is avoided by transforming equation with the help of n-step ahead predictor. Unknown functions are approximated by RBFNN along with some adaptation laws with the application of appropriate lyapunov functions descriptive stability analysis of MIMO NL system has been done.

References

1 H. G. Han and J. F. Qiao, "An adaptive computation algorithm for RBF neural network," IEEE Trans. Neural Netw. Learn. Syst., vol. 23, no. 2, pp. 342–347, Feb. 2012.
2 C. L. P. Chen, G. X. Wen, Y. J. Liu, and F. Y. Wang, "Adaptive consensus control for a class of nonlinear multiagent time-delay systems using neural networks," IEEE Trans. Neural Netw. Learn. Syst., vol. 25, no. 6, pp. 1217–1226, Jun. 2014.

3   Y. N. Li, C. G. Yang, S. S. Ge, and T. H. Lee, "Adaptive output feedback NN control of a class of discrete-time MIMO nonlinear systems with unknown control directions," IEEE Trans. Syst. Man, Cybern. B, Cybern., vol. 41, no. 2, pp. 507–517, Apr. 2011.

4   W. S. Chen, L. C. Jiao, J. Li, and R. Li, "Adaptive NN backstepping output-feedback control for stochastic nonlinear strict-feedback systems with time-varying delays," IEEE Trans. Syst. Man, Cybern. B, Cybern., vol. 40, no. 3, pp. 939–950, Jun. 2010.

5   A. Y. Alanis, E. N. Sanchez, and A. G. Loukianov, "Real-time discrete backstepping neural control for induction motors," IEEE Trans. Control Syst. Technol., vol. 19, no. 2, pp. 359–366, Mar. 2011.

6   J. Na, X. M. Ren, and D. D. Zheng, "Adaptive control for nonlinear pure- feedback systems with high-order sliding mode observer," IEEE Trans. Neural Netw. Learn. Syst., vol. 24, no. 3, pp. 370–382, Mar. 2013.

7   S. S. Ge, J. Zhang, and T. H. Lee, "Adaptive neural network control for a class of MIMO nonlinear systems with disturbances in discrete- time," IEEE Trans. Syst. Man, Cybern. B, Cybern., vol. 34, no. 4, pp. 1630–1645, Aug. 2004.

8   Yan-Jun Liu, Li Tang, Shaocheng Tong and C. L. Philip Chen, "Adaptive NN controller design for a class of nonlinear MIMO discrete-time systems," IEEE transactions on neural networks and learning systems, vol. 26, no. 5, pp. 1007-1018, May 2015.

9   V. K. Deolia, S. Purwar, and T. N. Sharma, "Backstepping control of discrete-time nonlinear system under unknown dead-zone constraint," in Proc. Int. Conf. Commun. Syst. Netw. Technol., pp. 344–349, Jun. 2011.

10   C. L. P. Chen, Y. J. Liu, and G. X. Wen, "Fuzzy neural network-based adaptive control for a class of uncertain nonlinear stochastic systems," IEEE Trans. Cybern., vol. 44, no. 5, pp. 583–593, May 2014.

11   Zhongsheng Hou and Shangtai Jin, "Data sriven model free adaptive control for a class of MIMO nonlinear discrete-time systems," IEEE transactions on neural networks, vol. 22, no. 12, pp. 2173-2188, Dec. 2011.

12   Yan-Jun Liu and Shaocheng Tong, "Adaptive NN tracking control of uncertain nonlinear discrete-time systems with nonaffine dead-zone input," IEEE transactions on cybernetics, vol. 45, no. 3, pp. 497-505, March 2015.

13   Mou Chen, Shuzhi Sam Ge and Bernard Voon Ee How, "Robust adaptive NN control for a class of uncertain MIMO nonlinear systems with input nonlinearities," IEEE transactions on neural networks, Vol. 21, No. 5, pp. 796-812, May 2010.

14   M. Gopal, "Digital control and state variable Methods: conventional and intelligent control systems", 4th edition, 2009.

15   V. K. Deolia, S. Purwar, and T. N. Sharma, "Adaptive Backstepping controller of discrete-time nonlinear system with input saturation," IJCSN, Vol. 1, No. 3.

Prerna shukla [1] and Gaurav kumar sharma [2]

Comparative Study of Comparator Topologies: A Review

Abstract: The Analog to digital converter (ADC) is widely used in many applications such as wireless communication and digital audio video fields to improve digital system performance with respect to analog solution. In design of ADC, comparator plays a significant role. The comparator compares two analog signals and produces the output in the digital form or in other word comparator is a 1-bit ADC. Comparators are used in zero crossing detector, switching power regulator, memory sense amplifier etc. This paper presents six different topologies of comparator. SIS based comparator has the advantage of reduced power consumption, Where as dynamic latched comparator is utilized to minimize the circuit complexity along with maintain the circuit accuracy.

Keywords: ADC, dynamic latch comparator, SIS comparator,VSV comparator,power consumption,area.

1 Introduction

In electronics world the second most widely used   component is comparator while first is operational amplifier [1].Therefore operational amplifier and comparator plays an important role in the electronics. As we know that the comparator is one bit analog to digital converter [2][3]. The conversion procedure is, firstly it samples the analog signal and then hold it and converts it into digital form. The conversion speed of comparator is limited by the decision making response time of the comparator. There are many applications where comparators are used like zero crossing detector, peak detectors, switching power regulators, analog to digital converter etc. In flash and pipeline ADC the preamplifier based comparator have been used in previous years [8],but disadvantage of this comparator is offset voltage. This drawback of preamplifier based comparator is overcome by the dynamic latched comparator. That's why

---

1 M.tech scholor,GLA university mathura
Prerna.shukla_mtec14@gla.ac.in
2 Assistant professor,GLA university Mathura
Gauravkr.sharma@gla.ac.in

today most widely used comparator is dynamic latched comparator [4]. However this comparator suffers from larger power dissipation than that of preamplifier based comparator.But increases the rate of conversion in analog to digital converter. By using the dynamic comparator in back to back form which provides the positive feedback. By this mechanism we can convert a smaller voltage difference in full scale digital level output.TIQ technique provides high conversion speed and makes ADC very fast .VSV based comparator reduces the power dissipation in comparison to TIQ comparator. Figure (1) shows the schematic of comparator and figure (2) shows the ideal voltage transfer characteristic of comparator.

This paper is organized as follows.In section II the main three architectures comparator topologies are explained.In section III the different topologies of comparators with there advantages and disadvantages are discussed and the related equations also explaine.In section IV the comparison table of different comparator topologies are present which explains the power consumption and area requirement of each topology.In section V summaries the work.

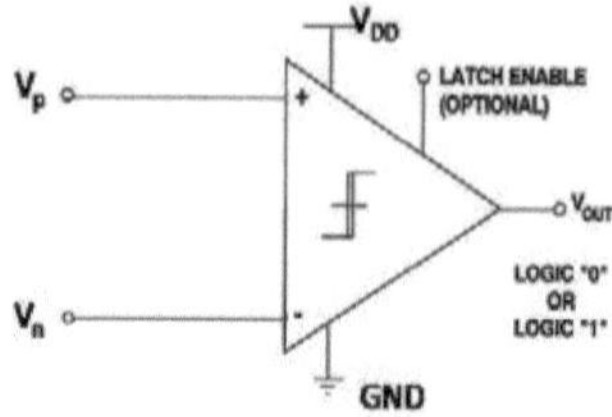

Fig 1. Schematic of comparator

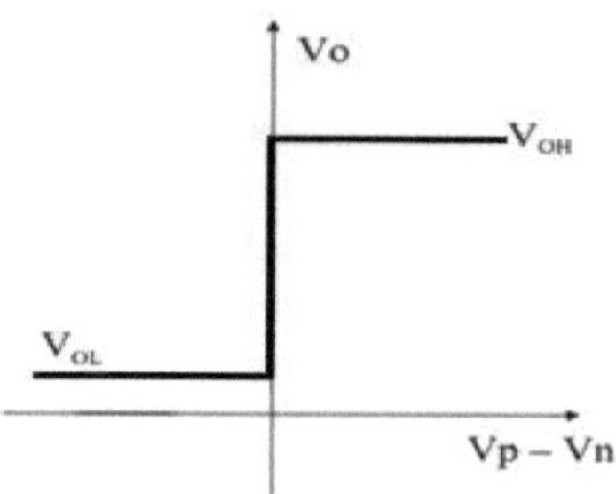

Fig 2. Ideal voltage transfer characteristic of comparator

In figure (1) at positive input terminal vp is applied and at negative input terminal or at non-inverting terminal vn voltage is applied.

If vp >vn then vo is logic "1"

If vn>vp then vo is logic "0"

2 Comparator architectures

There are basically main three different architectures of comparator

A) Open loop comparator

B) Regenerative comparator

C) High speed comparator

A) Open loop comparator: These comparators are basically operational amplifier without compensation for improving the time response and to obtain the largest bandwidth. Due to the limited bandwidth product open loop comparators are too slow.But on the other hand if these are connected in cascaded form than it provides a large gain bandwidth product in comparison with single stage amplifier. However it increases the area and power consumption.

B) Regenerative comparator: Positive feedback is used in regenerative comparators to compare signals. This comparator is also known as latch in which the two cross coupled MOSFETs.The advantage of using latches having a faster switching speed.

C) High speed comparator: It consists of the above two types of comparators which gives faster response.

3 Different Topologies Of Comparator

3.1 Preamplifier based comparator

This comparator consist of 3 stages, the input preamplifier stage, a latch stage and output buffer stage. The preamplifier stage is basically a differential amplifier with active load. To improve the comparator sensitivity the preamplifier stage amplifies the input signal, and it also isolates the input from switching noise coming from the positive feedback stage [1].Input referred latch offset voltage can also be reduced by using this comparator. To find which one of the input signal is larger and to amplify their difference the latch stage can be used, whereas to get the rail to rail output, the noise present in the circuit can be removed by amplification. Output buffer stage is used to converter the latch stage output to a full scale digital level output. This topology has huge static power consumption.

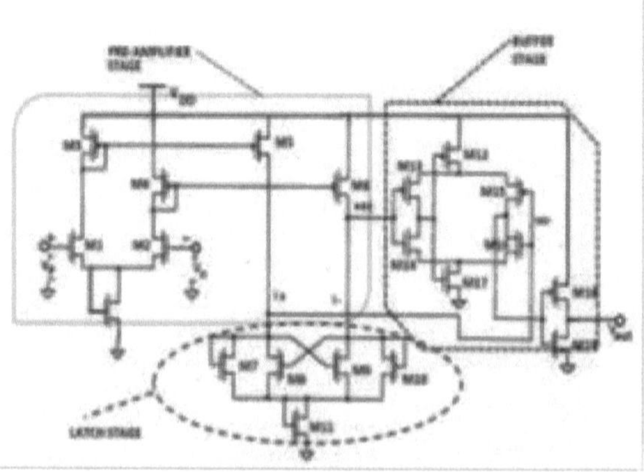

Fig 3. preamplifier based comparator

3.2 Switched inverter scheme (SIS) based comparator

This technique is also known as TIQ (threshold inverter quantization) technique. In this technique two inverters are connected in cascaded form to generate the range of internal reference voltages [3].The first inverter is having a unique threshold voltage for every step of input voltage. At first stage the form factor for CMOS inverter is selected and corresponding threshold voltage for all TIQ comparator is set.Whereas the second inverter is designed with a constant form factor. Second inverter is used to increase the voltage gain and present unbalanced propagation delay of all comparator. Advantage of using this topology, it makes the ADC very fast because it provides high conversion rate along with this it reduces power dissipation of the data converter.Calculating the threshold voltage for every transistor is the disadvantage of this technique. Here the Vm is internal voltage of a comparator and is fixed for each comparator. Vm is represented by the mathematical equation,

$$v_m = \left(\frac{\left(v_{dd} - |v_{tp}| + v_{pn}\right)\sqrt{\dfrac{k_p}{k_n}}}{1 + \sqrt{\dfrac{k_p}{k_n}}} \right)$$

Where wp and wn are the width of PMOS and NMOS, μn and μp are electron and hole mobility. Cox is the oxide thickness. L is the channel length of devices.

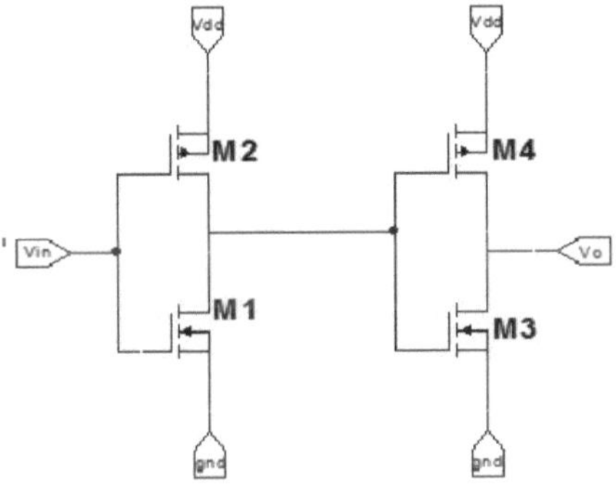

Fig 4. SIS based comparator

3.3 Clocked SIS comparator

In this topology the two cascaded inverters are used as SIS and two set of NMOS and PMOS connected in parallel at the pull up and pull down network. One pair is connected at the side of pull up network while another one is connected to the pull down network of SIS comparator[3]. The clock pulse is provided to both NMOS and PMOS, in which NMOS is connected to clock while PMOS is connected to the clockbar. This topology reduces the power dissipation upto 60.37% to that of the SIS comparator.

3.4 SIS comparator with sleep transistor

In this topology the SIS comparator is modified by adding the high threshold PMOS at pull up and NMOS at pull down. The static power dissipation is reduced by using this topology region being it increases the resistance of high threshold PMOS and NMOS, by the addition of header and footer [3]. Every

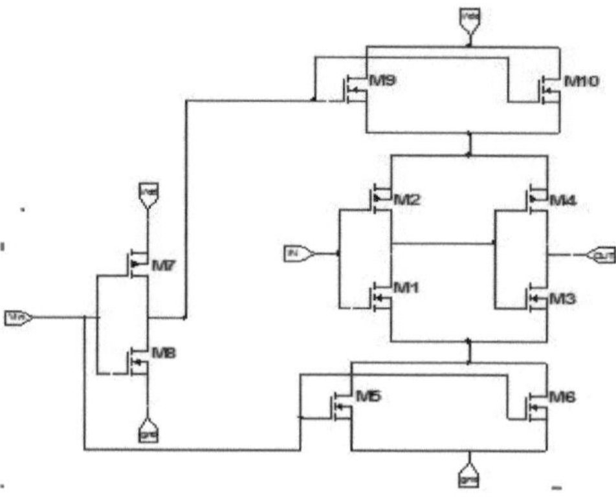

Fig 5. clocked SIS comparator

comparator is differently sized and hence the current through each comparator section is also different.

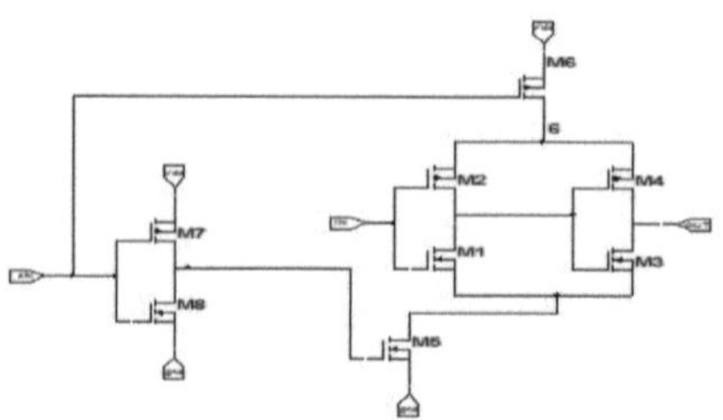

Fig 6. SIS comparator with sleep transistor

The calculation for the sizing of a sleep transistor is done by using equation,

$$\left(\frac{W}{L}\right)_{sleep} = \frac{I_{sleep}}{0.00281\,\mu_n C_{ox}(V_{DD} - V_T)(V_{DD} - V_{TH})}$$

By simulating the circuit without sleep transistor network, I_{sleep} can be calculated and also finding a maximum current that flows through ground or V_{DD}. V_{th} is the threshold voltage for short channel device and is V_{th} equal to threshold voltage of high k device.

This technique reduces the power dissipation approximately 77.3% to that of the SIS comparator.

3.5 Variable switching voltage (VSV) based comparator

This comparator consists of the eight number of transistors. However the number of transistor has been doubled as compared to the traditional TIQ comparator. There is a addition of two transistor M1 and M2 in the circuit which provides negative feedback in the comparator due to which there is a reduction in the drain current Mn1 and Mp1 of the first stage [4]. This modification decreases the power consumption of the first stage. The same concept is applied to the second stage of the comparator. Which provides sharper switching of the logical voltages and it is used to invert the output of the first stage. The drain and gate terminal of all the devices M1,M4,M5 and M8 are at same potential. Which causes all these are always in saturation (Vg=Vd).Using the transconductance (gm) the active resistance can be determined which is offered by the saturated device. For the NMOS device, transconductance is given by the following equation,

$$g_m = \frac{\partial I_D}{\partial V_{gs}} = \sqrt{\left(\frac{W}{L}\right)\mu_n c_{ox} I_D}$$

Where,µn is electron mobility and cox is gate oxide capacitance and Id is the drain current.

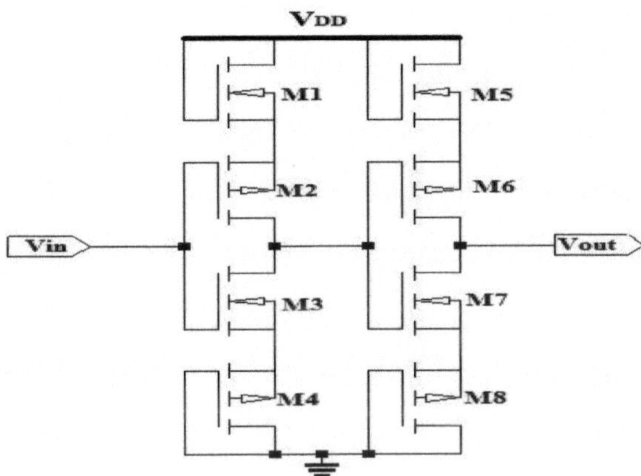

Fig 7. VSV based comparator

The disadvantage of this VSV based comparator is that it increases the area requirement in comparision to TIQ based technique.

3.6 Double tail dynamic latched comparator

This comparator is improved version of the conventional dynamic comparator the disadvantage of double tail dynamic latched comparator is that it requires the more number of transistor for that high supply voltage is needed for proper delay time [2]. In the present double tail dynamic latched comparator the two control transistor have been added in cross coupled manner with the first stage in parallel to M3 and M4. When clock=0, in this case both f_n and f_p nodes have been charged to v_{DD} and hence both switches (Mc1 and Mc2) are closed than it (fn and fp) starts to drop with different discharging rates. If any one of them discharge faster than comparator detects it and control by the control transistor (Mc1 and Mc2) which increases the voltage difference between them. But when the control transistor turns ON in other case than, current drawn from vdd to a ground via input transistor and the tail1 transistor.Which causes static power consumption. To minimize this effect, below the transistor M1 and M2 two NMOS switches are used (Msw1 and Msw2).

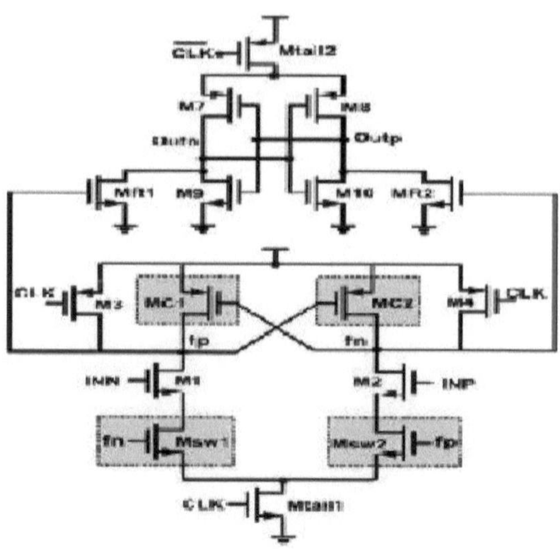

Fig 8. Dynamic latched comparator

The advantage of using dynamic latch comparator is that it has an excellent speed along with accuracy and also has zero static power dissipation.

4 Comparison table

Comparator topologies	Power consumption	Area
Double tail dynamic latched comparator	Low static power consumption	increase
SIS based comparator	low	minimum
Clocked SIS based comparator	60.37% reduced to that of SIS	increase
SIS comparator with sleep transistor	77.37% to that of SIS	increases
VSV based comparator	low	Increases to that of SIS
Preamplifier based comparator	More static power consumption	------

Conclusion

In this paper the different topologies of comparator is present. The comparison has been done on the basis of power, area. We found that dynamic latched comparator minimizes the static power dissipation and required lesser area in comparison to preamplifier based comparator but it produces the kick back noise. While TIQ technique is very useful in Flash ADC because they require a lesser area so it replaces the resistor ladder in flash ADC and provides high speed. As the technology is going to scale down there is further need to improve the comparator topologies.

References

1   R.jacob Baker,harry W.Li,David E. Boyce, "CMOS circuit Design,Layout,And simulation",IEEE press series on microelectronic system,IEEE press,prentice Hall of india private limited,Eastern Economy Edition,2002.
2   Samaneh babayan mashhadi et al, "Analysis and design of a low voltage low power double tail comparator" IEEE 2014.
3   Arun kumar sunaniya and kavita khare, "Design of improved resistor less 45nm switched inverter scheme (SIS) Analog to Digital converter, international journal of VLSI design & communication systems(VLSICS),3 june 2013.
4   Gulzer Ahmed and Rajendra Kumar Baghel,"Design of 6-bit flash analog to digital converter using variable switching voltage CMOS comparator,"international journal of VLSI design & communication system,3 june 2014.

5   neha bansal and kuldeep singh et al. Presented",comparative analysis of CMOS comparator in various Topologies for ADC applications,5 may 2014.

6   H.J.Jeon, Y.B. Kim, "a CMOS low-power low-offset and high speed fully dynamic latched comparator" IEEE, 2010.

7   Shakil mastan vali et. al "A 3GHz low offset fully dynamic latched comparator for   high speed low power ADCs," International Journal of Emerging Technology and Advanced Engineering (ISSN 2250-2459, ISO 9001:2008 Certified Journal, Volume 3, Issue 6, June 2013).

8   S.sheikhaei,S.mirabbasi and A.inanov, "A 0.35µm CMOS comparator comparator circuit for high speed ADC m on circuit and system appilication",IEEE international symposium PP.6134-6137,2005.

9   Amir zjajo jose,Pineda degyvez(2011), "low-power high resolution analog to digital converters,design test and calibration" ISBN 978-90-481-9724-8,First edition, springer new York.

10 dharmendra mani verma(2011) "reduced comparator low power flash ADC using 35nm CMOS",IEEE conference on electronics computer technology(ICECT),vol.1,pp.385-388.

11 G.torfs,Z.Li,j.Bauwelinck,X. yin,G. van der plas and j. vandewege, "low-power 4-bit flash analogue to digital converter for ranging applications",IEEE electronics letters,vol.47 no.1,pp 20-22.

12 T.esther rani,dr. rameshwar rao(2011) "area and power optimized multipliers with minimum leakage" , IEEE international conference on electronic computer technology-ICECT, ISBN no.978-1-4244-8679 vol.3, pp.284-287.

13 Pedro m. figeireodo, Joao c.vital, "Kickback noise reduction techniques for CMOS latched comparator", IEEE transactions on circuits and system,vol.53,no. 7,pp.541-545,july 2006.

14 Philip E. Allen and Douglas R.holberg, "CMOS analog circuit design",2ⁿᵈ Ed. New York, NY: Oxford,2002.

15 Jun He "Analysis of static and dynamic random offset voltages in dynamic comparators", IEEE transactions on circuits and system-I regular papers,vol.56,No.5,may2009.

16 De-Shiuan chiou,Yu-ting chen, Da-cheng juan, and shih-chieh,(2010) "Sleep transistor sizing for leakage power minimization considering temporal correlation",IEEE transactions on computer-aided design of integrated circuits and systems,vol.29,no.8,pp.1285-1290.

17 P.Iyappan,P.Jamuna and S. Vijayasamundiswary,(2009), "Design of analog to digital converter using CMOS logic", IEEE international conference on advances in recent technologies in communication and computing.ISBN no. 978-0-7695-3845-7,pp.74-76.

18 P.Nuzzo,F.D.Bernardinis,p.terreni,and G.van der plas, "Noise analysis of regenerative comparators for reconfigurable ADC architecture," IEE Trans. Circuits syst. I ,reg. papers,vol.55,no. 6,pp.1441-1454,jul.2008.

19 B.Goll and H.Zimmermann, "A 65nm CMOS comparator with modified latch to achieve 7GHz/1.3mW at 1.2v and 700MHz/47uW at 0.6V",in proc.IEEE int.solid-state circuits conf.Dig.Tech.papers,feb.2009,pp.328-329.

20 B.Wicht,T.Nirchl,and D.Schmitt-Landsiedel, "Yield and speed optimization of a latch-type voltage sence amplifier," IEEE J.solid state circuits,vol.39,no. 7,pp.1148-1158,jul.2004.

Dr.P.Poongodi [1], N.Prema [2] and R.Madhu Sudhanan [3]

Implementation of Temperature Process Control using Conventional and Parameter Optimization Technique

Abstract: The objective of this paper is to design controllers for a temperature process using conventional and parameter optimization technique. PID (Proportional Integral Derivative) controller is the well known and most widely used in the industries. Ant Colony Optimization (ACO) has become the leading form of advanced algorithm used in temperature control in the process industry. The proposed ACO-PID controller uses the combination of a PID and ACO technique. The comparative performance analysis has been done for PID and ACO-PID controllers.

Keywords: PID, ACO, ACO-PID, FOPDT

1 Introduction

The aim of the project is to control the temperature with in a desired limit. The temperature controller can be used to control the temperature of the any system. The temperature process is a non linear system. Temperature process typically contains a controller unit, temperature input unit and controller output unit. The temperature sensor form the temperature input unit and Solid State Relay drive forms the control output unit. Reference [4] narrates the implementation of temperature control using Supervisory control and data acquisition (SCADA). The system is based on a first order model and the process model also contains time delay. The system contains a first order transfer function representing a time constant .The block diagram for temperature process is shown in Fig. 1.

1 Professor, Dept of ECE, Karpagam College of Engineering, Coimbatore- 641 032, India
poongodiravikumar@gmail.com
2 PG Scholar, Dept of ECE, Karpagam College of Engineering, Coimbatore- 641 032, India
premanrpr@gmail.com
3 Research Scholar, Dept of ECE, Karpagam College of Engineering, Coimbatore- 641 032, India
c_r_mathusuthanan@yahoo.co.in

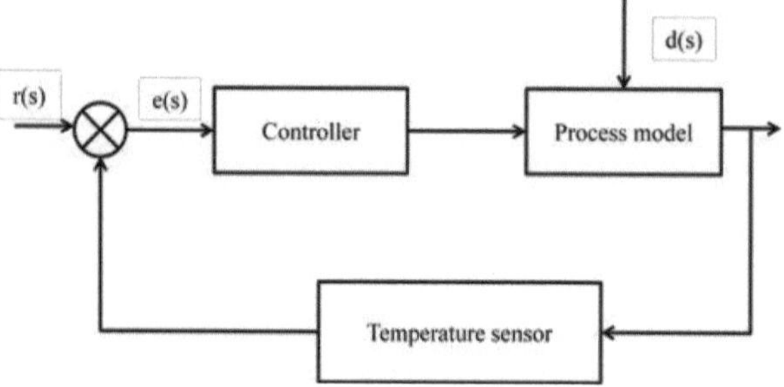

Fig 1. Block diagram of temperature process control

2 System Modeling

The most commonly industrial process can be modeled by first order plus dead
time models. The system can be done mainly in two ways one is the
mathematical modeling and another one is the empirical modeling.
Temperature process was modeled using the empirical modeling. The process
model of the transfer function is obtained from the step response. The First
Order Plus Delay Time (FOPDT) parameters that are to be determined are the
process gain (K), delay time (t_d), time constant (τ) units of minutes or seconds.
The First order plus time delay process equation is given by,

$$G_p(s) = \frac{K}{\tau s+1} e^{-t_d s} \qquad (1)$$

Where, K =Process gain, t_d = Delay time, τ = Time constant.

The temperature process of the transfer function is experimentally obtained
from [5] as,

$$G_p(s) = \frac{7}{12s + 1} e^{-1s} \qquad (2)$$

3 Control Strategies

The control strategies used for temperature control are PID Control and ACO-
PID control.

3.1 PID Controller

The PID Controller has Proportional, Integral and Derivative actions inbuilt. The output is proportional to the error at the instant time't'. It is reducing the rise time of process. The output is proportional to the integral error at the instant time't'; it can be interpreted at the accumulation of the past error. The controller removes the steady state error but maintain the transient response. The output is proportional to the derivative error at the instant time't'; it can be interpreted at the prediction of the future error. The controller affects the system by increasing stability and by reducing overshoot, improving the transient response.

$$P = K_p e(t) \quad ; I = K_i \int_0^t e(t)d(t) ; D = K_d \frac{de(t)}{dt}$$

The transfer function of the PID controller,

$$G(s) = \left(K_p + \frac{K_i}{s} + K_d s\right) \tag{3}$$

K_p= Proportional gain, K_i= Integral gain, K_d=Derivative gain

The open loop response of plant and the closed loop response of plant with PID Controller for unit step input is also shown in Fig 2.

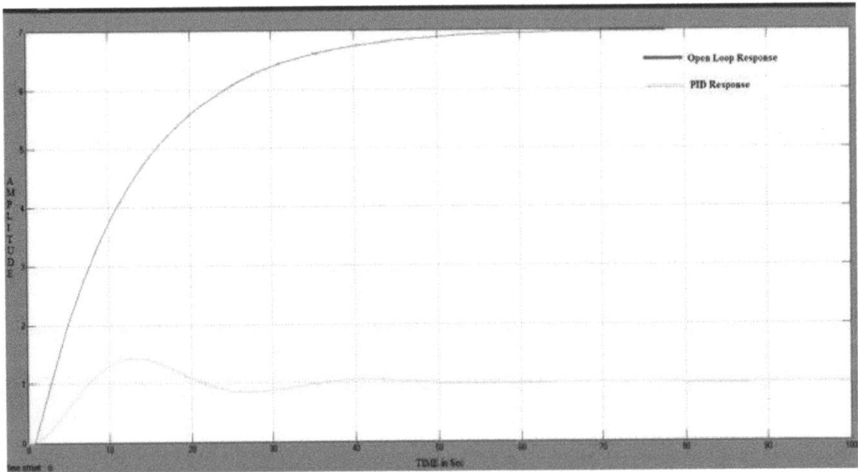

Fig 2. Open loop response and closed loop Response of Plant with PID Controller

The conventional PID controller has high settling time and rise time of 47.8 sec and 5.01 sec respectively with 1.5% overshoot.

3.2 ACO- PID Controller

The application of ACO algorithm for conical tank level control is detailed in [1].
The GA based PID controllers [6], IMC based PID controller [7, 9] and Fuzzy PID
based Controllers [8] has also been a new arena in temperature control process.
ACO's are especially suited for finding solutions to different optimization problems. A colony of artificial ants cooperates to find good solutions, which are an
emergent property of ant's co-operative interaction. Based on their similarities
with ant colonies in nature, ant algorithm are adaptive and robust and can be
applied to different versions of the same problem as well as to different
optimization problems. The main traits of artificial ants are taken from their
natural model. These traits are artificial ants exist in colonies of cooperating
individuals, they communicate indirectly by depositing pheromone they use a
sequence of local moves to find the shortest paths from the starting position, to
a destination point they apply a stochastic decision policy using local information only to find the best solution. If necessary in order to solve a particular
optimization problem, artificial ants have been enriched with some additional
capabilities not present in real ants. An ant searches collectively for a good
solution to a given optimization problem. Each individual ant can find a solution or at least part of a solution to the optimization problem on its own but only
when many ants work together they can find the optimal solution.

Since the optimal solution can only be found through the global
cooperation of all the ants in the colony, it is an emergent result of such this cooperation. While searching for a solution the ants do not communicate directly
but indirectly by adding pheromone to the environment. Based on the specific
problem an ant is given a starting state and moves through a sequence of
neighboring states   trying to find the shortest path. It moves based on a
stochastic local search policy directed by its internal state. The pheromone
trails, and local information encoded in the environment. Ants use this private
and public information in order to decide when and where to deposit pheromone. In most application the amount of pheromone deposited is proportional to
the quality of the move ant has made. Thus more pheromone, the better the
solution found. After an ant has found a solution, it dies; it is deleted from the
system. Ant Colony Optimization (ACO) algorithm is used to optimize the gains
and the values are applied in to the controller for the given plant. The proportional gain makes the controller respond to the error .The integral gain help to
eliminate steady state error and derivative gain prevent overshoot respectively.
ACO is depending upon the pheromone matrix $\tau = \{\tau_{ij}\}$ for the construction of
good solutions. The values of τ are

$$\tau_{ij} = \tau_0 \forall(i,j) \text{ Where } \tau_0 > 0$$

The probability $P_{ij}^A(t)$ of choosing a node j at node i. At each generation of the algorithm, the ant constructs a complete solution using this equation, starting at source node.

$$P_{ij}^A = \frac{[\tau_{ij}(t)^\alpha][\eta_{ij}]^\beta}{\sum_{ij\in T^A}[\tau(t)]^\alpha[\eta_{ij}]^\beta}; i,j \in T^A \tag{4}$$

Where $\eta_{ij} = \frac{1}{kj}$, j=[p,i,d] represents the heuristic function.

α and β are constant that determine the relative influence of the pheromone values and the heuristic values on the decision of the ant.

T^A is the path effectuated by the ant A at a given time.

The quantity of pheromone on each path may be defined as

$$\nabla\tau_{ij}^A = \begin{cases} \frac{L^{min}}{L^A} & \text{if i,j} \in T^A \\ 0 \end{cases} \tag{5}$$

Where L^A is the value of the objective function found by the ant A

L^{min} is the best solution carried out by the set of ants until the current iteration.

The pheromone evaporation is the way to avoid unlimited increase of pheromone trials and also it allows the forgetfulness of the bad choices.

$$\tau_{ij}(t) = \rho\tau_{ij}(t-1) + \sum_{A=1}^{NA}\Delta\tau_{ij}^A(t) \tag{6}$$

Where NA is the number of ants ρ is the evaporation rate $0< \rho \leq 1$.

3.3 ACO –PID Controller Design

The PID Controller is a well known technique in the process control area. The design of this controller requires three main parameters, proportional gain (K_p), Integral time constant (K_i), and derivative time constant (K_d).Design of ACO algorithm is used to optimize the parameters of the PID Controller for the temperature controller process. A plethora of ACO algorithm [2,3,10] has been studied for PID parameter optimization. Fig. 3 Shows the ACO-PID process control loop.

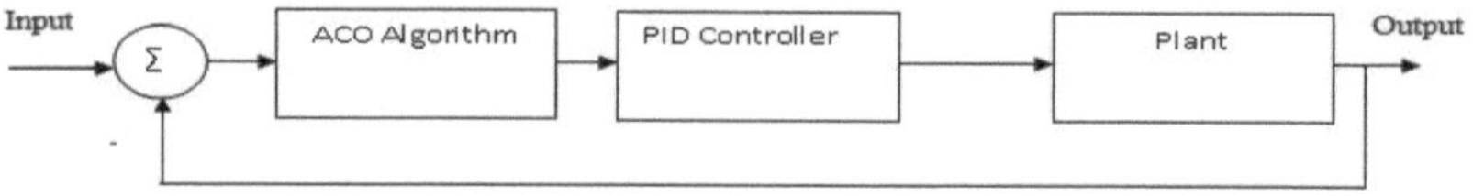

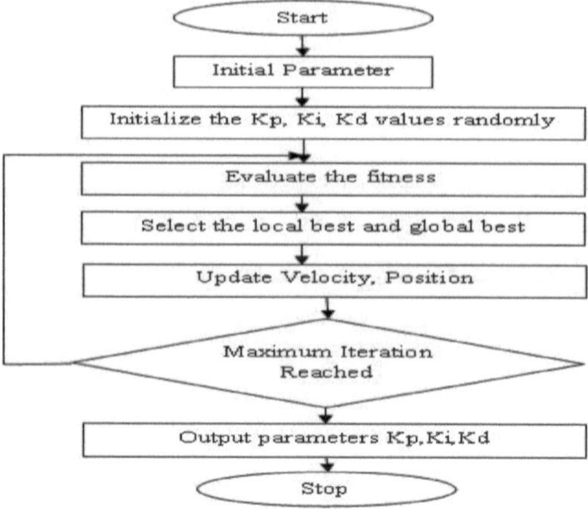

Fig 3. Process Loop with ACO-PID Controller

3.4 ACO Algorithm

Step 1: Initialize randomly a potential solution of the solution of the parameters (K_p ,K_i , K_d) by using uniform distribution. Initialize the pheromone trial and the heuristic value.

Step 2: Place the Ath ant on the complete heuristic value associated in the objective (minimize the error).

Step 3: Use pheromone evaporation to avoid unlimited increases of pheromone trials and allow the forget fitness of bad choices

Step 4: Evaluate the obtained solution according to the objective.

Step 5: Display the optimum value of the optimization parameters.

Step 6: Globally updates the pheromone, according to the optimum solution calculated at step 5. Iteration from step 2 until the maximum of iteration reached.

The Fig 4. Shows the flow chart of ACO –PID algorithm.

Fig 4. Flow Chart of ACO-PID

The parameters chosen are $\alpha = 1$, $\beta = 2$, $\rho = 0.8$, Number of ants = 500 for the plant transfer function $\frac{7}{12s+1}e^{-1s}$. The proportional gain makes the controller respond to the error. The integral gain helps to eliminate steady state error. The derivative gain to eliminate .The step response of ACO-PID was shown in Fig 5. The settling time of IMC controller is 0.6sec with a small overshoot.

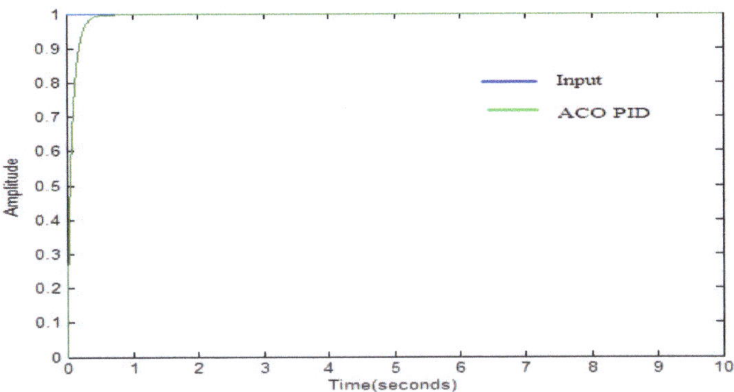

Fig 5. Response of Plant with ACO-PID Controller

4 Results And Discussion

Conventional PID controller has high settling time and rise time of 47.8 sec and 5.01 sec respectively with 1.5% overshoot. The settling time and rise time of ACO-PID controller is 0.6 sec and 0.5 sec respectively with no overshoot.

Table 1. Performance indices of PID & ACO-PID controllers

S.No	Controller Type	Settling Time (sec)	Rise Time (sec)	Over Shoot (%)
1	Conventional PID	47.8	5.01	1.5
2	ACO-PID	0.6	0.5	0

Conclusion

First Order plus Delay Time model is obtained for temperature process through system identification. Conventional PID and ACO-PID responses are obtained for the FOPDT temperature process model. The simulation is carried out using MATLAB/ Simulink for both the conventional PID and ACO-PID controller.

Acknowledgement

The authors are thankful to the institution for providing good infrastructure facilities. The authors are also thankful to the project co-coordinator Dr.C.Aravind for lending his support for the progress of this project.

References

1   Giri Rajkumar.S.M, Ram kumar.K & Sanjay Sarma O.V. ,“ Real Time Application of Ants Colony Optimization”, International Journal of Computer Applications, 3, 2010, 34-46.
2   Duan Hai-bin, Wang Dao-bo & Yu Xiu –fen, “Novel Approach to Nonlinear PID Parameter Optimization Algorithm ”, Journal of Bionic Engineering, 3, 2006,73-78
3   Hong He, Fang Liu , Li Li, Jin-Rong Yang, Le su and Yi Wu, “ Study of PID Control for Ant Colony Algorithm”, Global Congress on Intelligent System, IEEE Computer Society, 2009,204-207
4   Laware, A. R., Bandal,V. S., & Talange, D.B., “Real Time Temperature Control System Using PID Controller and Supervisory Control and Data Acquisition System (SCADA)”. International Journal of Application or Innovation in Engineering & Management, 2 (2), 2013,88-95.
5   Nithyarani, N., & GirirajKumar, S. M, “Model Identification of Temperature Process and Tuning with Advanced Control Techniques”. IJIREEICE, 1(9), 2013,443-447.
6   Nithyarani, N., Dr.GirirajKumar, S. M., & Dr.Anantharaman, N.“Modeling and control of temperature process using genetic algorithm”, IJAREEIE, 2 (11), 2013, 5355-5364.
7   Nithyarani, N., & Ranganathan, S. “Advances in Control Techniques and Process Analysis with LabVIEW and DCS”. International Journal of Electronics, Communication & Instrumentation Engineering Research and Development, 3(2), 2013,137-148.
8   Pamela, D., & Jebarajan, T. “ Intelligent Controller for Temperature Process”. International Journal of Control and Automation, 6(5), 2013, 191-198.
9   Shamsuzzoha, M.(2007), “IMC-PID controller design for improved disturbance rejection of time-delayed process”, Ind. Eng. Chem. Res., 46, 2007, 2077-2091.
10   Ying-Tung Hsiao & Cheng-Long Chuang. “Ant Colony Optimization for Designing of PID Controllers”, IEEE International symposium on Computer Aided Control Systems Design, 2004, 321-326

Dr.S.Rajkumar [1] and Dr.M.Ramkumar Prabhu [2]

A Novel Compact Dual-band slot Microstrip Antenna for Internal Wireless Communication

Abstract: This paper presents a compact, wideband microstrip patch antenna with a E-shape slot for dual frequency. The proposed antenna operates at 2.15 GHz and 2.98 GHz bands. The antenna size is very compact (40 mm x 60 mm x 1.6 mm) and is fed from a 50 Ω microstrip line. Using IE3D, according to the set size, the antenna is simulated. The computer simulation results show that the antenna can realize wide band characters. The two operating bands exhibit broad impedance bandwidths (VSWR ≤ 1.8) of about 25% and 15%.

Keywords: Microstrip antenna, E-shape slot, dual frequency, microstrip line, IE3D

1 Introduction

The rapid development of wireless communication technology has increased the demand for compact microstrip antennas with high gain and wideband operating frequencies. Microstrip patch antennas are very advantageous because of their low cost, low profile, light weight and simple realization process. However, the general microstrip patch antennas have some disadvantages such as narrow bandwidth etc. Enhancement of the performance to meet the demanding bandwidth is necessary [1]. There are numerous and well-known methods to increase the bandwidth of antennas, including increase of the substrate thickness, the use of a low dielectric substrate, slotted patch antenna, the use of various impedance matching and feeding techniques [2-17].

This paper presents a new E-shaped slot loaded patch antenna that is investigating for enhancing the impedance bandwidth. By choosing the suitable slot shape, selecting a proper feed and tuning their dimensions, a large operating bandwidth is obtained. The design employs 50 Ω microstrip

---

1 Professor, ECE Dept, Nehru college of Enginring & Research Center,Pampady,Kerala
2 Professor, ECE Dept, Dhaanish Ahmed College of Engineering, Chennai, India

line feeding. The antenna is simulated using IE3D software package of Zealand. The results show that the impedance bandwidth has achieved a good match.

2 Antenna Geometry

The dielectric constant of the substrate is closely related to the size and the bandwidth of the microstrip antenna. Low dielectric constant of the substrate produces larger bandwidth, while the high dielectric constant of the substrate results in smaller size of antenna. A trade-off relationship exists between antenna size and bandwidth [18].

The resonant frequency of microstrip antenna and the size of the radiation patch can be similar to the following formulas [19].

$$f \cong \frac{c}{2L\sqrt{\varepsilon_r}} \tag{1}$$

$$W = \frac{2}{f_r}\left(\frac{\varepsilon_r + 1}{2}\right)^{\frac{1}{2}} \tag{2}$$

$$L = \frac{c}{2f_r\sqrt{\varepsilon_r}} - 2\Delta l \tag{3}$$

Where f is the resonant frequency of the antenna, c is the free space velocity of the light, L is the actual length of the current, ε_r is the effective dielectric constant of the substrate and Δl is the length of equivalent radiation gap.

The geometries of the E-shaped slot antenna are shown in figure 1. The antenna is built on a glass epoxy substrate with dielectric constant 4.2 and height h of 1.6 mm. A substrate of low dielectric constant is selected to obtain a compact radiating structure that meets the demanding bandwidth specification. For the given design the slot and the feeding line are printed on the same side of the dielectric substrate. The geometry of the top view and side view of the proposed antenna is shown in figure 1 (a) and (b) respectively. The dimensions of the slotted patch are shown in figure (c). Here L_G and W_G represent the length and width of the ground plane which is finite whereas Lp and Wp are those of the patch. The lengths of the horizontal arms of the E-shape slot are defined by L_1, L_2 and L_3 where as W_1, W_2 and W_3 define the widths. The length and width of the vertical arm are defined by L_4 and W_4 respectively. The patch is also loaded with a horizontal slot at center coordinates (20, 50). The length and width of the slot are shown by L_5 and W_5 respectively. An additional horizontal slot is cut on the ground plane, whose length and width are shown by L_6 and W_6 respectively. The patch is fed by a 50 Ω microstrip line whose length and width are defined by L_7 and W_7 respectively. The use of microstrip line feeding technique provides the bandwidth enhancement. Table 1 shows the optimized design parameters for the proposedantenna.

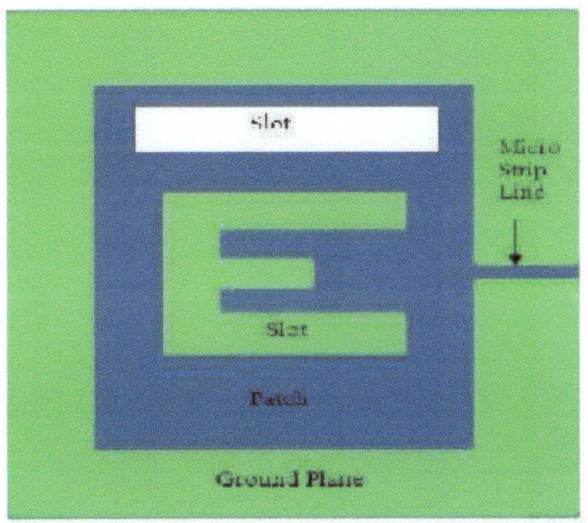

(a)   Top view of the antenna.

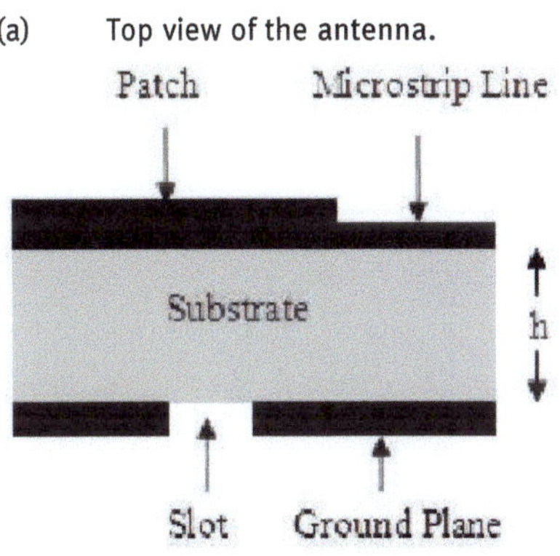

(b)   Side view of the antenna.

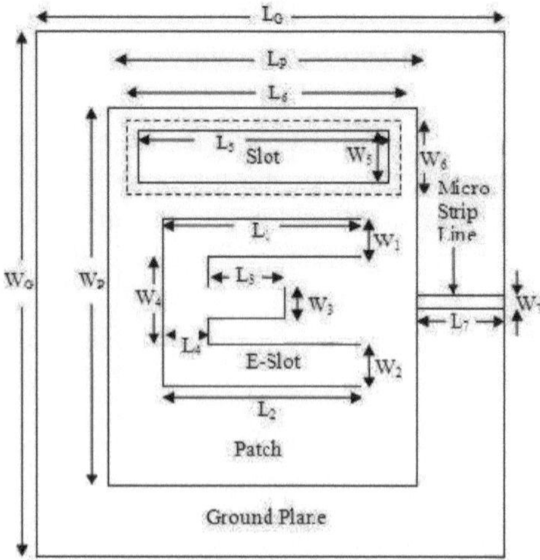

(c) Dimensions of the antenna.

Fig 1. Geometry of the proposed antenna

Table 1. The proposed patch antenna design parameters

Parameter	Value [mm]	Parameter	Value [mm]
W	5	L	15
L	4	W	10
W	6	L	10
L	3	W	10
W	5	L	25
L	2	W	9
W	5	L	30
L	2	W	2
W	5	L	5

Reducing the size of the antenna is one of the key factors to miniaturize the wireless communication devices. However, reducing the antenna size will usually reduce its impedance bandwidth as well. Therefore designing a small antenna operating with a wide impedance bandwidth which satisfies future generation wireless application is a challenging work, especially having stable radiation patterns across the operating frequency band [19]. In this paper microstrip line feeding, slots on the patch as well as on ground plane provide the wide bandwidth and gain enhancement.

3 Results And Discussions

The proposed antenna was simulated and optimized by IE3D software package of Zealand. This was used to calculate the return loss, impedance bandwidth and radiation pattern.

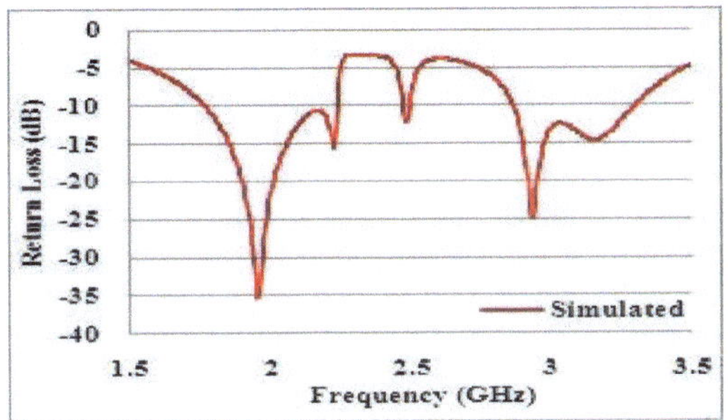

Fig 2. Simulated return loss of the proposed antenna

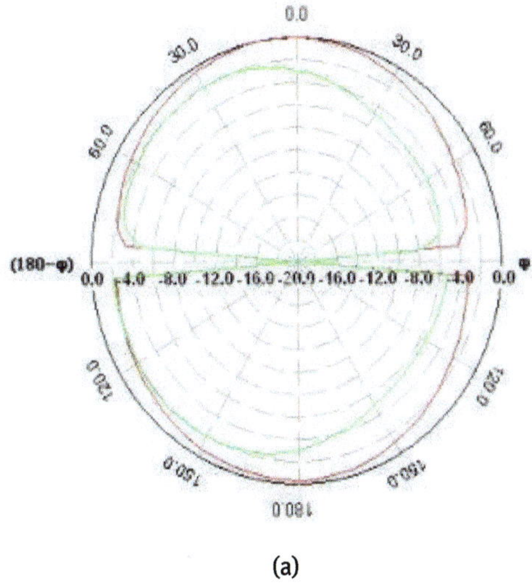

(a)

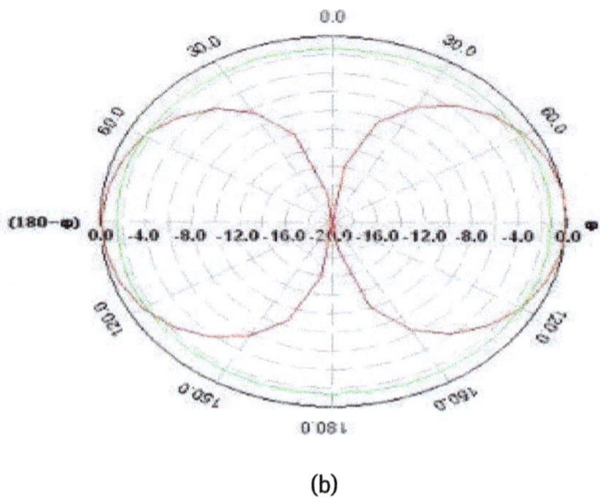

(b)

Fig 3. Radiation pattern of the proposed antenna. (a) Elevation and (b) Azimuth

The simulated -10 dB return loss of the proposed antenna is shown in the figure 2. The simulated result shows that the first band resonant frequency locates at about 2.15 GHz with the -10 dB impedance bandwidth from about 1.965 GHz to 2.8 GHz. The second band resonant frequency locates at about 2.98 GHz with the impedance bandwidth from about 2.8 GHz to 3.5 GHz. The -10 dB return loss impedance bandwidths for first and second band are 35% and 25% respectively. The simulated radiation patterns of the elevation and azimuth of the proposed antenna are shown in figure 3. It can be observed that the proposed antenna have the same radiation patterns over the entire frequency band.

Conclusion

A novel compact dual-band slot microstrip antenna for 1.965/3.5 GHz is presented. The proposed antenna has a compact size of (40 mm x 60 mm x 1.6 mm) and it can effectively cover the AMPS, GSM and WLAN applications. Good antenna performance and impedance matching can be realized by adjusting the length and width of microstrip line. The two operating bands exhibit broad impedance bandwidths (VSWR ≤ 1.8) of about 35% and 25%. It can be concluded from the results that the designed antenna has satisfactory performance and hence can be used for indoor wireless applications.

References

1   Ramadan, A., K. Y. Kabalan, A. El-Hajj, S. Khoury, and M. Al-usseini, *"A reconfigurable U-Koch microstrip antenna for wireless applications,"* Progress In Electromagnetics Research, PIER 93, 355-367, 2009.

2   Kawei Qian and Xiaohong Tang, *"Compact LTCC dual-band circularly polarized perturbed hexagonal microstrip antenna,"* IEEE Antennas and Wireless Propagation Letters, Vol. 10 pp. 1212-1215,2011.

3   K. Kumar and N. Gunasekaran, *"A novel wideband slotted mm wave microstrip patch antenna,"* Proc. IEEE Vol. 987-1, pp. 10-14, 2011.

4   Aliakbar Dastranj and Habibollah Abiri, *"Bandwidth enhancement of printed E-shaped slot antennas fed by CPW and microstrip line,"* IEEE Transactions on Antennas and Propagation, Vol. 58 No. 4, 2010.

5   Ayesha Aslam and F. A. Bhatti, *"Novel inset feed design technique for microstrip patch antenna,"* Proc. IEEE Vol. 978-1, pp. 215-219, 2010.

6   A. Sen, J. S. Roy and S.R. Bhadra Chaudhuri, "Investigation on a dual-frequency microstrip antenna for wireless applications," Proc. IEEE. Vol. 978-1, 2009.

7   M. T. Islam, M. N. Shakib and N. Misran, *"Broadband E-H shaped microstrip patch antenna for wireless systems,"* Progress In Electromagnetics Research, PIER 98,163-173, 2009.

8   Kasabegoudar, V. G. and K. J. Vinoy, *"A broadband suspended microstrip antenna for circular polarization,"* Progress In Electromagnetics Research, PIER 90, 353-368,2009.

9   Bai-wen Tian, Jin-ming, *"A Design of Dual-Band H- shaped Microstrip-Line-Fed Printed Wide-Slot Antenna,"* Proc. IEEE, vol. 978-1, pp. 201 - 203, 2008.

10 Lin S-, Row J-, *"Bandwidth enhancement for dual- frequency microstrip antenna with conical radiation,"* Electronics Letters. 2008; 44(1): 2-3.

11 Albooyeh, M. N. Kamjani, and M. Shobeyri, *"A novel cross-slot geometry to improve impedance bandwidth of microstrip antennas,"* Progress In Electromagnetics Research Letters, Vol. 4, 63-72, 2008.

12 G. M. Zhang, J. S. Hong and B. Z. Wang, *"Two novel band-notched UWB slot antennas fed by microstrip line,"* Progress In Electromagnetics Research, PIER 78, 209-218, 2008.

13 Ren, W., J. Y. Deng, and K. S. Chen, *"Compact PCB monopole antenna for UWB applications,"* Journal of Electromagnetic Wave and Applications, Vol. 21, No. 10, 1411-1420, 2007.

14 B. K. Ang and B. K. Chung, *"A Wideband E-shaped microstrip patch antenna for 5–6 GHz wireless Communications,"* Progress In Electromagnetics Research, PIER 75, 397-407, 2007.

15 Ravi Pratap Singh Kushwaha," *Bow-shape single layer with single probe feed microstrip patch antenna,"* Proc. IEEE, 2005.

16 Y. F. Liu, K. L. Lau, Q. Xue, C. H. Chan. *"Experimental Studies of Printed Wide-Slot Antenna for Wide-Band Applications,"* IEEE Antennas and Wireless Propagation Letters, Vol. 3, pp. 273-275, 2004.

17 Yu, A. and X. X. Zhang, *"A method to enhance the bandwidth of microstrip antennas using a modified E- shaped patch,"* Proceedings of Radio and Wireless Conference, 261–264, Aug. 10–13, 2003.

18 D. M. Pozar, *"Microstrip Antennas,"* Proc. IEEE, vol.80, No. 1, pp. 79-81, January 1992.

19 Xiaofei Shi, Zhihong Wang, Hua Su, Yun Zhao, *"A H- type Microstrip Slot Antenna in Ku-band Using LTCC Technology with MultipleLayerSubstrates,"* Proc. IEEE, Vol. 978-1, pp. 7104 - 7106, 2011.

Annushree Bablani[1] and PrakritiTrivedi[2]

Comparing Data Of Left And Right Hemisphere Of Brain Recorded Using EEGLAB

Abstract: Non-invasive type of brain computer interaction records brain signal by simply placing electrodes on scalp. It does not need any surgery. Electroencephalography (EEG) is a non-invasive type of technique that capture brain signals using an EEG machine consisting electrodes to be placed on scalp using 10-20 international system of electrode placement. Raw signals of EEG always have some noise with them. Eye movement causes great change in EEG waves (alpha, beta gamma, and theta) produced. For removing these noise it is necessary to identify them that in which part of brain they are more active and where not. So we have analysed Alpha EEG wave recorded from various electrodes from brain scalp during open eye of subjects. Results of our work show that alpha wave is dominant in occipital lobe of brain and a variation in readings collected from left and right hemisphere is also analysed.

Keywords: BCI, EEG, EEG waves, frontal lobe, left hemisphere, right hemisphere

1 Introduction

Inspired by the social recognition of people who suffer from severe neuromuscular disabilities, an interdisciplinary field of research has been created to offer direct human computer interaction via signals generated by the brain itself that technique of interaction is known as *brain computer interaction*. Brain computer interface can be called as a bridge between person's thoughts (as input) and a computer device (as output). It maps one's brain signals and provide specific output. BCI records brain signals via different techniques, in our work we have used the non-invasive method of recording signals i.e. brain

1 Department of Computer Science and Engineering, Government Engineering College, Ajmer, Rajasthan
annushree.bablani@gmail.com
2 Department of Computer Science and Engineering, Government Engineering College, Ajmer, Rajasthan
prakrititrivedi27@gmail.com

activities recorded from scalp via electrodes. Electroencephalography, is one of the non-invasive method to record brain data. EEG measure the combined electrical activity of neurons. This combined activity of brain is captured by placing electrodes on various lobes of brain and then these activity in form of signals is visualized using any interface. When we record data, with brain signals noise in data is also recorded. In brain signals these artifacts are due to muscle movement, eye movement, some voice heard by subject etc.

Our brain is divided in four lobes [1] that are (a) frontal lobe, responsible for thinking, planning, emotion recognition, problem solving etc. it is the front part of brain, (b) parietal lobe, integrate sensory information and manipulate objects, middle top of brain (c) temporal lobe, senses smell and sound, temple region i.e. above ears and (d) Occipital lobe, sense of sight, at extreme back of head. 64 channel electrodes are placed on these lobes and the brain activity is recorded by EEG machine. This EEG activity is shown in EEG waveform. For different activities different waveforms are generated.

2 Literature survey

First ever EEG machine was given to the world by Hans Berger ('he was a NeuropsychiatristatUniversity of Jena in Germany'), in 1929. To describe graphical map of the brain signals he used the German term *"elektrenkephalogramm"*. He was the first to suggest that- *"Depending on the functional status of the brain, brain waves change"* such as sleep, awake, thinking, anaesthesia, and epilepsy.

A method based on linear discriminant analysis has been proposed by Denis Delisle et al. [3] to detect events associated to eyes opening and closing, by capturing alpha waves measured from the occipital lobe.

Another method PSD (Power spectral density), helps identify the frequency domain where strength of signal is more or less. PSD can be seen as frequency response of periodic or random signals. The PSD for certain types of random signals is independent of time and hence it is deterministic. PSD is useful for analysis of random signals especially in BCI because brain signals are highly variable. SamanehValipour et al. in their paper [2] have recorded EEG signals only for 10 seconds because these signals are stationary for a time period of less than 12 seconds. MATLAB with EEGLAB has been used for computation of PSD for EEG signals. Authors have observed PSD on three subjects on channels P_z, P_3, P_4, F_z and C_z with subjects opened and closed eyes.

ICA is a powerful method used to separate independent data mixed linearly using various channels. For instance, when recording electroencephalograms (EEG) on the scalp, ICA can separate out artifacts embedded in the data (since they are usually autonomous of each other). The number of channels directly influence the performance of separation between artifact and pure EEG components. Before applying ICA on EEG data pre-processing is performed i.e. "whitening of data". In this pre-processing step simple linear change is performed over the coordinates of mixed data. Some of the ICA properties are: ICA can only separate the data that are linearly mixed, changing order of data points that are plotted cannot affect the result of ICA algorithm applied, change in position of electrode will also not affect the result of ICA, if data is dependent than also ICA algorithms finds the maximum separation between the sources and give the outcome.

While recording data from brain scalp via electrodes the electrical activity of neurons are displayed in waves form on the computers. These waves represents different mental state of the subject. Frequency is best way to represent the varying state of any EEG activity as it is easy to asses it. Brain neurons oscillations are observed in form of waves (defined on basis of their frequency) from a person mind are:

a) Delta (0.1 to 3.5 Hz): This rhythm is dominant in infants and during deep sleep of adults and when a person is suffering from serious brain disorder.

b) Theta (3.5 to 7.5 Hz): This rhythm is found while a person is in sleeping state (or drowsy). It is also found in children when they are awake. Mainly observed at frontal, temporal and parietal region.

c) Alpha (7.5 to 13 Hz): It is dominant when a person is awake performing daily tasks. Mostly found at occipital and parietal lobes of brain and stronger over the right hemisphere. Present when a subject is mentally inactive, alert, with eyes closed. Blocked or weakened by deep sleep, attention, especially visual, and mental effort.

Beta (13 to 30 Hz): Beta waves, with lower frequencies disappear during mental activity and Beta waves, with higher frequencies appear while a person is in tension and under intense mental activity. Under intense mental activity beta can extend up as far as 50Hz.

3 Working

EEGLAB: EEGLAB is a tool that collaboratively work with MATLAB toolbox for processing continuous/single and event-related EEG and other

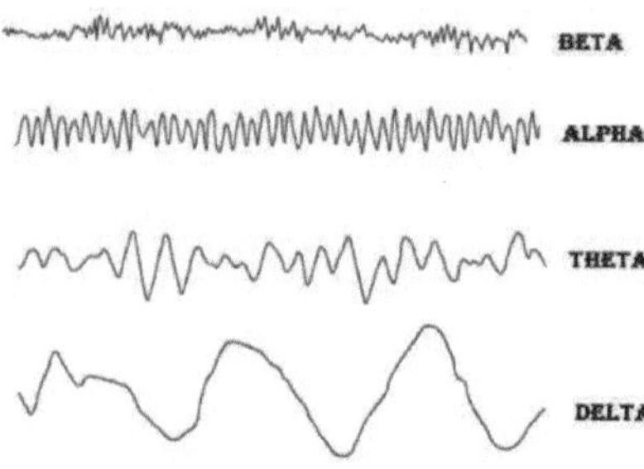

Figure 1: EEG rhythms

electrophysiological data. It includes methods like independent component analysis (ICA), time analysis, frequency analysis and some other artifact removal methods. To process complex EEG data, EEGLAB provides its users an interactive graphical user interface (GUI) which allow easy and interactive processing of EEG data and other available methods. EEGLAB also integrates general tutorial and help windows and a command history function which allow users to easily switch from one mode to another i.e. from GUI-based to script-based. EEGLAB has treasure of methods using those one can visualise and models the event-related brain data. Users who have a knowledge of MATLAB for them EEGLAB provides a programming environment, which offers storing, measuring, editing, updating, accessing and visualising the EEG data. EEGLAB is an open source platform, it allows researchers to create and share their new methods with the world.

4 Results

In our paper we have compared and analyzed activity of brain during open eye of subjects.

For our work we have downloaded data, the downloaded dataset created and contributed to PhysioNet by the developers of the BCI2000 instrumentation system, which they used in making theserecordings. The system is described by Goldberger ALet al. in their work [6]. We have collected data for 23 channels from left and 23 channels from right hemisphere of brain. Channel number and their respective electrodes name are shown in table 1. Data was recorded with subjects having open eyes [5].

For analyzing and comparing our results we have selected alpha wave having frequency range from 8 to 13 Hz, for all channels.Channel power and frequency plot using GUI interface of EEGLAB has been performed (Figure 3 and Figure 4). In our paper we have only shown channel plot of electrodes from occipital lobes and other electrodes values have been tabulated (Table 2).

On observing the nature of alpha wave we concluded at two results, first, power values where higher for channel number 21,22 and 23 (Placed on occipital lobe) for maximum subjects and secondly these values where higher in left hemisphere than in right hemisphere. (Comparison is plotted in figure 5).

Table 1. Channel number and their respective electrodes

Left Hemisphere		Right Hemisphere	
Channel number	Electrode Name	Channel number	Electrode Name
1	FC5	1	FC2
2	FC3	2	FC4
3	FC1	3	FC6
4	C5	4	C2
5	C3	5	C4
6	C1	6	C6
7	CP5	7	CP2
8	CP3	8	CP4
9	CP1	9	CP6
10	FP1	10	FP2
11	AF1	11	AF4
12	AF3	12	AF8
13	F7	13	F2
14	F5	14	F4
15	F3	15	F6
16	F1	16	F8
17	P7	17	P2
18	P5	18	P6
19	P3	19	P4
20	P1	20	P8
21	PO7	21	PO8
22	PO3	22	PO4
23	O1	23	O2

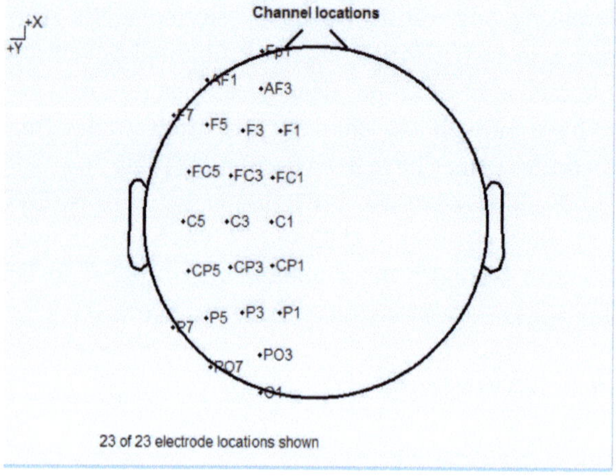

(a)

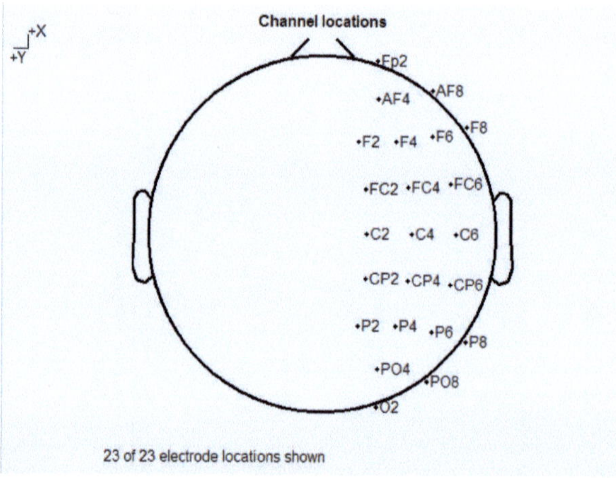

(b)

Figure 2. Electrodes placement in (a) left hemisphere and (b) right hemisphere

Figure 3 and Figure 4 shows power spectrum values for occipital lobe during eye open from left and right hemisphere respectively.

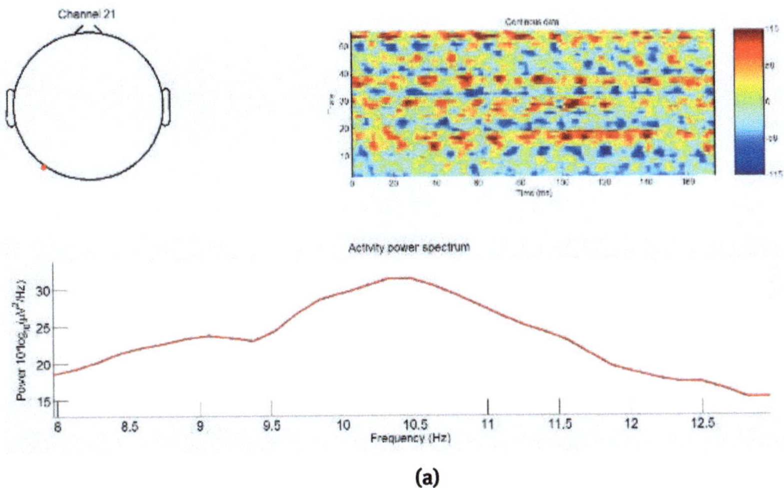

(a)

For channel 21 or electrode PO7 power value reaches 31.4 µV²/Hz at 10.48 Hz frequency

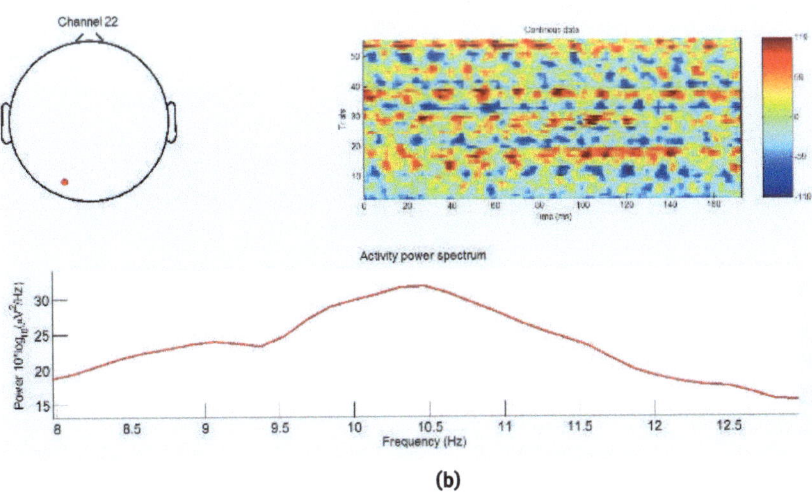

(b)

For channel 22 or electrode PO3 power value reaches 31.7 µV²/Hz at 10.48 Hz frequency

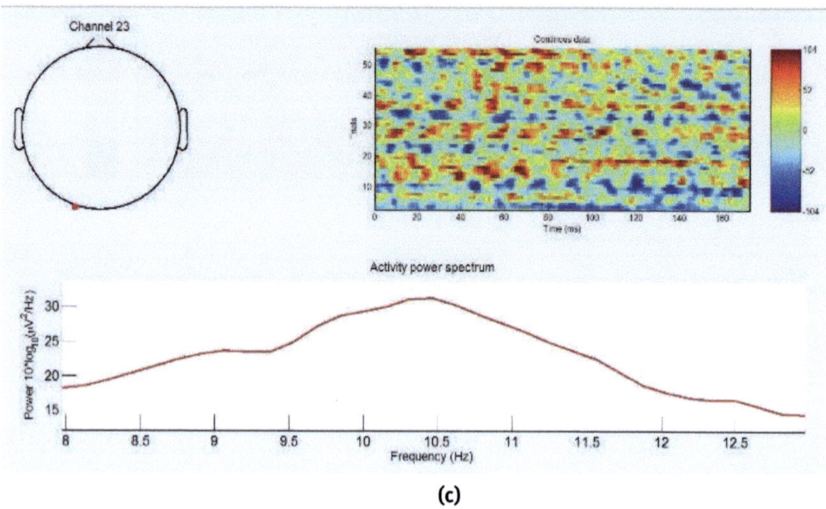

(c)

Figure 3: (a) to (c) shows Alpha wave power spectrum for subject 8 recorded from all channels of left hemisphere

For channel 23 or electrode O1 power value reaches 31.24 $\mu V^2/Hz$ at 10.47 Hz frequency

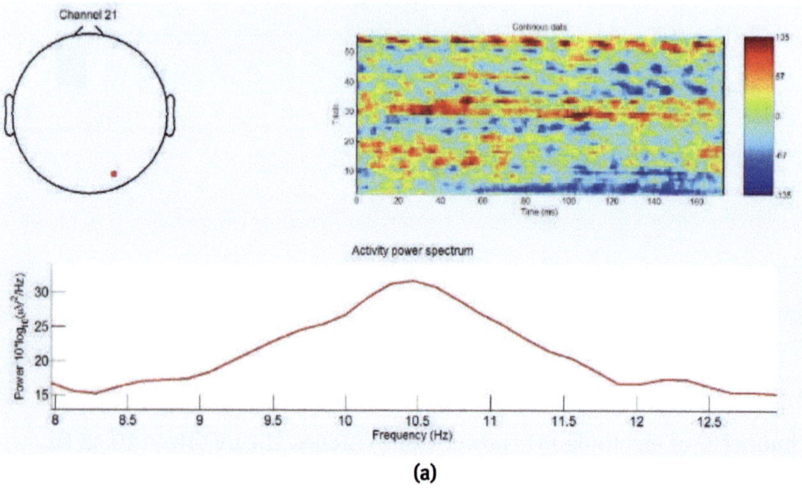

(a)

For channel 21 or electrode PO4 power value reaches 31.75 $\mu V^2/Hz$ at 10.48 Hz frequency

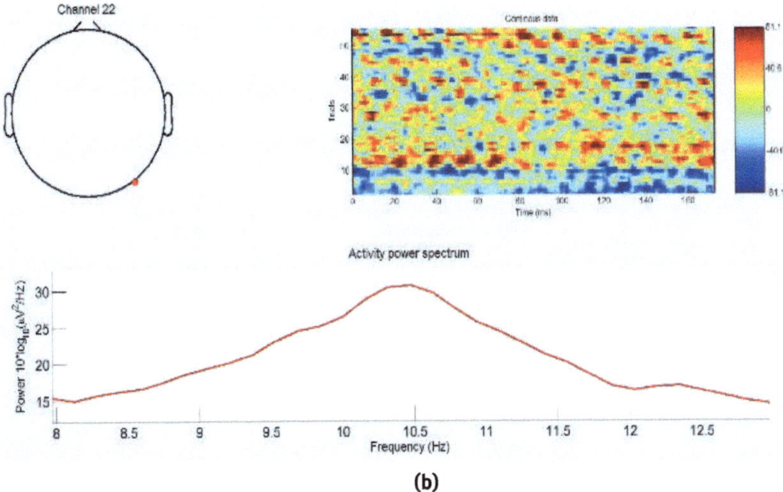

(b)

For channel 22 or electrode PO8 power value reaches 30.5 µV^2/Hz at 10.5 Hz frequency

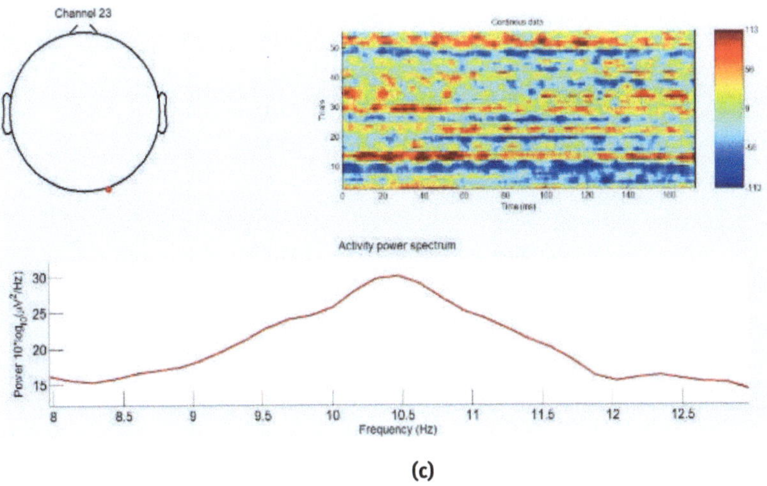

(c)

Figure 4. (a) To (c) Alpha wave power spectrum for subject 8 recorded from all channels of right hemisphere

For channel 23 or electrode O2 power value reaches 30.07 µV^2/Hz at 10.4 Hz frequency

From table 2 we can see that during eye open electrodes of occipital lobe have higher power values than electrodes of other lobe (Value for occipital lobe is shown in bold).

Table 2. Power value for subject 8 for left and right hemisphere

	Left hemisphere	Right hemisphere
Channel no.	Power (μV^2/Hz)	Power (μV^2/Hz)
1	24.1	23.37
2	24.7	22.62
3	24.2	20.77
4	24.5	23.6
5	25.7	23
6	24.4	20.02
7	26.6	27.4
8	27.6	27.5
9	27.8	25.43
10	24.3	23.49
11	24.1	22.95
12	24.4	22.8
13	24.2	23.51
14	24.4	23.13
15	24.6	23.10
16	24.5	21.45
17	27.3	31.00
18	30.3	30.40
19	30.32	29.66
20	30.8	26.75
21	*31.4*	*31.75*
22	*31.7*	*30.50*
23	*31.24*	*30.07*

Mean power value for both hemispheres is:

Mean (LH) = 26.65 μV^2/Hz

Mean (RH) = 25.40 μV^2/Hz

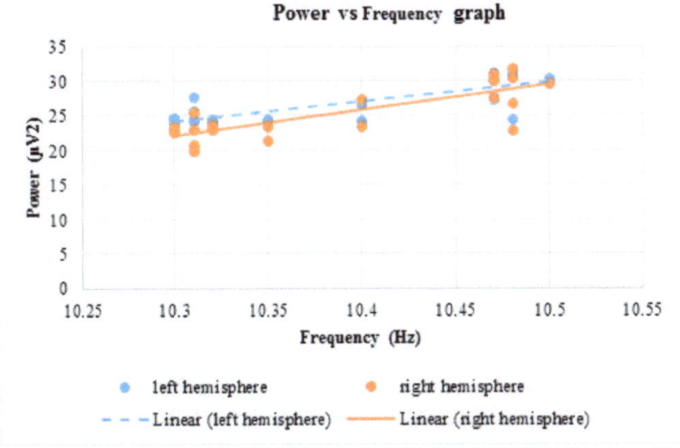

Figure 5: Power vs. frequency plot for left and right hemisphere

Result plotted show that power values of left hemisphere are more than that of right hemisphere for alpha wave, which shows that left part of brain is more active.Standard deviation values have been tabulated and plotted for the same subject.

Table 3. Standard deviation of subject 8 for left and right hemisphere

Channel number	Standard Deviation	
	Left Hemisphere	**Right Hemisphere**
1	101.9	60.09
2	96.61	65.43
3	67.27	85.79
4	78.76	55.08
5	82.24	56.23
6	61.97	53.13
7	88.17	58.54
8	68.5	60.35
9	67.09	78.49
10	110	97.26
11	148.9	86.46
12	96.15	117.9
13	138.6	68.06

14	103.9	65.67
15	90.26	95.29
16	82.03	121.8
17	82.59	60.57
18	101.1	74.35
19	87.15	48.26
20	66.96	42.47
21	73.43	82.5
22	75.63	57.83
23	72.63	80.92

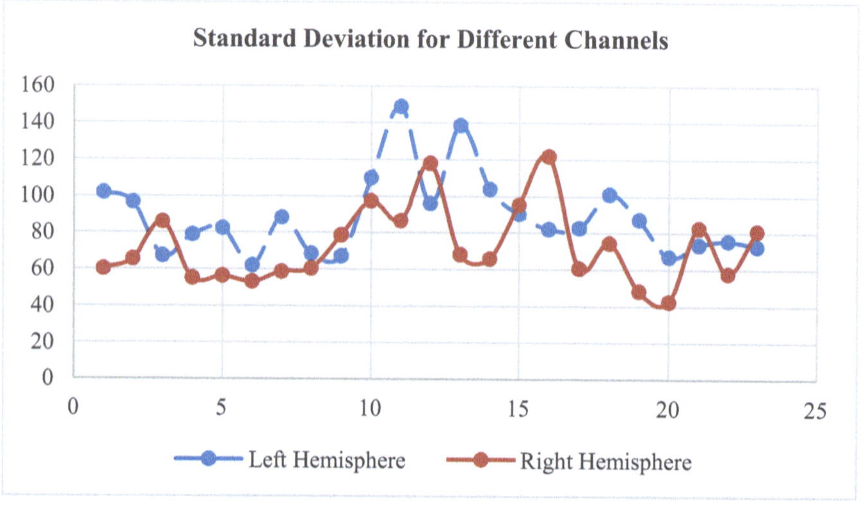

Figure 6: Plot of Standard Deviation

Root mean square value for each channel from both hemisphere is also calculated which is shown in table 3 for all the channels. Let, 'V$_n$', be the peak voltage, 'V$_{rms}$' be the rms voltage and 'n' be the channel number, than the value of V$_{rms}$ can be calculated as: $Vrms = \frac{1}{\sqrt{2}}.Vn; \forall\, n = 1\ to\ 23$

Table 4: Showing average min and max RMS value of all 23 channels for all 11 subjects

Subject no.	Left hemisphere		Right Hemisphere	
	Maximum (RMS in microvolt)	Minimum (RMS in microvolt)	Maximum (RMS in microvolt)	Maximum (RMS in microvolt)
Subject-1	133.26	-127.94	125.78	-131.39
Subject-2	269.69	-269.91	319.13	-316.34
Subject-3	94	-108.95	86	-11.04
Subject-4	166.13	-153.21	122	-122.34
Subject-5	121.30	-121.82	107	-112.39
Subject-6	327.39	-319.34	270.39	-278.73
Subject-7	182.95	-221.86	159.73	218.47
Subject-8	333.04	-292.73	269.21	-251.08
Subject-9	171.39	-164.95	159.78	-159.34
Subject-10	311.17	-292.82	310.52	-296
Subject-11	147.47	-143.08	122.30	-139.65

Conclusion

Brain computer interface is a way to make one's thoughts visible to the world. It does so by recording data from various channels that have been placed on brain scalp. In our paper, we have analysed and studied 23channels, from left and right hemisphere of brain each and found out the alpha wave nature on these electrodes i.e. 8-13 Hz. It was observed that open eye have higher value of alpha wave on occipital lobe compared to any other. Occipital lobe seems to be more active when subject is with eye open and from values we can observe that Lefthemisphere is more active than right hemisphere.

In future researchers can analyse individually if for right handed people left hemisphere is equally active as right hemisphere for left handed. Then we can minimize the number of electrodes and instead of using 64channel electrode placement we can also use half of the channels in future.

References

1   B.D. Chaurasia (2004, volume 3), *"Human Anatomy"* fourth edition page no. 345-365

2   SamanehValipour , A.D. Shaligram, G.R.Kulkarni, *Detection of an alpha rhythm of EEG signal based on EEGLAB,* Department of Electronic Science, Pune University, Pune, 411 007, Maharashtra,IndiaDepartment of Physics, Pune University, Pune, 411 007,Maharashtra,India.

3   Denis Delisle-Rodriguez, Javier F. Castillo-Garcia, TeodianoBastos-Filho, AnselmoFrizera-Neto, Alberto Lopez-Delis, *"Using Linear Discriminant Function to Detect Eyes Closing Activities through Alpha Wave "*

4   Motoki Sakai, Daming Wei, Wanzeng Kong, Guojun Dai, and Hua Hu," *Detection of Change in Alpha Wave following Eye Closure Based on KM2O-Langevin Equation",*International Journal of BioelectromagnetismVol. 12, No. 2, pp. 89 -93, 2010

5   [http://www.bci2000.org/]

6   Goldberger AL, Amaral LAN, Glass L, Hausdorff JM, IvanovPCh, Mark RG, MietusJE,Moody GB, Peng CK,Stanley HE. PhysioBank, PhysioToolkit, and PhysioNet: Components of a New Research Resource for Complex Physiologic Signals. *Circulation*101(23):e215e22 [Circulation Electronic Pages;[http://circ.ahajournals.org/cgi/content/full/101/23/e215]]; 2000 (June 13).

7   Oner, M.; Gongzhu Hu, "Analyzing One-Channel EEG Signals for Detection of Close and Open Eyes Activities," in *Advanced Applied Informatics (IIAIAAI), 2013 IIAI International Conference on* , vol., no., pp.318-323, Aug. 31 2013-Sept. 4 2013

8   Vargic, R.; Kacur, J., "Detection of Closed and Open eyes via brain control interface," in *ELMAR (ELMAR), 2014 56th International Symposium* , vol., no., pp.1-4, 10-12 Sept. 2014

9   Aziz, F.; Rahman, M.M.; Ahmad, T.; Jahan, M.S.; Tosrif, T.M.; Huq, M.M.; Reza, C.M.F.S.; Ahmed, A.U.; Day, P.; Badsha, S.; Al Mamoon, A.; Hasan, S.M.M., "Discrimination analysis of EEG signals at eye open and eye close condition for ECS switching system," in *Electrical Information and Communication Technology (EICT), 2013 International Conference on* , vol., no., pp.1-4, 13-15 Feb. 2014

10   Nishifuji, S.; Sugita, Y.; Hirano, H., "Event-related modulation of steady-state visual evoked potentials for eyes-closed brain computer interface," in *Engineering in Medicine and Biology Society (EMBC), 2015 37th Annual International Conference of the IEEE* , vol., no., pp.1918-1921, 25-29 Aug. 2015

11   Sosa, O.A.P.; Quijano, Y.; Doniz, M.; Quero, J.E.C., "Development of an EEG signal processing program based on EEGLAB," in *Health Care Exchanges (PAHCE), 2011 Pan American* , vol., no., pp.199-202, March 28 2011-April 1 2011

12   Ching-Chang Kuo; Lin, W.S.; Dressel, C.A.; Chiu, A.W.L., "Classification of intended motor movement using surface EEG ensemble empirical mode decomposition," in *Engineering in Medicine and Biology Society, EMBC, 2011 Annual International Conference of the IEEE* , vol., no., pp.6281-6284, Aug. 30 2011-Sept. 3 2011

13   Farina, D.; Jensen, W.; Akay, M., "Methods for Noninvasive Electroencephalogram Detection," in *Introduction to Neural Engineering for Motor Rehabilitation* , 1, Wiley-IEEE Press, 2013, pp.600

Robert Ssali Balagadde [1] and Parvataneni Premchand [2]

Non-Word Error Correction for Luganda

Abstract: Editing or word processing Luganda text has been an uphill task mainly because of lack of a system in this environment which could give a feedback on the spelling of Luganda words. In this context, this research paper presents a model for Non-Word error correction for Luganda (LugCorrect) which comes in handy to address this gap and consequently provide a more user friendly environment for editing Luganda text. To the best of our knowledge LugCorrect is the first of this kind of system developed for Luganda. *Jaccard Coefficient (JC)* - since it correlates well with *Damerau Levenshtein Distance (DLD)* and is computationally light (linear algorithm) - is used in LugCorrect for selection and ranking the *Correction Candidate List (CCL)*. Experimental results show that LugCorrect provided correction at an accuracy of 98% for erroneous words which are *Edit Distance of One (ED1)* through ED3 from their respectively targets at a correcting speed of 1365 Hz (words per second) with CCL size of 10.

Keywords: spelling error correction, non-word error correction for Luganda, Luganda corrector, Jaccard Coefficient, Damerau Levenshtein Distance, edit distance, and correction candidate list

1 Introduction

Editing or word processing Luganda text has been an uphill task mainly because of lack of a environmental system which could give a feedback on the spelling of Luganda words. In this context, this research work presents a model for Non-Word Spell Correction For Luganda (NWSCL- LugCorrect) which comes in handy to address this gap, and consequently, provide a more user friendly environment for editing Luganda text. To the best of our knowledge, LugCorrect is the first of this kind of system developed for Luganda.

---

1 Department of Computer Science & Engineering, University College of Engineering, Osmania University, Hyderabad, 500007, TS, India
baross.kla@gmail.com
2 Department of Computer Science & Engineering, University College of Engineering, Osmania University, Hyderabad, 500007, TS, India
Profpremchand.p@gmail.com

One challenge encountered while developing a model for Luganda spell checking is dealing with the infinite number of clitic-host word combinations (CHWCs) which make Luganda distinct from other foreign languages especially non-Bantu languages. In this research work, three types of CHWC are identified in respect to the use of inter-word apostrophe (IWA) defined as an apostrophe found within a word, rather than at the end or beginning of a word.

Type one - CHWC_1, bolded in Example 1 - are created as a result of compounding a modified monosyllabic word (MMW) or modified disyllabic word with an initial vowel (MDWIV) with another succeeding word which begins with a vowel. Compounding in Luganda may involve two to three words. MMW or MDWIV- referred to as clitic - is formed by dropping the ending vowel of the mono-syllabic word (MW) or disyllabic word with an initial vowel (DWIV) and replacing it with an apostrophe resulting into a long sound that is not represented by a double vowel.

Example 1
- ➤ omwenge **n'ennyama** (alcohol and meat) [conjunctive form]
- ➤ **n'otema** omuti (and you cut the tree) [narrative form]
- ➤ **ew'omuyizzi** (at the hunter's place) [locative form]
- ➤ Minisita omubeezi **ow'ebyobusuubuzi** ne tekinologiya (Minister of State for Trade and Technology)
- ➤ **n'obw'embwa** sibwagala (even that for the dog, I don't like)
- ➤ **n'olw'ensonga** eyo, sijja kujja (and for that reason, I will not come)

Example 2 shows some MWs and DWIVs (bolded) used in compounding to form CHWC_Is. Not all MWs and DWIVs are used in compounding, and Example 3 shows some of these.

Example 2
- ➤ ne, na, nga [conjunction]
- ➤ ne, nga [narratives]
- ➤ be, ze, ge. gwe , bwe, bye, lwe, lye, kye, bye [object relatives]
- ➤ kya, ya, za. lya ba, bya, ga, gwa, ka, lwa, wa [possessives]
- ➤ **ekya, eya, eza. erya, owa, aba** [possessives with initial vowel]
- ➤ kye, ye, be, ze, twe [copulatives]
- ➤ e [locatives]

Example 3
- ➤ era, ate, nti, so [conjunction]
- ➤ atya, oti [adverbs]
- ➤ ggwe, nze, ye, yo. bo, zo, bwo [emphatic pronouns]

> bba [nonn]
> si [negation]

Type II - CHWC_II, shown in Example 4 - are formed as a result of using clitic ng' and nng' to represent 'ŋ' character and double 'ŋ' character respectively in words which are not initially CHWC_1.

Example 4
> **ng'ang'**ala (to whimper like a dog)
> bbiri**ng'**anya (egg plant)
> agaku**ng'**anyizzaamu (to collect in it something)
> e**nng'**anda (relatives)
> **nng'**oma (dram)

Type III - CHWC_III, shown in Example 5 - are formed as a result of using clitic ng' and nng' to represent 'ŋ' character and double 'ŋ' character respectively in words which are initially CHWC_1.

Example 5
> **n'**agaku**ng'**anyizzaamu (and to collect in it something)
> **ng'eng'**anda (like relatives)
> **ng'enng'**oma (like a dram)

In view of this, the first task is disambiguating the three types of CHWCs. The approach adopted to disambiguate these words is capture in the Error Detection Mechanism (EDM) or LugDetect whose algorithm is shown in Figure 1.

It is worthwhile noting that the CHWC_1 akin to Luganda are also found in other Bantu languages like Runyankore-Rukiga, Kinyarwanda, among others. Example 6 and Example 7 shows samples of CHWC_I extracted from Runyankore-Rukiga and Kinyarwanda text respectively.

Example 6
> ky'okutunga
> nk'eihanga
> n'ekya
> g'ebyemikono

Example 7
> n'uw'Umukuru (and His Excellence)
> Nk'uko (like that)
> y'Umukuru (His Excellence)
> w'Igihugu (of a country)

Other Challenges include:

> ➢ Selection of the similarity matrix. The criteria used to address this challenge is elucidated in Section 3,
> ➢ Determining the number of element in the CCL. The approach adopted to address this is discussed in Section 3.

The scope of LugCorrect is limited to Luganda non-word correction. Named entities, abbreviations, e-mail addresses, uniform resource locator (URL) strings are not handled in LugCorrect, and neither are the real word errors.

2 Literature Survey

2.1 Errors

According to Peterson [19], spelling errors are basically of two types. Firstly, cognitive errors which are errors due to lack of knowledge of the language and are errors often ignored on the assertion that they are infrequent. In this context, their frequency is evaluated to be between 10% and 15% [21]. Secondly, typographical errors are 80% of "typos" (also called single character errors) and are one of the following four types: one exceeding character, one missing character, a mistaken character, or the transposition of two consecutive characters [5]. This means that 80% of errors are within edit distance of one and almost all errors - within edit distance of two.

2.2 Error Correction (EC)

EC involves generating recommendation or suggestion list (also called correction candidates) of similar real words to the erroneous one, from which the end user can choose. The similarity of two words can be measured in several ways. Ref [2] and [22] provide some of the ways by defining metrics for determining the distance between words, which in turn are used to find in the dictionary words with smallest distances with respect to the erroneous word.

Other techniques used in generating correction list including: Minimum Edit Distance Technique [14], and [5]; A Reverse Minimum Edit Distance Technique (RMEDT) used by [8], [9] and [12];The Longest Common Character Subsequence (LCCS) Technique [1]; The Hamming Distance (HD) Technique [11]; Jaccard Coefficient technique used by [17]; among others.

2.3 Lexicon

User-lexicon can be interactively enriched with new entries enabling the checker to recognize all the possible inflexions derived from them. A lexicon for a spelling correction or text recognition application must be carefully tuned to its intended domain of discourse. Too small a lexicon can burden the user with too many false rejections of valid terms; too large a lexicon can result in an unacceptably high number of false acceptances. The relationship between misspellings and word frequencies is not straightforward.

Peterson recommend that lexicon for spelling correction be kept relatively small based on the fact that approximately half a percent of all single error transformations of each of the words on a 350,000-item word list result in other valid words on the list [20]. However, Damerau and Mays challenge this recommendation by using a corpus of over 22 million words of text from various genres [6]; and they found that by increasing the size of their frequency rank-ordered word list from 50,000 to 60,000 words, they were able to eliminate 1,348 false rejections while incurring only 23 additional false acceptances. Since this 50-to-1 differential error rate represents a significant improvement in correction accuracy; therefore, they recommend the use of larger lexicons.

Dictionaries alone are often insufficient sources for lexicon construction. Walker and Amsle [23] observed that nearly two-thirds (61%) of the words in the Merriam-Webster Seventh Collegiate Dictionary did not appear in an eight million word corpus of New York Times news wire text, and conversely, almost two-thirds (64.9%) of the words in the text were not in the dictionary.

On the topic of construction of the lexicon for the spell program, McIlroy [16] provides some helpful insights into appropriate sources that may be drawn upon for general lexicon construction.

An article by Damerau [7] provides some insights into and guidelines for the automatic construction and customization of domain oriented vocabularies for specialized NLP (Natural Language Processing) applications.

3 LugCorrect

3.1 Introduction

LugCorrect, developed using Python programming language, provides correction candidate list (CCL) or an explanation or hint on the type of error committed and works hand in hand with LugDetect which assist in identifying the type of Luganda word which need be corrected, and therefore, using the correct lexicon for selecting the correction candidates, given that there are four lexicons used for Luganda detection and correction. In other words, LugDetect is required to perform the necessary preprocessing before LugCorrect can be invoked.

The two models form LugSpell, a model for an interactive spell checker providing spelling feedback while the end user is word processing their document in the editor. However, in this article, more emphasis is directed towards LugCorrect and therefore, LugDerect, whose mechanism is shown in Figure 1, is discussed briefly.

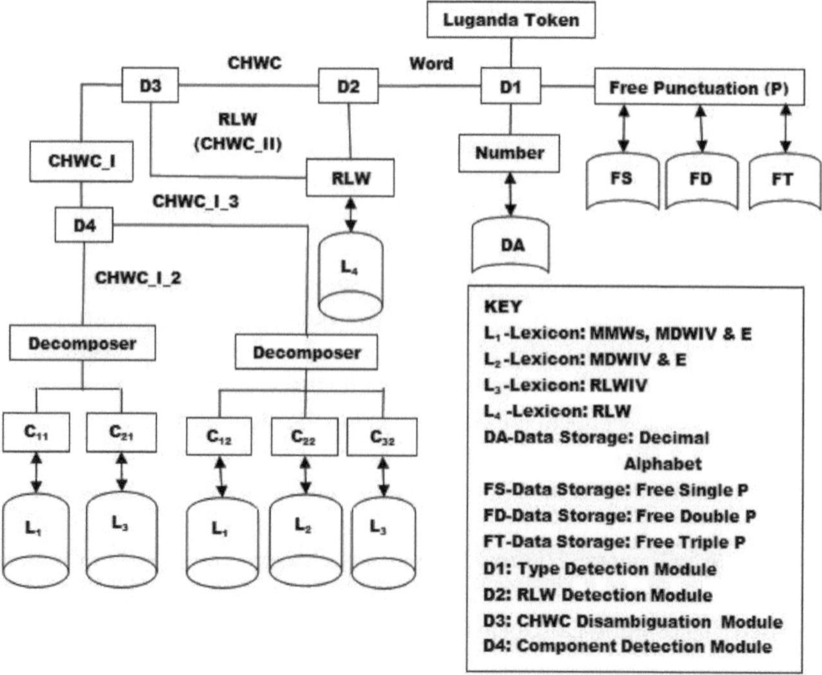

Figure 1. Error Detection Mechanism (LugDetect) which works hand in hand with LugCorrect

3.2 LugDetect

The *Error Detection Mechanism* (EDM) or LugDetect is invoked when a token is passed to the *Type Detection Module* (TDM) - labeled as D1 in Figure 1 which determines the type of token passes on to it as per the type rules elucidated below, and consequently, invokes the appropriate module from the three, namely, RLW (Real Luganda Word) Detection Module - labeled as D2 in the figure; Number Module; and Free Punctuation module.

The type rules used for identifying the different types of token are as follows:

i. A token is a word if it consists of only letter or a mixture of letters and punctuation marks

ii. A token is a number if it consists of any of the decimal alphabet. This is inclusive of a mixture of decimal digits and punctuation marks.

iii. A token is free punctuation if it consists of only punctuation marks.

The *Number Module* uses the decimal alphabet to detect and correct errors in numbers; while as the *Free Punctuation Module* uses three punctuation data storages for the same purpose.

The *RLW Detection Module (D2)* differentiates between CHWCs and RLWs by taking into consideration the following features which were captured through a compositional study of these two categories of words:

i. CHWCs are characterised by inter-word apostrophes

ii. RLW are characterised by absence of inter-word apostrophes

Where, the inter word apostrophes are apostrophes found within a word, rather than at the end or beginning of a word.

The *CHWC Disambiguation Module (D3)* and *RLW Module are* the two modules invoked by D2 in respect of whether the received token is a CHWC or RLW.

D3 determines the type of *CHWC* which has been passed to it, and then passes it on to the appropriate module as stipulated by the following rules:

i. If token is *CHWC_I* then pass it on to *CHWC_I Module*

ii. If token is *CHWC_II* then convert it to *RLW* and pass it on to *RLW Module* aforementioned.

Component Detection Module (D4), which determines the number of components in the CHCW_I, and Decomposer, which decomposes the CHCW_I into its constituent components, are sub modules in the *CHWC_1 Module.*

The LugDectect modules - which invoke LugCorrect and passes onto it the corresponding lexicon or data storage when an error is detected - Include: Free Punctuation Module; Number Module; RLW Module; C11 and C21 Modules of the "two component decomposer"; and finally, C12, C22, and C32 Modules of the "three component decomposer".

The content of the four lexicons used in detection and correction of errors in Luganda words include:

> L_1: contains MMWs, MDWIVs and 'e' (locative)
> L_2: contains MDWIVs and 'e'
> L_3: contains RLWIVs (Real Luganda Words with Initial Vowels)
> L_4: contains RLWs (Real Luganda Words)

3.3 LugCorrect Conceptualisation

The Error Correction Module (ECM) or LugCorrect, whose algorithm is shown in Figure 2 was developed putting into account the following two assumptions that:

> 93% to 95% of spelling errors are an edit distance of one (ED1) from the target [20]
> It is most unlikely for an end-user to commit an error on the first character [26]

These two assumptions have tremendously simplified the work of developing ECM, firstly, by narrowing the search scope, and thereby, reducing the number of correction candidates (CCs), and secondly, by reducing the need for more processing power These assumptions were implemented by:

> using Jaccard Coefficient (JC) [17] - a similarity metric evaluated using Equation 1 - for selecting and ranking the best CCs for presentation to the end user in descending order of JC. Note that CCs with ED1 are automatically selected because they tend to have the highest JC value as corroborated by Table 1 and Figure 3.

$$J_c = n(A \cap B) / n(A \cup B) \qquad (1)$$

where: A represents a set of constituent n-grams of string A

B denotes a set of constituent n-grams of string B.

> Prefixing the first letter in the searching process for generating CC list (CCL)

JC was chosen because, firstly - compared to other similarity metrics, for instance Damerau Levenshtein Distance (DLD) (quadratic algorithm) - JC is computationally light (linear algorithm); secondly, JC correlates well with DLD as demonstrated by Table 1 and Figure 3; and thirdly, JC produced good result as a selection technique in cluster plus, a genetic clustering algorithm used in IGC+ (Information Gain through Cluster Plus), a recommender system developed by Mohd Abdul Hameed et al, [17].

Apart from generating CCL, ECM or LugCorrect working in tandem with LugDetect, provides feedbacks or hints on word level punctuation and numeric errors committed by the end user; however, in this paper this is just mentioned.

In a bid to obtain maximum accuracy, the next task is to determine the optimal size of the CCL - that is, the minimum number of choices presented to the end user for selection - given that the bigger the size of CCL the higher the probability of finding the user's desired word in the CCL, and therefore, the bigger the *accuracy (A_P)* of LugCorrect.

In this context, an experiment was conducted to investigate the effect of increasing CCL size on A_P. Three dictionaries, namely ED1, ED2 and ED3 described in figure 6, were selected and passed onto LugCorrect. Figure 8 show that unknown in the experimental dictionaries lower the accuracy, therefore in this experimentation, dictionaries with unknown were discarded - namely, ED1U, ED2U, and ED3U. The accuracy of ECM for each dictionary for each given size of CCL was determined using Equation 2. The results of the experimentation are presented in Table 2 and Figure 4.

The result show that the optimum CCL size of six is required to obtain the maximum A_P of 98% for ED1 and ED2 while the optimum value of 10 is required to obtain the same accuracy for ED1 through ED3. A choice of 10 was adopted for LugCorrect.

Algorithm : Error Correction Module
    Get Erroneous Word
    IF{punctuation Error}
    Give feedback on punctuation error
    ELSE IF{Numeric Error }
    Give feedback on numeric error
    ELSE IF (Word Error)
    Generate correction candidate list (GCCL)
    ENDIF

    GCCL
    SubDic=Get all words beginning with 1st character of erroneous word from lexicon
    L=Number of words in SubDic
    IF {L>10}
    Get first 10 words from SubDic load in Candidate Dictionary (D_C) with their corresponding Jaccard Coefficient (JC)
    ELSE

Get all words in SubDic load in D_C with their corresponding JC
ENDIF

REPEAT
swap if JC of word in SubDic is greater than minimum JC of word in D_C
UNTILL all words in SubDic have been checked

Sort words in D_C in descending order of JC
Present the word in D_C to the user for selection
IF{user selects a word in D_C and gives consent}
Change word in text editor
ELSE

Figure 2. Algorithm for Error Correction Module

Case	Target	Erroneous word	ED	JC
1	Abakungu	abakuungu	1	0.889
	(Officials)	abakkuungu	2	0.800
		abokuunggu	3	0.636
2	Ttiimu	tiimu	1	0.833
	(Team)	timu	2	0.667
		time	3	0.429

Table 1. The relationship between ED and JC

CCL_LENGTH	1	2	3	4	5	6	7	8	9	10	12	14	20
ED1	90	94	96	96	96	98	98	98	98	98	98	98	98
ED2	90	98	98	98	98	98	98	98	98	98	98	98	98
ED3	94	90	96	96	96	96	96	96	96	98	98	98	98
Average $A_{P1,2}$ (ED1-ED2)	90	96	97	97	97	98	98	98	98	98	98	98	98
Average $A_{P1,3}$ (ED1-ED3)	91.3	94	96.7	96.7	96.7	97.3	97.3	97.3	97.3	98	98	98	98

Table 2. The effects of CCL Size on Accuracy (AP)

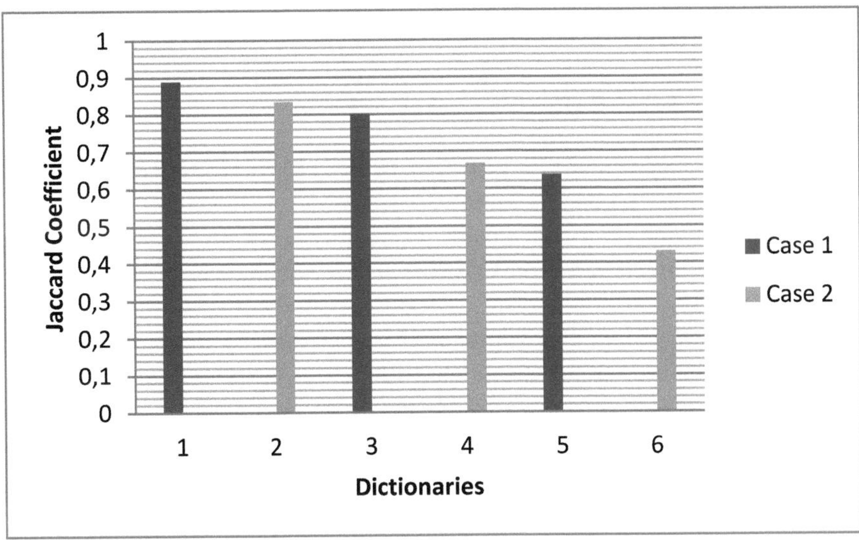

Figure 3. Graph visualizing information in table 1 which shows the relationship between Jacccard Coefficient (JC) and ED. ED1: bar 1&2; ED2: bar 3&4; and ED3: bar 5&6

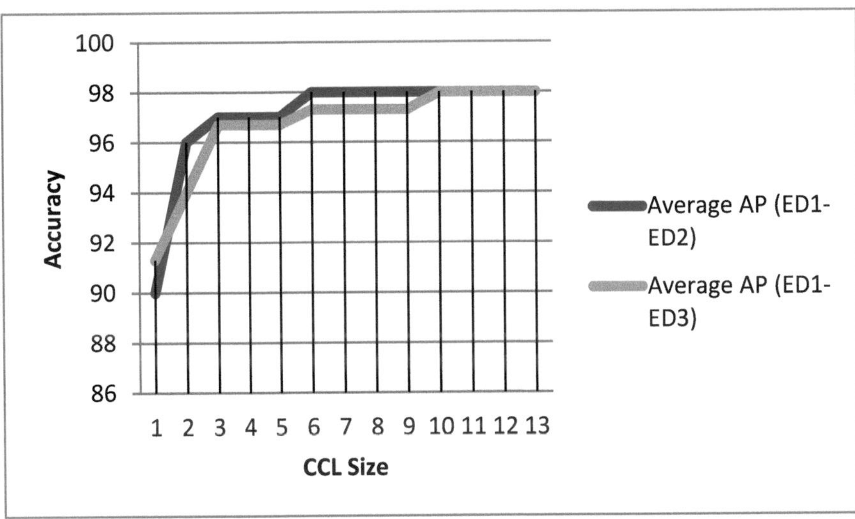

Figure 4. Graph showing the effect of CCL size on accuracy(AP)

4 Evaluation

4.1 Experimental Setup

Investigating the Effect of Unknown Targets on Accuracy and Speed of Correction:
The evaluation process is initiated by passing, as input to LugCorrect, a dictionary containing key - value pairs, where the key is representing the correct Luganda word (target) and value - a string of corresponding erroneous words or non-words.

For this purpose, six dictionaries were prepared each containing 50 tokens (non-words). Figure 5 shows a sample dictionary, while Figure 6 presents the characteristics of each dictionary used in the experimentation.

For each non-word in the dictionary, a corresponding *correction candidate list (CCL)* is generated and if the target exists in the *CCL* than LugCorrect has accurately provided a correction.

The metrics used for evaluation are *accuracy (A_P)* in percentage and *frequency of correction (F_C)* in hertz (Hz). A_P, evaluated using Equation 2, is a measure of how accurate LugCorrect provides correct suggestions or corrections.

$$A_P = 100((N - U_{CCL})/(N - U_L)) \qquad (2)$$

where:

> ➢ N denotes total number of tokens processed in this case, non-words or erroneous words
> ➢ U_{CCL} represents the total number of inaccurate corrections, that is, number of targets not in CCL
> ➢ U_L denotes the total number of unknown targets, that is, targets not in the lexicon

---

{'eriisa':'erisa', 'ttiimu' : 'tttiimu timu tiimu', 'abadde' : 'abade', 'abakulira' : 'abakurira abakulila', 'abakungu' : 'abakuungu', 'abalagira' : 'abalagila abaragira', 'abalyetaba' : 'abaryetaba abalyeetaba'}

Figure 5. Sample dictionary used in the experimentation for non-words which are edit distance of one (ED1) from their corresponding targets (real words)

1. ED1 - Dictionary containing non-words which are *edit distance of one* (ED1) from their corresponding targets (real words).
2. ED1U - Dictionary containing non-words which are ED1 from their corresponding targets (real words) and also contains unknown targets (not in the lexicon)
3. ED2 - Dictionary containing non-words which are ED2 from their corresponding targets (real words).
4. ED2U - Dictionary containing non-words which are ED2 from their corresponding targets (real words) and also contains unknown targets
5. ED3 - Dictionary containing non-words which are ED3 from their corresponding targets (real words).
6. ED3U - Dictionary containing non-words which are ED3 from their corresponding targets (real words) and also contains unknown targets

Figure 6. Characteristics of the dictionaries used in the experimentation

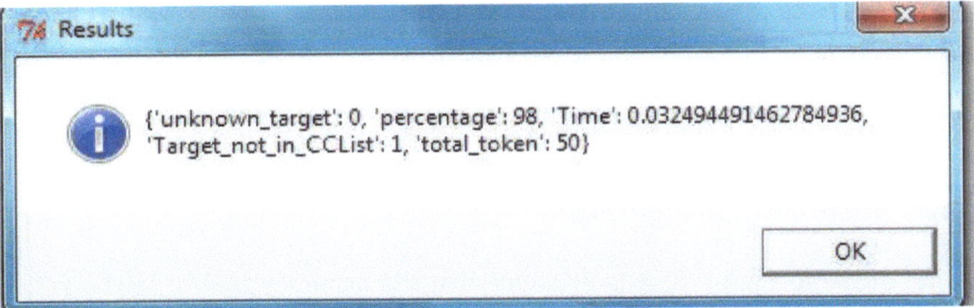

Figure 7. Sample of results collected from LugCorrect for dictionary ED1

Table 3. Summary of Results of the experimentation on LugCorrect

	1 (ED 1)	2 (ED1U)	3 (ED2)	4 (ED2U)	5 (ED3)	6 (ED3U)
U_L	0	10	0	6	0	9
T(ms)	35.2002	37.1603	35. 8806	36.3901	36.7947	38.4117
U_{CCL}	1	11	1	7	2	10
N	50	50	50	50	50	50
F_C	1420	1346	1394	1374	1358	1302
A_P	98	97	98	97	98	97

FC, evaluated using Equation 3, is a measure of how fast LugCorrect is in terms of generating CCL. F_C which depends on the processing power of the computer, defines how many words can be checked in one second.

$$F_C = N \, / \, T \tag{3}$$

where:

- ➤ N denotes total number of tokens
- ➤ T presents time taken to process the tokens

The computer used in the experimentation uses an ADM FX(tm) - 8150 eight core 3.6 GHz Processor running 64 bit Windows 7 Ultimate Operating System.

Sample results captured from LugCorrect during experimentation are shown in Figure 7. All the results captured during experimentation are summarised in Table 3. The graphs for A_P and F_C are shown in Figure 8 and Figure 9 respectively.

Investigating the Effect Of CCL Size on Accuracy: An experiment was conducted - refer to Section 3.3 - to investigate the effect of increasing CCL size on A_P. The main objective of this experimentation was to determine the optimal CCL size for obtaining the maximum permissible accuracy. The results are tabulated in Table 2 and visualized in Figure 4.

4.2 Discussion of Results

In reference to Figure 8, it is evidence that the accuracy (A_P) of LugCorrect is stable for ED1 through ED3 at 98% for input data (dictionaries) with only known targets (the dark grey bars); while for input data (dictionaries) with unknown targets (the light grey bars) A_P stabilises at 97%.

The effect of introducing unknown targets on A_P is that A_P slightly falls by 1% and this is true for all input data, that is to say 1 versus (vs) 2; 3 vs 4; and 5 vs 6. We can conclusively say that introduction of unknown targets negatively affects the accuracy of LugCorrect.

In reference to Figure 9 it is evident that the F_C falls gradually with respect to increase in ED for input data with only known target; however, for input data with unknown targets F_C rises than fall; therefore, there is no conclusive pattern on the impact of increasing ED on F_C, but we can conclude that the average F_C of 1365 Hz for LugCorrect is good enough for interactive correction, given that the minimal speed of 10 Hz is required for interactive correction, [18]. In reference to Table 2 and Figure 4, it is observed that, firstly, for ED1 and ED2 the optimal CCL size is six while for ED1 through ED3 the optimal size is ten.

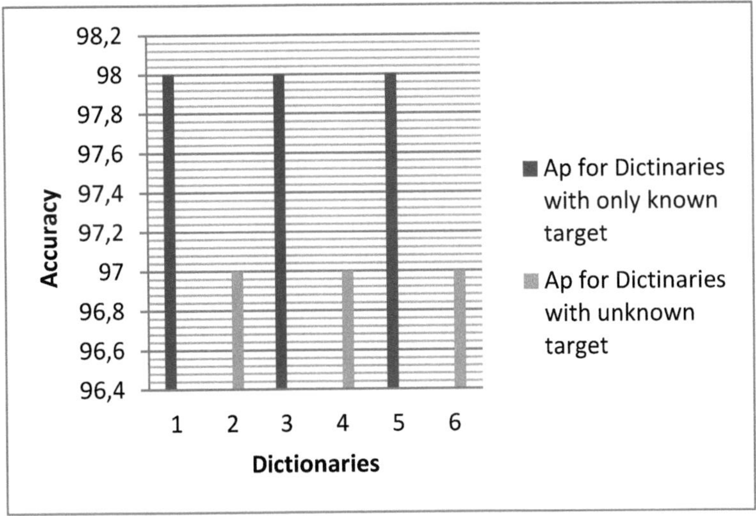

Figure 8. A graph showing Accuracy in percentage (AP) of the various input dictionaries

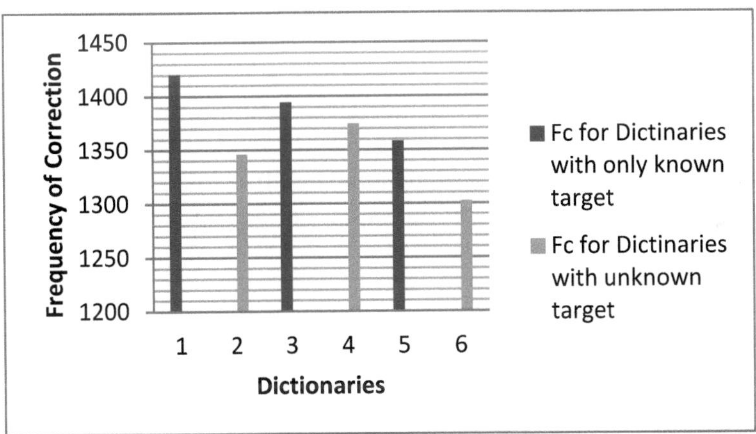

Figure 9. A graph showing the Frequency of correction (Fc) in hertz (Hz) for the various input dictionaries

Secondly, there is no conclusive pattern of the effect of ED on accuracy for the various CCL sizes.

4.3 Comparison with Other Systems

To the best of our knowledge there is no Luganda corrector to effect meaningful comparison, however the comparison in this section is focused mainly on methodology employed and accuracy obtained.

LugCorrect has an average accuracy of 98% for ED1 through ED3 which is much higher in comparison to 70.5% (74% and 67%) achieved by a statistical collector for English developed by Peter Norvig [18]. Peter's corrector uses a Reverse Minimum Edit Distance (RMED) technique combined with Dictionary Lookup Technique (DTL) to generate correction candidates, as well as statistical or probabilistic approach for selection. Another system which uses RMED for correction is AKHAR, the first Punjabi spell checker and part of the commercial Punjabi word processor AKHAR. The accuracy obtained for AKHAT is 83.5% [13].

The low accuracy in Peter's corrector is also attributed to provision of one choice in comparison to LugCorrect which at least gives 10 choices. A choice of ten is the optimal value (minimum CCL size) for obtaining maximum value of accuracy as demonstrated in Figure 4.

The rationale behind this is that the probability of the target word being in a CCL is higher when the number of elements in the CCL (choice) is bigger.

The average F_c for Peter's corrector is 16 Hz (15 Hz and 17 Hz) which is quite low due to the fact that Peter's corrector is computationally heavy. In comparison, LugCorrect, which is computationally light, has an edge in that it operates at average F_c of 1365 Hz.

The average accuracy (AV) of 98% achieved for ED1 through ED3 (Figure 8) is comparable to that for Punjabi corrector with AV of 98% but for ED1 [10].

5 Related Work

Aspell .31developed by Atkinson [3], is a Free and Open Source spell checker designed to eventually replace Ispell. Test results show that Aspell achieved a correction accuracy of 93.1% with a CCL size of 50 at a correcting speed of 137 Hz (words per second).

Youssef Bassil [27], proposes a new parallel shared-memory spell-checking algorithm that uses rich real-world word statistics from Yahoo! N-Grams Dataset (YND) [25] to correct non-word and real-word errors in electronic text. Youssef's system is divided into three sub-algorithms that run in a parallel fashion: The error detection algorithm that detects misspellings, the candidates generation algorithm that generates correction suggestions, and the error correction algorithm that performs contextual error correction. Experimentation conducted on a set of text articles containing misspellings, showed that a correction accuracy of 99% on non-word and 65% on real word error was achieved.

Kaur and Garg [13] proposed a hybrid system which checks both spelling and grammatical errors for Punjabi language. The CCL is generated by prefixing the first two characters of the erroneous word followed by searching the lexicon for words which beginning with these two characters. The elements in CCL are ranked alphabetically and presented to the end-user. The accuracy reported for the system is 83.5%.

Integrated Scoring for Spelling error correction, Abbreviation expansion and Case restoration (ISSAC) is a mechanism developed by Wilson Wong et al [24] for cleaning dirty texts from online sources and the idea of which was first conceived as part of the text preprocessing phase in an ontology engineering project. Evaluations of ISSAC using 400 chat records reveal an improved accuracy of 96.5% over the existing 74.4% based on the use of Aspell only.

Conclusion

This research work provides a methodology of dealing with non word error correction for Luganda, a Bantu language and as a result LugCorrect was developed.

CHWC_1 (Clictic-host Word Combination type 1) which are akin to Luganda text are also found in other Bantu Languages - for instance, Runyankore-Rukiga and Kinyarwanda; therefore, the approach adopted to deal with CHWC in LugCorrect can be adopted to deal with the same in these languages.

In addition, LugCorrect - although has been initially developed for Luganda - can easily be adopted to spell check or generate CCL for other Bantu languages, as most of them have a lot in common.

As per experimental results, LugCorrect provides correction for erroneous words which are *edit distance one* (ED1) through ED3 from the target (correct version of the word) at a consistent average accuracy (Av) of 98%, which covers 95% to 98% of the error cases. Literature claims that there are few error cases (less than 2%) which are greater than ED2 from the target. The accuracy achieved by LugCorrect for ED1 is comparable to that achieved by the Punjabi corrector with Av of 98% but for ED1 [10].

The results of the experimentation show that introduction of unknown targets negatively affects the accuracy of LugCorrect; and there is no conclusive correlation or pattern between ED and accuracy.

LugCorrect provide correction at an average speed of 1365 Hz (number of words per second) which is over and above the minimal speed of 10 Hz required for interactive correction.

It is worthwhile noting that the use of Jacard Coefficient (JC) - which is computationally lighter in comparison with *Damerau Levenshtein Distance (DLD)* - in the selection and ranking process has produced good results in LugCorrect as aforementioned.

Future work

The scope of LugCorrect has been limited to non-word errors, however this scope can be extended to cover real word errors. A number of approaches are in existence most of them use statistical approaches and therefore require an annotated corpus, in this case, for Luganda. The approach proposed by Mays et al., [15] is a statistical method which uses word-trigram probabilities for detecting and correcting real word errors.

Another approach, proposed by Kernighan, et al, [12] is a real word error detector based on a noisy channel model.

Word Sense Disambiguation (WSD) algorithm is another approach applied to correct real word errors in context. WSD algorithm uses the context of the sentence to resolve semantic ambiguity.

Consequently, there is a need to investigate the aforementioned approaches in bid to identify the most appropriate for the highly inflected Luganda language.

Another investigation on language and error models - statistical selection model for selecting correction candidate - is required in order to ascertain whether they can improve the performance of LugCorrect.

Luganda is an agglutinative language which means that it is highly inflected language (that is, it has many form of a given word). This corresponds to a large lexicon to capture these forms; and therefore, a call and need for a language-dependent algorithm for handling the language morphology, in a bid to reduced on the size of the required lexicon.

The use of a language and error models in the selection process can be investigated upon in order to access the impact of the two models on the selection process.

References

1   L. Allison and T.I. Dix., "A bit-string longest Common-Subsequence Algorithm", Information Processing Letters, 23:305–310, 1986.
2   Angell R.; Freund G.; and Willety P. "Automatic Spelling Correcting Using a Trigram Similarity Measure", Information Processing and Management, vol. 19, 1983.
3   Atkinson, K. (2006), "Gnu Aspell 0.60.4", http://aspell.sourceforge.net/.
4   Church, K. W.; and Gale, W. A. "Enhanced Good-Turing and Cat-Cal Two New Methods for Estimating Probabilities of English Bigrams", Comput. Speech Lang, 1991.
5   Damerau F. J., "A technique for Computer Detection and Correction of Spelling Errors", Communication ACM, 7:171–176, 1964.
6   Damerau F J.; and Mays. E., "An Examination of Undetected Typing Errors", Information Process Management, 25, 6, 659-664, 1989.
7   Damerau. F. J.' "Evaluating Computer-Generated Domain - Oriented Vocabularies", Information Process Management, 26, 6, 791–801, 1990.
8   Durham I.; Lamb, D A.; and Saxe, J. B., "Spelling Correction in User Interfaces", Communication ACM 26, 10 (Oct.), 764–773, 1983.
9   Gorin, R. E, "SPELL: A Spelling Checking and Correction Program", On line documentation for the DEC-10 computer, 1971.
10   Harpreet Kaur, Gurpreet Kaur, Manpreet Kaur, "Punjabi Spell Checker Using Dictionary Clustering". International Journal of Science, Engineering and Technology Research (IJSETR), Volume 4, Issue 7, July 2015 2369 ISSN: 2278 – 7798, 2015.
11   Hamming, Richard W., "Error detecting and error correcting codes", Bell System Technical Journal 29 (2): 147–160, 1950.
12   Kernighan; Mark D.; Kenneth W. Church; and William A. Gale, "A Spelling Correction Program Based on a Noisy Channel Model", Proceedings of COLING 1990, 205-210, 1990.
13   J. Kaur and K. Garg," Hybrid Approach for Spell Checker and Grammar Checker for Punjabi", International Journal of Advanced Research in Computer Science and Software Engineering, Volume 4, Issue 6, June 2014
14   Levenshtein, V. I., "Binary codes capable of correcting deletions, insertions, and reversals", Soviet Physics Doklady, 10:707–710, 1966.
15   Mays Eric; Fred J. Damerau; and Robert L. Mercer, "Context Based Spelling Correction", Information Processing and Management, 27(5):517–522, 1991.
16   MCIlroy, M. D; "Development of a Spelling List", IEEE Transaction Communication COM-30, 1 (Jan.), 91–99, 1992.
17   Mohd Abdul Hameed; S. Ramachandram; Omar Al Jadaan, "Genetic Algorithm Based Recommender System - IGC Plus: A Novel Clustering Dependant Recommender System", International Journal of Communication Systems and Network, ISBN: 2234-8018, Volume 4, issue 1., 2014.
18   Peter Norvig, "How to Write a Spelling Collector", an essay, 2007
19   Peterson J.L., "Computer Programs for Detecting and Correcting Spelling Errors", Comm. of ACM vol.23, issue 12, 1980.
20   Peterson, J L, "A Note on Undetected Typing Errors", Commun. ACM 29, 7 (July), 633-637, 1986.
21   Pollock J.; and Zamora A., "Automatic Spelling Correction in Scientific and Scholarly Text", Comm. of ACM vol.27 358- 368, 1984.

22 Tanaka E., and Kojima Y.. "A High Speed String Correction Method Using a Hierarchical File", IEEE Transactions on Pattern Analysis and Machine Intelligence vol. 9, 1987.

23 Walker, D. E.; and Amsler, R. A., "The use of Machine-Readable Dictionaries in Sublanguage Analysis", In Analysing Language in Restricted Domains: Sub-language Description and Processing. Lawrence Erlbaum, Hillsdale, N. J., 69-83, 1986.

24 Wilson Wong, Wei Liu and Mohammed Bennamoun, "Integrated Scoring for Spelling Error Correction, Abbreviation Expansion and Case Restoration in Dirty Text", Proc. Fifth Australasian Data Mining Conference (AusDM2006), 2006

25 Yahoo! Webscope dataset Yahoo! N-Grams, ver. 2.0, http://research.yahoo.com/Academic_Relation

26 Yannakoudakis, E.J. and Fawthrop, D., "The rules of spelling errors," Information Processing and Management, vol. 19, no. 2, pp. 87-99, 1983.

27 Youssef Bassil, "Parallel Spell-Checking Algorithm Based on Yahoo! N-Grams Dataset", International Journal of Research and Reviews in Computer Science (IJRRCS), ISSN: 2079-2557, Vol. 3, No. 1, February 2012.

Pawan Kumar Singh [1], Supratim Das [2], Ram Sarkar [3] and
Mita Nasipuri [4]

A New Approach for Texture based Script Identification At Block Level using Quad-Tree Decomposition

Abstract: A considerable amount of success has been achieved in developing monolingual OCR systems for *Indic* scripts. But in a country like India, where multi-script scenario is prevalent, identifying scripts beforehand becomes obligatory. In this paper, we present the significance of Gabor wavelets filters in extracting directional energy and entropy distributions for 11 official handwritten scripts *namely, Bangla, Devanagari, Gujarati, Gurumukhi, Kannada, Malayalam, Oriya, Tamil, Telugu, Urdu* and *Roman*. The experimentation is conducted at block level based on a quad-tree decomposition approach and evaluated using six different well-known classifiers. Finally, the best identification accuracy of 96.86% has been achieved by Multi Layer Perceptron (MLP) classifier for 3-fold cross validation at level-2 decomposition. The results serve to establish the efficacy of the present approach to the classification of handwritten *Indic* scripts.

Keywords: Handwritten script identification, *Indic* scripts, Gabor wavelet filters, Quad-tree decomposition, Optical Character Recognition, Multiple classifiers

1 Introduction

Script is defined as the graphic form of writing system which is used to express the written languages. Languages throughout the world are typeset in many different scripts. A script may be used by only one language or shared by many

1 Department of Computer Science and Engineering, Jadavpur University, Kolkata, India
pawansingh.ju@gmail.com
2 Department of Computer Science and Engineering, Jadavpur University, Kolkata, India
supratimdas21@gmail.com
3 Department of Computer Science and Engineering, Jadavpur University, Kolkata, India
raamsarkar@gmail.com
4 Department of Computer Science and Engineering, Jadavpur University, Kolkata, India
mitanasipuri@gmail.com

languages, with slight variations from one language to other. For example, *Devanagari* is used for writing a number of Indian languages like *Hindi, Konkani, Sanskrit, Nepali*, etc., whereas *Assamese* and *Bengali* languages use different variants of the *Bangla* script. India is a multilingual country with 23 constitutionally recognized languages written in 12 major scripts. Besides these, hundreds of other languages are used in India, each one with a number of dialects. The officially recognized languages are *Hindi, Bengali, Punjabi, Marathi, Gujarati, Oriya, Sindhi, Assamese, Nepali, Urdu, Sanskrit, Tamil, Telugu, Kannada, Malayalam, Kashmiri, Manipuri, Konkani, Maithali, Santhali, Bodo, Dogari* and *English*. The 12 major scripts used to write these languages are: *Devanagari, Bangla, Oriya, Gujarati, Gurumukhi, Tamil, Telugu, Kannada, Malayalam, Manipuri, Roman* and *Urdu*. Of these, *Urdu* is derived from the *Persian* script and is written from right to left. The first 10 scripts are originated from the early *Brahmi* script (300 BC) and are also referred to as *Indic* scripts [1-2]. *Indic* scripts are a logical composition of individual script symbols and follow a common logical structure. This can be referred to as the "script composition grammar" which has no counterpart in any other scripts in the world. *Indic* scripts are written syllabically and are visually composed in three tiers where constituent symbols in each tier play specific roles in the interpretation of that syllable [1].

Script identification aims to extract information presented in digital documents *viz.*, articles, newspapers, magazines and e-books. Automatic script identification facilitates sorting, searching, indexing, and retrieving of multilingual documents. Each script has its own set of characters which is very different from other scripts. In addition, a single OCR system can recognize only a script of particular type. On the contrary, it is perhaps impossible to design a single recognizer which can identify a variety of scripts/languages. Hence, in this multilingual and multi-script environment, OCR systems are supposed to be capable of recognizing characters irrespective of the script in which they are written. In general, recognition of different script characters in a single OCR module is difficult. This is because of features which are necessary for character recognition depends on the structural property, style and nature of writing which generally differ from one script to another. Alternative option for handling documents in a multi-script environment is to use a pool of OCRs (different OCR for different script) corresponding to different scripts. The characters in an input document can then be recognized reliably by selecting the appropriate OCR system from the said pool. However, it requires a priori knowledge of the script in which the document is written. Unfortunately, this information may not be readily available. At the same time, manual identification of the documents' scripts may be monotonous and time consuming. Therefore, it is necessary to identify the script of the document before feeding the document to the corresponding OCR system.

In general, script identification can be achieved at any of the three levels: (a) Page level, (b) Text-line level and (c) Word level. Identifying scripts at page-level can be sometimes too convoluted and protracted. Again, the identification of scripts at text-line or word level requires the exact segmentation of the document pages into their corresponding text lines and words. That is, the accuracy of the script identification in turns depends on the accuracies of its text-line and word segmentation methods respectively. In addition, identifying text words of different scripts with only a few numbers of characters may not always be feasible because at word-level, the number of characters present in a single word may not be always informative. So, in the present work, we have performed script identification at block level, where the entire script document pages are decomposed gradually into its corresponding smaller blocks using a quad-tree based segmentation approach.

The literature survey on the present topic reveals that very few researchers had performed script identification at block level other than three accustomed levels. P. K. Singh *et al.* [3] presented a detailed survey of the techniques applied on both printed and handwritten *Indic* script identification. M. Hangarge *et al.* [4] proposed a handwritten script identification scheme to identify 3 different scripts *namely, English, Devanagari*, and *Urdu*. A set of 13 spatial spread features were extracted using morphological filters. Further, *k*- Nearest Neighbor (*k*-NN) algorithm was used to classify 300 text blocks attaining an accuracy rate of 88.6% with five fold cross validation test. G.G. Rajput *et al.* [5] proposed a novel method towards multi-script identification at block level on 8 handwritten scripts including *Roman*. The recognition was based upon features extracted using Discrete Cosine Transform (DCT) and Wavelets of Daubechies family. Identification of the script was done using *k*-NN classifier on a dataset of 800 text block images and yielded an average recognition rate of 96.4%. G. D. Joshi *et al.* [6] presented a scheme to identify 10 different printed *Indic* scripts. The features were extracted globally from the responses of a multi-channel log-Gabor filter bank. This proposed system achieved an overall classification accuracy of 97.11% on a dataset of 2978 individual text blocks. M. B. Vijayalaxmi *et al.* [7] proposed a script identification of *Roman, Devanagari, Kannada, Tamil, Telugu* and *Malayalam* scripts at text block level using features of Correlation property of Gray Level Co-occurrence Matrix (GLCM) and multi resolutionality of Discrete Wavelet Transform (DWT). Using Support Vector Machine (SVM) classifier, the average script classification accuracy achieved in case of bi-script and tri-script combinations were 96.4333% and 93.9833% respectively on a dataset of 600 text block images. M. C. Padma *et al.* [8] presented a texture-based approach to identify the script type of the documents printed in 3 prioritized scripts *namely, Kannada, Hindi* and *English*. The texture features were extracted from the Shannon entropy values, computed from the sub-bands of the wavelet packet decomposition. Script classification performance was analyzed using *k*-NN classifier on a dataset of 2700 text images and the average success rate was found to be 99.33%. Sk. Md. Obaidullah *et al.* [9] proposed an automatic handwritten script identification technique at block level for document images of six popular *Indic* scripts *namely, Bangla, Devanagari, Malayalam, Oriya, Roman* and *Urdu*. Initially, a 34-dimensional feature vector was constructed applying different transforms (based on Radon Transform, Discrete Cosine Transform, Fast Fourier Transform and Distance Transform), textural and statistical techniques. Finally, using a GAS (Greedy Attribute Selection) method, 20 attributes were selected for learning process. Experimentation performed on a dataset of 600 image blocks using Logistic Model Tree which produced an accuracy of 84%. It is apparent from the literature survey that only a few works [4-5, 9] have been done on handwritten *Indic* scripts. Estimation of features of individual scripts from the whole document image involves a lot of computation time because of large image size. Apart from this, the major drawback of the existing works is that the recognition of scripts is mainly performed on individual fixed-sized text blocks, which are selected manually from the document images. Again, the size of text blocks has a remarkable impact on the accuracy of the overall system. This has motivated us to design an automatic script identification scheme based on texture based features at block level for 11 officially recognized *Indic* scripts *viz., Bangla, Devanagari, Gujarati, Gurumukhi, Kannada, Malayalam, Oriya, Tamil, Telugu, Urdu* and *Roman.* The block selection from the document images has been done automatically by using quad-tree decomposition approach.

2 Proposed Work

The proposed model is inspired by a simple observation that every script defines a finite set of text patterns, each having a distinct visual appearance [10]. Scripts are made up of different shaped patterns in forming the different character sets. Individual text patterns of one script are assembled together to form a meaningful text word, a text-line or a paragraph. This collection of the text patterns of the one script exhibits distinct visual appearance. A uniform block of texts, regardless of the content, may be considered as distinct texture patterns (a block of text as single entity) [10]. This observation implies that one may devise a suitable texture classification algorithm to perform identification of text language. In the proposed model, the texture-based features are extracted from the Gabor wavelets filters designed at multiple scales and multiple orientations. These features are estimated at different levels of the document images decomposed by the quad-tree based approach which is described in the next subsection.

2.1 Quad-Tree Decomposition Approach

The quad-tree based decomposition approach is designed by recursively segmenting the image into four equal-sized quadrants in top-down fashion. The decomposition rule can be formally defined as follows: The root node of the tree represents the whole image. This method starts dividing each image into four quadrants if the difference between maximum and minimum value of the inspection image is larger than the decomposition threshold, which means that the image is non-homogeneous. This rule is recursively applied to subdivided images until the image cannot be divided. So, at 1st Level, the whole document image is divided into four equal blocks- L_1, L_2, L_3 and L_4. For 2nd level, L_1 block is again sub-divided into another four sub-blocks *namely*, L_{11}, L_{12}, L_{13} and L_{14}. Similarly, each of the other blocks i.e., L_2, L_3 and L_4 are also sub-divided into four sub-blocks. Thus, for 2nd level decomposition, a total of 16 blocks is realized. Therefore, the total number of sub-blocks at l-th level can be written as 4^l where l denotes the level of decomposition. For an image of size $M \times N$, then the size of each sub-block at l-th level is $(M/2^l) \times (N/2^l)$. For the present work, the maximal value of l is chosen to be 4. Fig. 1 shows the quad-tree decomposition for a handwritten *Telugu* document image.

2.2 Gabor Wavelets

The different wavelet transform functions filter out different range of frequencies (i.e. sub bands). Thus, wavelet is a powerful tool, which decomposes the image into low frequency and high frequency sub-band images. Among various wavelet bases, Gabor functions provide the optimal resolution in both the time (spatial) and frequency domains, and the Gabor wavelet transform seems to be the optimal basis to extract local features. Besides, it has been found to yield distortion tolerance for pattern recognition tasks. The Gabor kernel is a complex sinusoid modulated by a Gaussian envelope. The Gabor wavelets have filter responses similar in shape to the respective fields in the

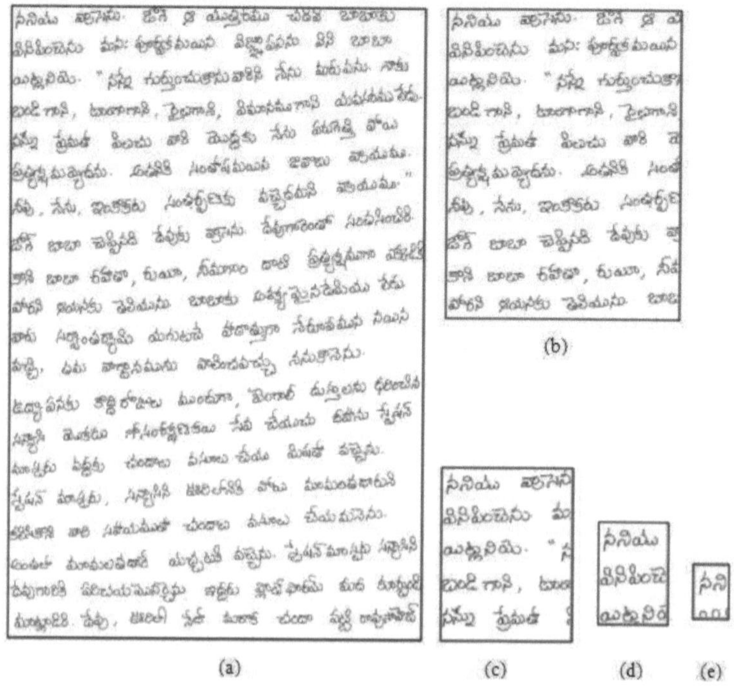

Figure 1. Illustration of: (a) handwritten Telugu document page, (b) 1st level, (c) 2nd level (L₁), (d) 3rd level (L₁₁), and (e) 4th level (L₁₁₁)decomposition

primary visual cortex in mammalian brains [11]. The kernel or mother wavelet in the spatial domain in given by [12]:

$$\psi_j(\vec{x}) = \frac{k_j^2}{\sigma^2} exp\left(-\frac{k_j^2 x^2}{2\sigma^2}\right)\left[exp(j\vec{k_j}\vec{x}) - exp\left(-\frac{\sigma^2}{2}\right)\right] \tag{1}$$

where,

$$\vec{k_j} = \begin{pmatrix} k_v \sin\phi_\mu \\ k_v \cos\phi_\mu \end{pmatrix}, k_v = 2^{-\frac{v+1}{2}}\pi, \phi_\mu = \mu\frac{\pi}{8}, \vec{x} = (x_1, x_2)\ \forall\ x_1, x_2$$
$$\in \mathbb{R}^2 \tag{2}$$

σ is the standard deviation of the Gaussian, $\vec{k}$ is the wave vector of the plane wave, ϕ_μ and k_v denote the orientations and frequency scales of Gabor wavelets respectively which are obtained from the mother wavelet. The Gabor wavelet representation of a block image is obtained by doing a convolution between the image and a family of Gabor filters as described by Eqn. (3). The convolution of image $I(x)$ and a Gabor filter $\psi_j(\vec{x})$ can be defined as follows:

$$J_j(\vec{x}) = I(\vec{x}) * \psi_j(\vec{x}) \qquad (3)$$

Here, $*$denotes the convolution operator and $J_j(\vec{x})$ is the Gabor filter response of the image block with orientation ϕ_μ and scale k_v. This is referred to as wavelet transform because the family of kernels are self-similar and are generated from one mother wavelet by scaling and rotation of frequency. The transform extracts features oriented along ϕ_μ and for the frequency k_v. Each combination of μ and v results in a sub-band of same dimension as the input image I. For the present work, $\mu \in \{0,1,....,5\}$ and $v = \{1,2,.....,5\}$. Five frequency scales and six orientations would yield 30 sub-bands. Fig. 2 shows the directional selectivity property of the Gabor wavelet transform used in the present work whereas Fig. 3 shows the resulting image blocks at five frequency scales and six orientations for a sample handwritten *Kannada* block image obtained at 2nd level of image decomposition. For the feature extraction purpose, we proposed to compute the energy and entropy values [13] from each of the sub-bands of the Gabor wavelet transform which makes the size of our feature vector to 60. Finally, this set of 60 multi-scale and multi-directional oriented features has been used for the block level recognition of 11 official scripts.

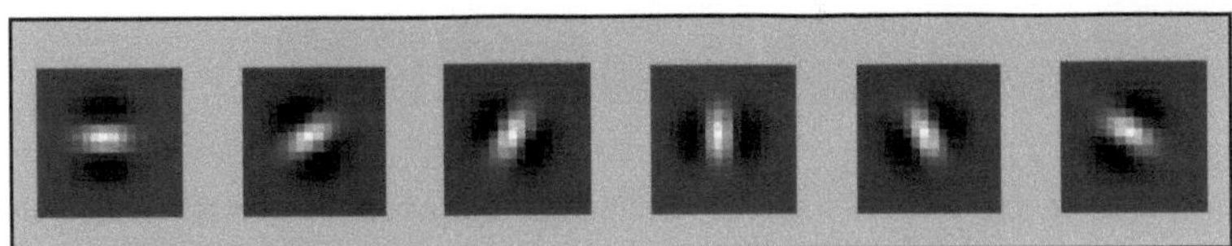

Figure 2. Illustration of six kernels of size 16X16 used in the present work

The algorithm of the proposed methodology is described as follows
Step 1: Input a RGB document image and convert it into a gray scale image $I(x, y)$ where, x and y are the spatial coordinates of the image.
Step 2: Select an integer value l where l indicates the level of decomposition. For the present work, the optimal value of l is varied from 2 to 4.
Step 3: Perform l-level quad-tree decomposition to generate 4^l blocks. Feature values are generated from each of these blocks.
Step 4: Convolve the block image of Step 3 with Gabor wavelet kernel with five different scale values ($v = 1, 2, 3, 4, and\ 5$) and six different orientations

$(\phi_\mu = 0^0, 30^0, 60^0, 90^0, 120^0, and\ 150^0)$. Thus, for each block we get a total of 30 convolved filter outputs.

Step 5: Acquire the absolute value of these filtered output block images and calculate the energy and entropy values.

Step 6: Repeat Step 2 to Step 5 to extract a total of 60 features from each of the script block images.

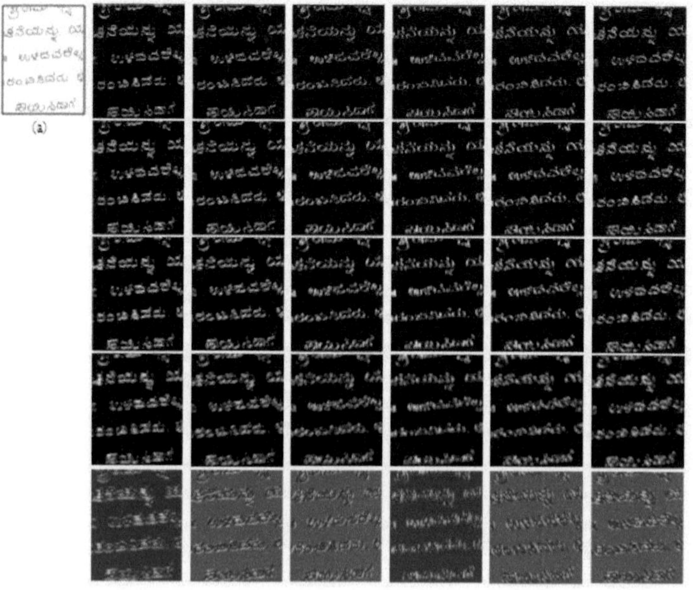

Figure 3. Illustration of the output images of Gabor filter responses at five scales and six orientations for a sample handwritten *Kannada* script block image shown in (a). (The first row shows the output for $k_v = 1$ and six orientations, the second row shows the output for $k_v = 2$ and six orientations, and so on)

3 Experimental Study and analysis

The experiments are carried out on eleven official scripts *namely, Bangla, Devanagari, Gujarati, Gurumukhi, Kannada, Malayalam, Oriya, Tamil, Telugu, Urdu* and *Roman*. A total dataset of 110 handwritten document pages (10 pages from each script) is collected from different people with varying age, sex, educational qualification etc. These documents are digitized using a HP flatbed scanner and stored as bitmap (BMP) image. Otsu's global thresholding approach [14] is used to convert them into two-tone images (0 and 1) where the label '1' represents the object and '0' represents the background. Binarized block images may contain noisy pixels which have been removed by using Gaussian filter [13]. A set of 160, 640 and 2560 text blocks for each of the scripts are prepared by applying 2-level, 3-level and 4-level quad-tree based page segmentation approach. Sample handwritten text block images of 2-level decomposition for all the above mentioned scripts are shown in Fig. 4. A 3-fold cross-validation approach is used for testing the present approach. That is, for 2-level decomposition approach, a total of 1175 text blocks are used for the training purpose and the remaining 585 text blocks are used for the testing purpose. Similarly, for 3-level and 4-level decomposition approaches, a set of 4695 and 18775 text blocks are taken for the training purpose and the remaining 2345 and 9285 text blocks are used for testing the system respectively. The proposed approach is then evaluated using six different classifiers *namely*, Naïve Bayes, MLP, SVM, Random Forest, Bagging, MultiClass classifier. The graphical comparison of the identification accuracy rates of the said classifiers for 2-level, 3-level and 4-level approaches are shown with the help of a bar chart in Fig. 5.

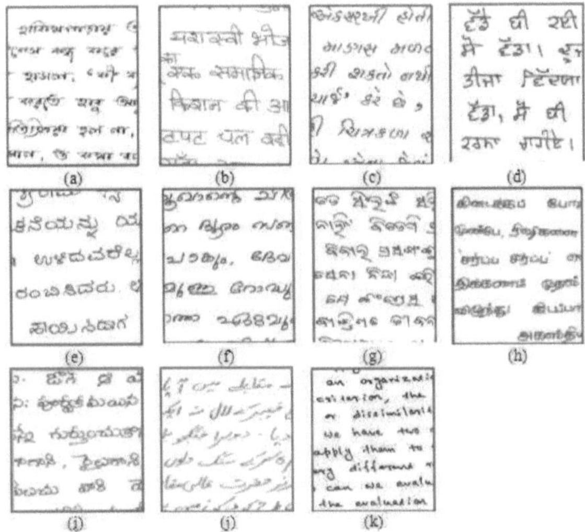

Figure 4. Samples of text block images obtained at 2-level quad-tree based decomposition scheme written in: (a) *Bangla*, (b) *Devanagari*, (c) *Gujarati*, (d) *Gurumukhi*, (e) *Kannada*, (f) *Malayalam*, (g) *Oriya*, (h) *Tamil*, (i) *Telugu*, (j) *Urdu*, and (k) *Roman* scripts respectively

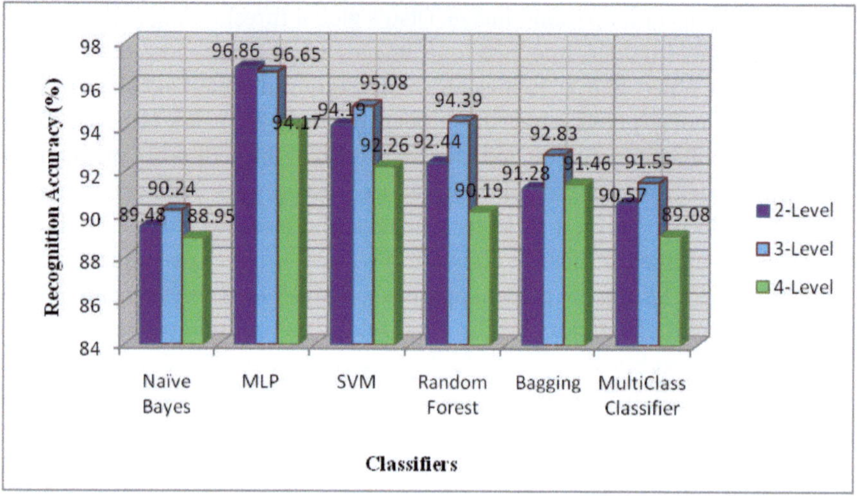

Figure 5. Graph showing the recognition accuracies of the proposed script identification technique for 2-level, 3-level and 4-level decomposition approaches using six different classifiers

It can be seen form Fig. 5 that the best identification accuracy is found to be 96.86% for 2-level decomposition approach by MLP classifier. In the present work, detailed error analysis of MLP classifier with respect to different well-known parameters *namely*, Kappa statistics, mean absolute error (MAE), root mean square error (RMSE), True Positive rate (TPR), False Positive rate (FPR), precision, recall, F-measure, and Area Under ROC (AUC) on eleven handwritten scripts are also computed. Table 1 provides a statistical performance analysis of MLP classifier for text blocks written in each of the aforementioned scripts.

Although the recognition accuracy achieved by the present technique is quite impressive considering the inherent complexities of scripts involved, but still some of the block images written in one script are misclassified into another script. The main reasons for this misclassification are due to presence of noise, distortion, and peculiar writing styles of different writers. Apart from this, sometimes the sub-block images (decomposed from the document image) may contain more background part than the foreground ones which in turn estimate the feature values misleading enough for the present script identification technique.

Table 1. Statistical performance measures of MLP classifier along with their respective means (shaded in grey and styled in bold) achieved by the present technique on 11 handwritten scripts at block level

Scripts	Kappa Statistics	MAE	RMSE	TPR	FPR	Precision	Recall	F-measure	AUC
Bangla				0.938	0.008	0.938	0.938	0.938	0.983
Devanagari				0.956	0.003	0.975	0.956	0.965	0.983
Gujarati				0.988	0.001	0.994	0.988	0.991	1.000
Gurumukhi				1.000	0.005	0.964	1.000	0.982	0.999
Kannada				0.981	0.001	0.994	0.981	0.987	0.996
Malayalam	0.9647	0.1917	0.305	0.988	0.005	0.958	0.988	0.972	1.000
Oriya				0.969	0.002	0.981	0.969	0.975	0.992
Tamil				0.938	0.003	0.974	0.938	0.955	0.988
Telugu				0.961	0.007	0.942	0.961	0.951	0.982
Urdu				0.969	0.004	0.971	0.969	0.995	1.000
Roman				0.971	0.002	0.968	0.971	0.948	0.989
Mean	**0.9647**	**0.1917**	**0.305**	**0.969**	**0.004**	**0.969**	**0.969**	**0.969**	**0.992**

Conclusion

Script identification, a challenging research problem in any multilingual environment, has got attention to the researchers few decades ago. Research in the field of script identification aims at conceiving and establishing an automatic system which would be able to discriminate a certain number of scripts. As developing a common OCR engine for different scripts is near to impossible, it is necessary to identify the scripts correctly before feeding them to corresponding OCR engine. In this work, we assessed the effectiveness of using Gabor wavelets filters on 11 officially recognized handwritten *Indic* scripts along with *Roman* script. The scheme is applied at block level using a quad-tree based decomposition approach. The present technique attains an identification accuracy of 96.86% for 2-level decomposition. As the present technique is evaluated on a limited dataset, we have achieved satisfactory result. This work is first of its kind to the best of our knowledge as far as the number of scripts is concerned. As the key feature used in this technique is mainly texture based, in future, the technique could be applicable for recognizing other non-*Indic* scripts in any multi-script environment. Future scope may also include comparing different wavelet transforms with the Gabor wavelet transform in terms of computational accuracy.

Acknowledgment

The authors are thankful to the Center for Microprocessor Application for Training Education and Research (*CMATER*) and Project on Storage Retrieval and Understanding of Video for Multimedia (SRUVM) of Computer Science and Engineering Department, Jadavpur University, for providing infrastructure facilities during progress of the work. The current work, reported here, has been partially funded by Technical Education Quality Improvement Programme Phase–II (TEQIP-II), Jadavpur University, Kolkata, India.

References

1   H. Scharfe, "*Kharosti and Brahmi*", J. Am. Oriental Soc., vol. 122, no. 2, pp. 391-393, 2002.

2   A. S. Mahmud, "*Crisis and Need: Information and Communication Technology in Development Initiatives Runs through a Paradox*", ITU Document WSIS/PC-2/CONTR/17-E, World Summit on Information Society, In: International Telecommunication Union (ITU), Geneva, 2003.

3   P. K. Singh, R. Sarkar, M. Nasipuri: "*Offline Script Identification from Multilingual Indic-script Documents: A state-of-the-art*", In: Computer Science Review (Elsevier), vol. 15-16, pp. 1-28, 2015.

4   M. Hangarge, B. V. Dhandra, "Offline Handwritten Script Identification in Document Images", In: International Journal of Computer Applications, vol. 4, no. 6, pp. 6-10, 2010.

5   G. G. Rajput, Anita H. B., "*Handwritten Script Recognition using DCT and Wavelet Features at Block Level*", In: IJCA Special Issue on "Recent Trends in Image Processing and Pattern Recognition", pp. 158-163, 2010.

6   G. D. Joshi, S. Garg, J. Sivaswamy, "*Script Identification from Indian documents*", In: DAS, LNCS 3872, pp. 255-267, 2006.

7   M. B. Vijayalaxmi, B. V. Dhandra, "*Script Recognition using GLCM and DWT Features*", In: International Journal of Advanced Research in Computer and Communication Engineering, vol. 4, issue 1, pp. 256-260, 2015.

8   M. C. Padma, P. A. Vijaya, "*Entropy Based Texture Features Useful for Automatic Script Identification*", In: International Journal on Computer Science and Engineering, vol. 2, no. 2, pp. 115-120, 2010.

9   Sk. Md. Obaidullah, C. Halder, N. Das, K. Roy, "*Indic Script Identification from Handwritten Document Images-An Unconstrained Block-level Approach*", In: Proc. of 2nd IEEE International Conference on Recent Trends in Information Systems, pp. 213-218, 2015.

10 T. N. Tan, "*Rotation Invariant Texture Features and their use in Automatic Script Identification*", In: IEEE Transactions on Pattern Analysis and Machine Intelligence, vol. 20, no. 7, pp. 751-756, 1998.

11 J. G. Daugman, "*Uncertainty relation for resolution in space, spatial-frequency, and orientation optimized by two-dimensional visual cortical filters*", In: J. Optical Soc. Amer., 2(7), pp. 1160-1169, 1985.

12 T. S. Lee, *"Image representation using 2D Gabor wavelets"*, In: IEEE Transactions on Pattern Analysis and Machine Intelligence, vol. 18, no. 10, 1996.

13 R. C. Gonzalez, R. E. Woods, *"Digital Image Processing"*, vol. I. Prentice-Hall, India (1992).

14 N. Ostu *"A thresholding selection method from gray-level histogram"*, In: IEEE Transactions on Systems Man Cybernetics, SMC-8, pp. 62-66, 1978.

Preeja Babu [1] and Dhanya Sethumadhavan [2]

Ligand based Virtual Screening using Graph Wavelet Alignment Kernel based on Boiling Point and Electronegativity

Abstract: Cost of drug discovery is mainly due to the failure rate in the clinical testing. The failure rate can be reduced by the proper utilization of data generated during the previous drug discovery processes. Amount of medical data is increasing drastically and efficient machine learning helps to extract more useful information for reducing the failures in the later stages of the drug discovery. Extraction of information from the data requires efficient representation and mining algorithm. Molecular graphs are very expressive which allows productive implementation of the machine-learning algorithms. In the present work, molecular graphs are used for the efficient representation of chemical data, and wavelets are used to capture local topology as the feature for molecular graph comparison to predict the activity. The model is tested for two classification problems to predict mutagenicity and toxicity on two publically-available datasets.

Keywords: Drug discovery, virtual screening, SVM, graph kernel, wavelet

1 Introduction

Drug discovery is a complex multi-step process for finding a new chemical compound with desired chemical and biological properties. Two types of screenings are involved in the drug discovery process – in silico/virtual screening (VS), followed by experimental in vivo /high-throughput screening (HTS). In computer aided drug design, in silico screening like virtual screening, is used to find the promising candidates for drug discovery from large compound database. In silico method is faster, flexible, low cost and less time-consuming than in vivo method [1].

1 NMIMS University
Preeja.Babu@nmims.edu
2 Amrita University
dhanyasethumadhavan@gmail.com

In the HTS, large number of compounds from various sources like in-house compound databases, commercial vendors, combinatorial libraries, natural product libraries, etc. - are quickly assayed using automated platforms; but the cost is very high. The VS can complement the HTS by reducing the number of compounds to be screened, with increased probability of success. The cost-factor for VS is substantially low [1].

Virtual screening methods are classified into two, ligand-based and structure-based methods. In ligand-based strategy, the existing information about the ligand molecules is used for virtual screening compounds from the chemical database or from the set of newly designed molecule. For structure-based strategy, the information about both target and ligand is used for the 'design cycle'. It includes mainly the binding affinity between ligand and the target. Ligand-based VS can be further classified into empirical filters-based, similarity-based and quantitative-structure activity relationship (QSAR)-based methods.

In QSAR-based method, the accuracy of screening depends on the way of representation of the molecules (molecular descriptors) and the metrics used to find the similarity or dissimilarity between molecules. Machine-learning methods are used for the development of accurate QSAR models. Recent research shows that kernel methods such as support vector machines (SVM) give more accurate QSAR models for non-linear models. SVM gives very good result for both predicting the accuracy of complex as well as simple relation between the activity and structure of a molecule. In VS, the relation between molecular structures and activity is highly non-linear. In case of SVM, accuracy on the decision of classifier is based on the accuracy on similarity measure which in turn, depends on the representation as well as the similarity calculation. Various types of representations like One-dimensional (1D), 2D and 3D representations of molecules (with increasing accuracy and computational complexity) can be used for virtual screening [2-6]. Higher dimensions are more expressive than lower dimensions but the computational cost is also high for 3D representation [7-12]. The 2D representation such as molecular graph is simple and expressive due to which it is well suited for the faster and is an accurate method for finding the similarity between molecules [3].

Structural data can be represented as a feature vector with numerical entries or with local structures. Representation of structured data using numerical values require huge amount of pre-processing. Pre-processing is required to find the corresponding properties of each molecule from the structure analysis. In graph kernels, substructures like random walk, subtree, shortest path etc. are compared to find similar molecular graphs. The substructure similar graphs should be more or less similar. Based on the type of substructures used for the comparisons, graph kernels are classified into Walk kernel, Wavelet kernel, Tree kernel, graphlet kernel etc.

Graph kernel between two graphs G_1 and G_2 is defined as

$$k(G_1, G_2) = \sum_{i \in substr_1} \sum_{j \in substr_2} k(substr_i, substr_j) \tag{1}$$

where $substr_1$ and $substr_2$ are the set of substructures of graph G_1 and G_2

Recent studies show that wavelets can be used for the multi-resolution analysis of graphs [7-11]. Wavelet functions can be used for the feature extraction for the graph kernel computation. Smalter et al., [9] proposed wavelet alignment kernel for virtual screening in drug discovery.

In the present work, boiling point and electronegativity based Graph Wavelet Alignment Kernel was used for ligand-based VS. The models were developed and then the implementation of the model and the performance of the algorithm on several different benchmark datasets were evaluated.

2 Ligand-based Virtual Screening

In this work, two types of Graph Wavelet Alignment Kernels for ligand-based VS screening were implemented. In the first type, electronegativity of atoms was used for the classification of compounds into two sets, active and inactive. In the second type, atoms are labelled using boiling points. This was based on the QSAR modeling. The accuracies of two types are compared and found that the electronegativity of atoms is more efficient compared to that of the boiling point for increasing the prediction accuracy for classifying the compounds into active and inactive.

3 QSAR-based Virtual Screening

Normally machine learning methods are used to create models for QSAR. In machine learning artificial neural learning (ANN), decision tree and support vector machines are mainly used. In which SVM plays main role because of its ability to handle nonlinear QSAR problem. In machine learning based QSAR can use different types of representation like 1D descriptors, molecular graphs and 3D pharmacophore. In this work, molecular graphs are used and wavelet is used for extracting features from the molecular graphs.

4 Graph Wavelet Alignment Kernel for Classification

In wavelet based kernels, feature extraction can be done using wavelet analysis and features are the representatives of local topology. Wavelet functions can have different levels of analysis. Each levels of analysis will give different levels of information. Wavelet function helps to include the information about a node and its neighbourhood into a single entry of the feature vector of a node. Each entry in the feature vector of a node represents different level of information. These feature vectors of nodes can be used to compute the alignment of the molecular graph. Wavelet functions are used to represent functions and signals as its constituents. Wavelet decomposing and representation is possible for mathematical functions, matrices and images. A lot of research work is going on for the representation and analysis of graph using wavelet function. The advantage of this method over others is the representation of various scales and resolutions at a reasonable computational complexity. Different wavelet functions are available for the multi resolution analysis of graph [8-10].

In this method, different properties of nodes and its neighbouring nodes are used for kernel computation. Nodes and edges can be labelled using boiling point and electronegativity of atoms. All these features give different types of information for finding the structural similarity. Using wavelet functions the functions and signals can be decomposed into its constituent parts. The same method is applicable for the multi-resolution analysis of graphs. Using wavelet analysis, graphs can be represented as topologies between its data elements [7-9]. This method works well with finite and infinite dimensional feature space. In this work, infinite dimensional feature space is considered. Information about the local and non-local environment was considered as features for the comparison. After finding these features for each node, the next step was the alignment of one graph over the other. In this step, all possible alignment between graphs had to be considered and the kernel value was based on the optimal alignment where the similarity was maximized. This kind of comparison is not only based on the similarity between nodes but also similarity between its neighbours is included in this kernel computation. This kernel computation is like subgraph kernel computation. Classification is done in four steps. Step 1 deal with the extraction of features using various atomic properties. Best possible alignment between graphs is found in step 2. In step 3, the similarity between graphs is calculated for the best possible alignment. Best possible alignment between graph G_1 and graph G_2 is the alignment in which the similarity is maximum. Classification of active and inactive compounds is done in step 4. Classification is done using the kernel matrix computed in step 3.

4.1 Feature Extraction

Smalter et al., [9] introduced $h - hop$ neighborhood and discrete wavelet function for feature extraction. Using $h - hop$ distance, authors introduced wavelet function to discrete structures. Using these two concepts they extracted information about the local region and constructed features using this local region information. The $h- hop$ distance helps to project graphs in high dimensional space with arbitrary topology to Euclidian space where wavelet operation is possible. The $h - hop$ gives the distance metric based on the shortest distance between the vertices. Discrete wavelet function operates on the $h - hop$ euclidian space.

Definition 1: h-hop neighbourhood of node b in graph G' is the set of nodes that are exactly h-hop distance away from node b. It is denoted as $N_h(b)$ [9].

$f(b)$ is the feature vector associated with node b and $f_i(b)$ can be calculated from different properties like boiling point and electronegativity of atoms.

$$\overline{f_i}(b) = \frac{1}{|N_i(v)|} \sum_{u \in N_i(b)} f_u \qquad (2)$$

where f_u is the property of the atom and $\overline{f_i}(b)$ is the average value of the property of the atom b. $\overline{f_i}(b)$ is used to represent the local information of the atom.

Wavelet can be used to extract nodes and its neighbourhood information. Continuous wavelet function cannot be applied on discrete structures. Wavelet analysis for discrete structures like graph requires discrete wavelet function. Smalter et al., [1] used wavelet measurement for finding the similarity between molecular graphs. The wavelet measurement for vertex v for h-hop distance is defined as

$$\Gamma_h(v) = C_{h,v} * \sum_{j=0}^{h} \psi_{j,h} * \overline{f_j}(v) \qquad (3)$$

where $C_{h,v}$ is the normalization factor and it is defined as

$$C_{h,v} = \left(\sum_{j=0}^{h} \frac{\psi_{j,h}^2}{|N_j(v)|} \right)^{-1/2} \qquad (4)$$

Wavelet measurement vector for node v is defined as

$$\Gamma^h(v) = \{\, \Gamma_0(v), \Gamma_1(v) \ldots \ldots \Gamma_h(v) \,\} \qquad (5)$$

Each entry represents different levels of analysis.

4.2 Graph Alignment and Kernel Computation

In graph alignment kernel, smaller graph is aligned on the larger graph in such a way that the total similarity is maximized. Finding the similarity between vertices and edges of graphs is an NP-Hard problem.

$$K_A(G, G') = \max_\pi \left[\sum_{v\,\in\,V_{[G]}} k_n(v, \pi(v)) + \sum_{u,v} k_n((u,v),(\pi(u),\pi(v))) \right] \qquad (6)$$

Graph alignment kernel requires the computation of similarities between all possible alignments between graphs G and G'. Smalter et al., [9] simplified the kernel in Equation (6). The simplified version contains the similarity between vertices only. Based on vertex similarity, authors proposed graph alignment kernel between graphs G and G' as

$$K_A(G, G') = \max_\pi \sum_{v\in V(G_1)} k_a(v_1, \pi(v_1')) \qquad (7)$$

where $\pi: V(G) \to V(G')$ is the alignment of vertices in graph G on the vertices in graph G'. In Equation (7) the kernel computation is based on the similarity between vertices of graph G and G'. Similarity measurement is based on the vertex labels and its properties.

The wavelet function is used to include the neighbourhood information of a node into a single feature of that node. Different levels of neighbourhood information forms the feature vector of a node. Based on these features Smalter et al., [9] modified the graph alignment kernel as

$$K_M(G, G') = \max_\pi \sum_{v\in V(G_1)} k_m\left(f(v_1), f(\pi(v_1')) \right) \qquad (8)$$

Using wavelet measurement vector, Smalter et al., (2009) defined wavelet alignment kernel as

$$K_\Gamma(G, G') = \max_\pi \sum_{v\in V(G_1)} k_l\left(\Gamma(v_1), \Gamma^h(\pi(v_1')) \right) \qquad (9)$$

where k_l is the linear or RBF kernel.

The computational complexity is less compared to other graph kernel as wavelet has been used for the feature extraction.

5 Ligand based Virtual screening using Graph Wavelet Alignment Kernel based on boiling point and electronegativity

In this method, two types of ligand based virtual screening are implemented. One is based on boiling point and other is based on electronegativity. In this method, graph wavelet alignment kernel is used for finding the similarity between molecular graphs based on their molecular properties. The kernel introduced by Smalter et al., [9] is used for this purpose. Using this kernel, the compounds are classified into two sets as active and inactive compounds. This is based on the QSAR modelling.

5.1 Graph Wavelet Alignment Kernel based on boiling point and electronegativity

This section describes the implementation of graph wavelet alignment kernel based on different properties and comparison of their accuracies. Different notions of similarity based on the properties were measured using kernel introduced by Smalter et al., [9]. Computational complexity of these graph wavelet alignment kernels are very less compared to other graph kernels.

Based on the electronegativity, similarity between molecular graphs G_1 and G_2 is calculated using

$$K_{wavelet\_EN}(G, G') = K_{\Gamma}(G, G') = \max_{\pi} \sum_{v \in V(G_1)} k_l \left(\Gamma(v_1), \Gamma^h(\pi(v_1')) \right) \quad (10)$$

where $\Gamma(v_1)$ and $\Gamma(v_1')$ is the wavelet measurement vector of vertex v_1 in graph G_1 and v_1' in graph G_2 respectively.

Normalization of the kernel in Equation (10) is done by using

$$k'_{wavelet\_EN}(G_1, G_2) = \frac{k'_{wavelet\_EN}(G_1, G_2)}{\sqrt{k'_{wavelet\_EN}(G_1, G_1) * k'_{wavelet\_EN}(G_1, G_1)}} \quad (11)$$

The decision of classifier depends on the value of the decision function

$$f(x) = sign(w^T x - \gamma) = sign(\sum_{i=1}^{m} u_i d_i k'_{wavelet\_EN}(G_i, G) - \gamma) \quad (12)$$

$$= sign(\sum_{i \in svindex} u_i d_i k'_{wavelet\_EN}(G_i, G) - \gamma)$$

where the value of $k'(G_i, G)$ was obtained from Equation 11.

6 Implementation

All the codes were written in Java Release 7, NetBeans IDE 7.3.1 and experiments were run on a 1.8 GHz Intel Core™ i3 processor with 4 GB of main memory running Ubuntu 13.04. After the computation of kernel matrix, the classification code was written in MATLAB Release 15. Classification was done with the help of LIBSVM version 2.88. The practical suitability of these kernels was tested on two real-world datasets, PTC (Helma *et al.*, 2001) and MUTAG (Debnath *et al.*, 1991). In this work, classification of active and inactive molecules was done by graph wavelet alignment kernel based QSAR model.

6.1 Implementation steps

Step 1: Feature Extraction
   a.   Creating an adjacency matrix using the bond information between atoms in a chemical compound. Bond information can be extracted from the molfile.
   b.   Calculation of different $h - hop$ neighbourhoods of each atom
   c.   Calculation of average properties of each atom at different levels. Properties at different $h - hop$ neighbourhoods are the elements of feature vector.
   d.   Applying Wavelet Analysis to the Graph Structure: After finding the h-hop neighbourhood, wavelet function was applied on the h-hop neighbourhood. In this work, Haar wavelet function was used. The support of this Haar wavelet function was [0, 1] with integral 0 and partitioned the function into h + 1 intervals. For each node of a graph, i.e. the wavelet measurement vector for each atom in molecular graph can be found out using this formula

$$\Gamma_h(v) = C_{h,v} * \sum_{j=0}^{h} \psi_{j,h} * \overline{f_j}(v)$$

$\Gamma_h(v)$ is the wavelet measurement vector. $C_{h,v}$ is the normalization factor

$$C_{h,v} = \left(\sum_{j=0}^{h} \frac{\psi_{j,h}^2}{|N_j(v)|} \right)^{-1/2}$$

where $\psi_{j,h} = \frac{1}{h+1} \int_{j/(h+1)}^{(j+1)/(h+1)} \psi(x) dx$

   e.   Finding out Feature Values Based on Physicochemical Properties

Instead of using neighbourhood information, physiochemical properties of atoms can also be included in the feature vector of atoms. Features such as electronegativity and boiling point are used.

Step 2: Graph Alignment

    a.   Find the size of graph G_1 and G_2.

    b.   Align the smaller graph on the larger graph.

Step 3: Kernel Computation- Graph Wavelet Alignment Kernel

    a.   Finding the similarity between two Graphs: Next step is finding out the similarity between the two graphs for all possible alignments and to choose the alignment which gives the maximum similarity.

$$
k_A(x, y) = \begin{cases} max_\pi \sum_{i=1}^{|x|} k_1(x_i, y_{\pi(i)}) & if\, |y| \geq x| \\ max_\pi \sum_{i-1}^{|y|} k_1(x_{\pi(j)}, y_j) & otherwise \end{cases}
$$

Similarity between feature vectors of nodes in graph G and G' can be found out using linear kernel or RBF kernel.

Step 4: Classification of active and inactive molecules using LIBSVM

    Training phase

    Testing phase

7 Classification Accuracy

This section illustrates the performance of graph wavelet alignment kernel for finding the similarity between vertexes labelled molecular graphs. Accuracy of the method was tested with PTC and MUTAG data sets and it is calculated using 10 fold cross validation. Different values of C were tried, in which C=0.01 gave more accurate result. Performance evaluation was done by calculating accuracy, specificity and sensitivity. Result of the proposed method is given in Table 1 and Table 2.

The Table 1 and Table 2 shows the test result of screening compounds

The accuracies of graph wavelet alignment kernel can be increased using more effective filters and the use of more relevant properties for feature extraction. The accuracy of model based on electronegativity is more compared to boiling point. Its shows activity is more dependent on electronegativity than boiling point.

Table 1: Test Results of Graph Wavelet Alignment Linear Kernel (based on Electronegativity)

	FM	FR	MM	MR	MUTAG
True Positive	65	57	70	75	104
True Negative	138	138	135	133	33
False Positive	68	92	72	59	30
False Negative	78	64	59	77	21
Specificity	67	60	65.3	69.3	52.4
Sensitivity	45.5	47.2	54.3	49.4	83.2
Accuracy	58.2	55.6	61.1	60.5	72.9

Table 2: Test Result of Graph wavelet alignment linear kernel (based on boiling point)

	FM	FR	MM	MR	MUTAG
True Positive	61	54	62	65	96
True Negative	136	135	131	128	30
False Positive	70	95	76	64	33
False Negative	82	67	67	87	29
Specificity	66.1	58.7	63.3	66.7	47.7
Sensitivity	42.7	44.7	48.1	42.8	76.8
Accuracy	56.5	53.9	57.5	56.2	67.1

Conclusions

In this paper, we have described the implementation and working of graph wavelet alignment kernel based on electronegativity and boiling point. Among these, electronegativity based wavelet alignment kernel gave more accurate results. We have tested the accuracy of proposed kernel using PTC dataset as well as MUTAG dataset. The predictive performance was measured using parameters like accuracy, sensitivity and specificity. Compared to other graph kernel the accuracy of graph wavelet alignment kernel is not that much good but this can be used as a relevant basic kernel for multiple kernel learning (MKL) because the accuracy of MKL depends on the number of relevant basic kernels.

The combination of different property based wavelet alignment kernels can also be used for the prediction of drug-likeness of a molecule.

Acknowledgment

We wished to thank Dr. K. P. Soman , Professor and Head ,CEN, Amrita Vishwavidyapeetham.

References

1   K P Soman, Preeja Ravishankar Babu, Hemant Palivela, Prashant S Kharkar 2015, LIGAND Based Virtual screening using random walk kernel and empirical filters

2   Kashima, Hisashi, and Akihiro Inokuchi. "Kernels for graph classification." In ICDM Workshop on Active Mining, vol. 2002. 2002.

3   Kashima, Hisashi, Koji Tsuda, and Akihiro Inokuchi. "Marginalized kernels between labeled graphs." In ICML, vol. 3, pp. 321-328. 2003.

4   Gärtner, Thomas, Peter Flach, and Stefan Wrobel. "On graph kernels: Hardness results and efficient alternatives." In Learning Theory and Kernel Machines, pp. 129-143. Springer Berlin Heidelberg, 2003.

5   Borgwardt, Karsten M., Nicol N. Schraudolph, and Svn Vishwanathan. "Fast computation of graph kernels." In Advances in neural information processing systems, pp. 1449-1456. 2006.

6   Shervashidze, Nino, Tobias Petri, Kurt Mehlhorn, Karsten M. Borgwardt, and Svn Vishwanathan. "Efficient graphlet kernels for large graph comparison." In International conference on artificial intelligence and statistics, pp. 488-495. 2009.

7   Crovella M, Kolaczyk E.(2003) Graph wavelets for spatial traffic analysis. *Infocom.* ;3:1848–1857.

8   Maggioni, M., Bremer, J.C., Coifman, R. R., & Szlam, A. D. (2005, August). Biorthogonal diffusion wavelets for multiscale representations on manifolds and graphs. In *Optics & Photonics 2005* (pp. 59141M-59141M). International Society for Optics and Photonics.

9   Smalter, A., Huan, J., & Lushington, G. (2009). Graph wavelet alignment kernels for drug virtual screening. *Journal of Bioinformatics and computational biology*, 7(03), 473-497.

10 Bai, L., Rossi, L., Zhang, Z., & Hancock, E. (2015). An Aligned Subtree Kernel for Weighted Graphs. In *Proceedings of the 32nd International Conference on Machine Learning (ICML-15)* (pp. 30-39)

11 Yanardag, P., & Vishwanathan, S. V. N. (2015, August). Deep graph kernels. In *Proceedings of the 21th ACM SIGKDD International Conference on Knowledge Discovery and Data Mining* (pp. 1365-1374). ACM.

12 Schölkopf, B., Smola, A., & Müller, K. R. (1998). Nonlinear component analysis as a kernel eigenvalue problem. *Neural computation*, 10(5), 1299-1319.

Pawan Kumar Singh[1], Shubham Sinha[2],
Sagnik Pal Chowdhury[3], Ram Sarkar[4] and Mita Nasipuri[5]

Word Segmentation from Unconstrained Handwritten Bangla Document Images using Distance Transform

Abstract: Segmentation of handwritten document images into text lines and words is one of the most significant and challenging tasks in the development of a complete Optical Character Recognition (OCR) system. This paper addresses the automatic segmentation of text words directly from unconstrained *Bangla* handwritten document images. The popular Distance transform (DT) algorithm is applied for locating the outer boundary of the word images. This technique is free from generating the over-segmented words. A simple post-processing procedure is applied to isolate the under-segmented word images, if any. The proposed technique is tested on 50 random images taken from *CMATERdb*1.1.1 database. Satisfactory result is achieved with a segmentation accuracy of 91.88% which confirms the robustness of the proposed methodology.

Keywords: Word Segmentation, Handwritten documents, *Bangla* Script, Distance Transform, Optical Character Recognition, *CMATERdb*1.1.1

1 Introduction

OCR system refers to a process of generating machine editable text when input is given by optical means, like scanning of any text documents either in printed

---

1 Department of Computer Science and Engineering, Jadavpur University, Kolkata, India
pawansingh.ju@gmail.com
2 Department of Computer Science and Technology, Indian Institute of Engineering Science and Technology, Shibpur, Howrah-711103, India
bitan1994@gmail.com
3 Department of Computer Science and Technology, Indian Institute of Engineering Science and Technology, Shibpur, Howrah-711103, India
sagnik.pc@gmail.com
4 Department of Computer Science and Engineering, Jadavpur University, Kolkata, India
raamsarkar@gmail.com
5 Department of Computer Science and Engineering, Jadavpur University, Kolkata, India
mitanasipuri@gmail.com

or handwritten form. Segmentation is the process of extracting the objects of interest from an image. Segmentation of a document image into its basic entities, *namely*, text lines and words, is considered as a critical dilemma in the field of document image processing. It is a common methodology to segment the text-line which immediately follows the word extraction procedure. The problem becomes more challenging when handwritten documents are considered due to its intrinsic complex nature. Firstly, the writing styles in handwritten documents may be cursive or discrete. In case of discrete handwriting, characters get combined in forming the words. Secondly, unlike machine printed text, handwritten text is not uniformly spaced. Thirdly, ascenders and descenders of the consecutive word images may get frequently connected and these word images can also be present at different orientations. Finally, noise and other artifacts are commonly seen more in handwritten documents than in printed ones. Therefore, if the page segmentation technique is implemented using the traditional two-step approach, the following problems need to be taken care of. In the case of text-line segmentation, major difficulties include the multi-level skewness, presence of overlapping and touching words among the successive text-lines. In the case of word segmentation, difficulties that arise include the appearance of skew and slant at the word-level, the existence of punctuation marks along the text-line and the non-uniform spacing of words which is a common enduring in handwritten documents. So, this two-stage approach to get the words may not be always useful in handwritten recognition because the problems encountered during word segmentation will multiply due to added intricacies in the text-line extraction process. But, if the words are extracted directly from the document page, then these errors can be avoided judiciously. Apart from this, there are situations where text-line segmentation is not essential for example, postal data, medical prescription data, etc.; word segmentation serves as the best alternative for achieving the maximum precision.

From the literature survey performed in [2], it can be seen most of the works addressing on segmentation of handwritten text-line, word or character have been focused on either *Roman, Chinese, Japanese* or *Arabic* scripts. Whereas a limited amount of work described in the state-of-the-art [3] has been done for unconstrained handwritten *Bangla* documents. S. Saha *et al.* [4] proposed a Hough transform based methodology for text line as well as word segmentation from digitized images. R. Sarkar *et al.* [5] used Spiral Run Length Smearing Algorithm (SRLSA) for word segmentation. The algorithm first segmented the document page into text lines and then smears the neighboring data pixels of the connected components to get the word boundaries which merges the neighboring components and thus word components in a particular text line is extracted. J. Ryu *et al.* [6] considered word segmentation of *English* and *Bangla* texts as a labeling problem. Each gap in a line was labeled whether it was inter-word or intra-word. A normalized super pixel representation method was first presented that extracted a set of candidate gaps in each text line. The assignment problem was considered as a binary quadratic problem as a result of which pairwise relations as well as local properties could be considered. K. Mullick *et al.* [7] proposed a novel approach for text-line segmentation where the image was first blurred out, smudging the gaps between words but preserving space between the lines. Initial segmentation was obtained by shredding the image based on white most pixels in between the blurred out lines. Touching lines were separated by thinning and then by finding most probable point of separation. In [8], initially the contour of the words present in a given text line were detected and then a threshold was chosen based on Median White-Run Length (MWR) and Average White Run-Length (AWR) present in the given text line. After that the word components were extracted from the text lines based on the contour and the previously chosen threshold value. At last, these words were represented in bounded boxes. A few more works [9-13] for word extraction was already done for other *Indic* scripts like *Oriya, Devanagari, Kannada, Tamil,* and *Gurumukhi* handwritten documents. N. Tripathy *et al.* [9] proposed a water-reservoir based scheme for segmentation of handwritten *Oriya* text documents. For text-line segmentation, the document was firstly divided into vertical stripes. Stripe-wise horizontal histograms were then computed by analyzing the heights of the water reservoirs and the relationship of the peak-valley points of the histograms was used for the final segmentation. Based on vertical projection profile and structural features of *Oriya* characters, text lines were segmented into words. For character segmentation, at first, isolated and connected (touching) characters in a word were detected. Using structural, topological and water-reservoir-concept based features, touching characters of the corresponding word images were then segmented. A. S. Ramteke *et al.* in [10] proposed a method based on Vertical Projection Profile to segment the characters from word and extract the base characters by identifying the empty spaces between them. This algorithm was implemented on bank cheques and concludes 97% of efficiency for isolated characters only. H. R. Mamatha *et al.* in [11] described a segmentation scheme for segmenting handwritten *Kannada* scripts into text-lines, words and characters using morphological operations and projection profiles. The method was tested on unconstrained handwritten *Kannada* scripts and an average segmentation rate of 82.35% and 73.08% for words and characters respectively was obtained. S. Pannirselvam *et al.* proposed a segmentation algorithm [12] for segmenting handwritten *Tamil* scripts into text-lines, words and characters using Horizontal and vertical profile. The methodology gave an average segmentation rate of 99%, 98.35% and 96% for text-lines, words and characters respectively. M. Kumar *et al.* described a strip based projection profile technique [13] for segmentation of handwritten *Gurumukhi* script documents into text-lines and a white space and pitch technique for segmentation of text-lines into words. The technique achieved accuracies of 93.7% and 98.4% for text-line and word segmentations respectively. It can be observed from the literature study that for the above mentioned word extraction approaches, the text lines were first considered and then words were extracted from them. But, till date there is no work available for extraction of words directly from unconstrained document images written in *Bangla* script.

2 Proposed Work

The proposed work consists of a word segmentation algorithm which directly extracts the text words from the document images without undergoing the error-prone text-line segmentation process. The well-known Gaussian filter [15] is applied to eliminate the noise present in the handwritten document images. The outline of the word images are initially traced by applying DT on the entire document image. This transformation generates an estimation of the word boundaries which is immediately followed by gray-level thresholding. As a result of thresholding, the individual word images are finally extracted from each of the smeared black regions. A suitable measurement is also taken care of to rectify the under-segmentation errors. The details of the present work are discussed in the following subsections.

2.1 Distance Transform (DT)

There are a lot of image analysis applications which require the measurement of images, the components of images or the relationship between images. One technique that may be used in a wide variety of applications is the DT or Euclidean distance transform (EDT). The DT maps each image pixel into its smallest distance to regions of interest [14]. Let the pixels within a two-dimensional digital image $I(x, y)$ be divided into two classes – object pixels and background pixels.

$$I(x,y) \in \{0,1\} \qquad (1)$$

The DT of this image, $I_d(x,y)$ then labels each object pixel of this binary image with the distance between that pixel and the nearest background pixel. Mathematically,

$$I_d(x,y) = \begin{cases} 0 & I(x,y) \in \{0\} \\ min(\|x-x_0, y-y_0\|, \forall I(x_0,y_0) \in 0) & I(x,y) \in \{1\} \end{cases} \qquad (2)$$

where, $\|x,y\|$ is some two-dimensional distance metric. Different distance metrics result in different distance transformations. From a measurement perspective, the Euclidean distance is the most useful because it corresponds to the way objects are measured in the real world. The Euclidean distance metric uses the L_2 norm and is defined as:

$$\|x,y\|_{L_2} = \sqrt{x^2 + y^2} \qquad (3)$$

This metric is isotropic in that distances measured are independent of image orientation, subject of course to the limitation that the image boundary is digital, and therefore in discrete locations. Fig. 1 shows an example of DT of a binary image. For each pixel in Fig. 1(a), the corresponding pixel in the DT of Fig. 1(b) holds the smallest Euclidean distance between this pixel and all the other black pixels. Illustration of sample handwritten *Bangla* document image and its corresponding EDT are shown in Fig. 2.

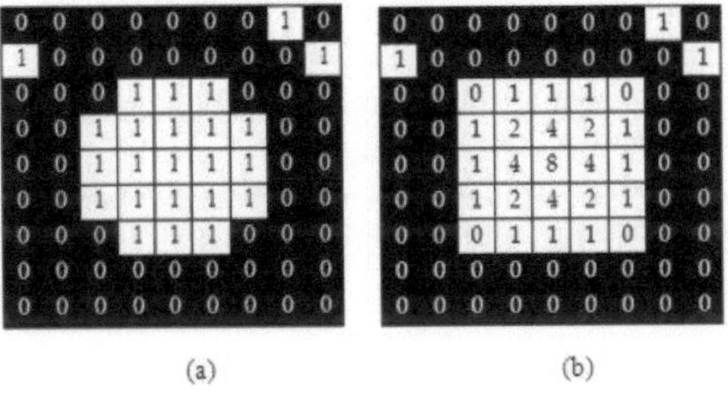

(a)   (b)

Figure 1. Numerical example of EDT. Here, (b) Euclidean distance of each pixel to the nearest black pixel of the input binary image shown in (a). (The distance values are squared so that only integer values are stored)

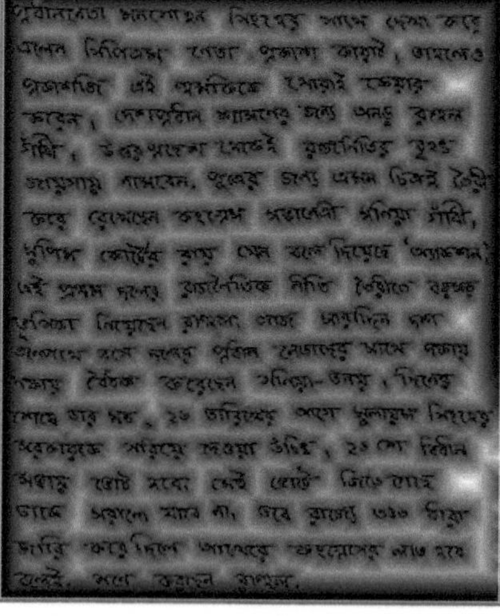

(a)   (b)

Figure 2. Illustration of: (a) sample handwritten *Bangla* document image, and (b) corresponding EDT of (a)

2.2 Word Extraction

After performing EDT on the original document image, gray-level thresholding is done to binarize the document images. The value of the threshold α is a user-defined parameter which heavily depends on the handwriting styles and quality of the document image. If the document image incorporates densely packed writing then, the threshold is to be decreased and vice versa. For the present work, the value of α is set after rigorously testing a variety of document images and the optimal value is found to vary between 120 and 180 on a scale of 0-255 gray-level intensities. This thresholding of the document image generates a set of smeared black regions, shown in Fig. 3, which wraps over each of the possible word images. Each smeared black region represents a single group of individual connected components indicating the word image. Now, the word images are formed by applying the connected component labelling (CCL) algorithm [15] in each of these smeared regions.

It is observed that the application of the EDT on the document image yields a non-realistic single connected component along the border region. This single large component includes all the word images which are close to the boundary of the document image. To overcome this, the largest connected component is

Figure 3. Sample handwritten *Bangla* document image showing a smeared black region wrapped over each of the possible word images

traversed throughout to locate all the horizontal valleys (present along the vertical outline of the component) and vertical valleys (present along the horizontal outline of the component). These valleys serve as slicing lines to get different words lying along the boundary.

2.3 Removal of Error cases

In general, there may occur two major errors in any word segmentation method: Over-segmentation and Under-segmentation of the words. If a single word component is erroneously broken down into two/more parts, then it is considered as over-segmentation error. Whereas if two/more words are recognized as a single word, it is considered as under-segmentation error. The present work is free from the over-segmentation errors. This is because of the dynamically selected threshold while binarizing each document image. So, this is one of the key advantages of our technique. But the technique generates some under-segmentation errors. Fig. 4 shows one such situation of under-segmentation for a portion of *Bangla* handwritten document image. Though, under-segmentation errors are found, but the way EDT produces the shape of the smeared regions, it makes the handling of the under-segmentation errors, a bit easier.

The under-segmentation errors are mainly seen of two types: (i) joining of two or more words on the same text-line (i.e., horizontally joined), and (b) joining of two or more words on two successive text-lines (i.e., vertically joined). For the first case, a vertical valley is observed in the word image and this valley is used to separate the joined words. For the second case, the word images protrude to form ascenders and descenders which in turn complicate the situation as the probability of finding an exactly horizontal valley decreases to a large extent. To overcome this, the minimum distance in the horizontal plane is considered and their corresponding boundary points are noted. From the line joining these two points, a user-defined parameter called β, is taken on both sides, which determines the area of the square through its trajectory. Every two pair of points inside this rectangle is measured and a perpendicular line is drawn from those two points traversing through the boundary of the square. This perpendicular line defines the slicing line between the two vertically under-segmented words. The value of β is determined as 0.2 times the word image height.

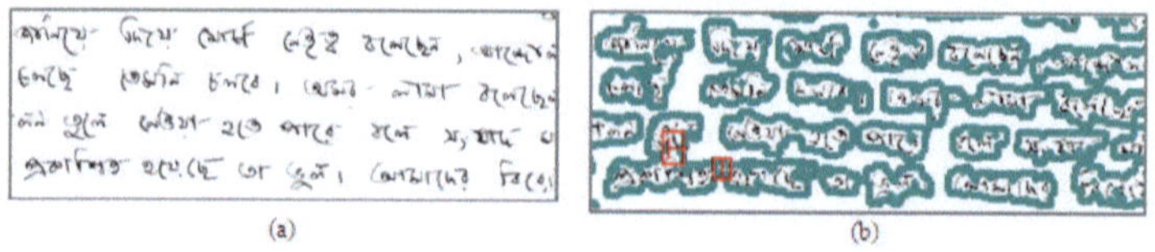

(a)   (b)

Figure 4. Illustration of: (a) a scrap of *Bangla* handwritten document image, and (b) under-segmentation error causing the amalgamation of two text words into a single one

The algorithm of the proposed word segmentation technique is as follows:

Step 1:	Read the original document image as RGB image and convert it to gray-scale image.
Step 2:	Apply Gaussian filter to remove the noise.
Step 3:	Perform EDT to the document image.
Step 4:	Select a threshold value α for each distance transformed images for performing gray-level thresholding.

Step 5:	Locate the boundary points for each binarized word images by examining the 8-connectivity neighborhood of each pixel and then store these boundary points in an array.
Step 6:	Apply CCL algorithm to label the connected components in the binarized image which, in this case, are the individual word images.
Step 7:	Accumulate the pixels that are present in each label into a new array to form an individual word image.
Step 8:	Perform a suitable post-processing technique to separate the under-segmented word images.

3 Experimental Results and Analysis

For the experimental evaluation of the present methodology, a set of 50 document pages are randomly selected from *CMATERdb*1.1.1 handwritten database [16]. *CMATERdb*1.1.1 contains 100 document images written entirely in *Bangla* script. For manual evaluation of the accuracy of word extraction technique, we have considered the errors produced due to under- and over-segmented word images. In both the cases, such extracted words are also treated as wrongly extracted words. The performance evaluation of the present technique is shown in Table 1. Fig. 5 illustrates the present word extraction technique on a sample document page.

Table 1. Performance evaluation of the present word extraction technique on *CMATERdb*1.1.1 handwritten database

Database	*CMATERdb*1.1.1
Number of document pages	50
Actual number of words present (T)	7503
Number of words extracted experimentally	6894
Number of over-segmented words (O)	0
Number of under-segmented words (U)	609
Success rate $[(T - (O + U) * 100)/T]$	**91.88%**

Figure 5. Illustration of successful word segmentation technique on a sample document page written in *Bangla* script

Conclusion

In the present work, we have designed a direct page-to-word segmentation algorithm for unconstrained *Bangla* handwritten document images. A general tendency of the researchers in this domain is to first identify the text-lines and then the extracted text-lines are fed to the word extraction module. Considering the typical complexities of the unconstrained handwritings, we have decided to avoid the text-line extraction module which otherwise could have been genera- ted unwanted errors, cumulative in nature; in due course which would lessen the segmentation accuracy of the word extraction module. Experimental result on the *CMATERdb*1.1.1 handwriting database has shown that the proposed technique yields reasonably satisfactory performance comparable to the state-of-the-art techniques. Along with that, this technique avoids the removal of skew present in the text-lines to some extent as we have extracted the words directly from the document images. Though the results are encouraging, still the proposed technique suffers from under-segmentation issues even if we have tried to rectify these errors to some extent. More appropriate post-processing modules need to be developed to cope up with these situations. Future scope of the work will be to deal with the connected component which is smeared across the border in a better way. Finally, we could say that with minor modifications, this technique could be successfully applied to other *Indic* scripts documents too.

Acknowledgment

The authors are thankful to the Center for Microprocessor Application for Training Education and Research (*CMATER*) and Project on Storage Retrieval and Understanding of Video for Multimedia (SRUVM) of Computer Science and Engineering Department, Jadavpur University, for providing infrastructure facilities during progress of the work. The current work, reported here, has been partially funded by University with Potential for Excellence (UPE), Phase-II, UGC, Government of India.

References

1   R. Sarkar, S. Malakar, N. Das, S. Basu, M. Kundu, M. Nasipuri, "*Word Extraction and Character Segmentation from Text Lines of Unconstrained Handwritten Bangla Document Images*", In: Journal of Intelligent Systems, vol. 2, Issue 3, pp. 227-260, 2011.
2   R. G. Casey, E. Lecolinet, "*A survey of methods and strategies in character segmentation*", In: IEEE Transactions on Pattern Analysis and Machine Intelligence, vol.18, pp. 690 - 706,1996.
3   C. J. Kumar, G. Singh, R. Rani, R. Dhir, "*Handwritten Segmentation in Bangla script: A Review of Offline Techniques*", In : International Journal of Advanced Research in Computer Science and Software Engineering, vol. 3, Issue 1, pp. 135-140, 2013.
4   S. Saha, S. Basu, M. Nasipuri, D. K. Basu, "*A Hough Transform based Technique for Text Segmentation*", In: Journal of Computing, vol. 2, Issue 2, pp. 134-141, 2010.
5   R. Sarkar, S. Moulik, N. Das, S. Basu, M. Nasipuri, D. K. Basu, "*Word extraction from unconstrained handwritten Bangla document images using Spiral Run Length Smearing Algorithm*", In: Proc. of 5th Indian International Conference on Artificial Intelligence (IICAI), pp. 71-90, 2011.

6 J. Ryu, H. Koo, N.Ik. Cho, *"Word Segmentation Method for Handwritten Documents based on Structured Learning"*, In: IEEE Signal Processing Letters, vol. 22, no. 8, pp. 1161-1165, 2015.

7 K. Mullick, S. Banerjee, U. Bhattacharya, *"An efficient line segmentation approach for handwritten Bangla document image"*, In: Proc. of 8th IEEE Advances in Pattern Recognition (ICAPR), pp. 1-6, 2015.

8 F. Kurniawan, A. R. Khan, D. Mohamad, *"Contour vs Non-Contour based Word Segmentation from Handwritten Text Lines: an experimental analysis"*, In: International Journal of Digital Content Technology and its Applications, vol. 3(2) pp. 127–131, 2009.

9 N. Tripathy, U. Pal, *"Handwriting segmentation of unconstrained Oriya text"*, In: Proc. of 9th IEEE International Workshop on Frontiers in Handwriting Recognition, pp. 306-311, 2004.

10 A. S. Ramteke, M. E. Rane, *"Offline Handwritten Devanagari Script Segmentation"*, In: International Journal of Scientific & Engineering Research, vol. 1, Issue 4, pp. 142-145, 2012.

11 H. R. Mamatha, K. Srikantamurthy, *"Morphological Operations and Projection Profiles based Segmentation of Handwritten Kannada Document"*, In: International Journal of Applied Information Systems (IJAIS), vol. 4, No. 5, pp. 13-19, 2012.

12 S. Pannirselvam, S. Ponmani, *"A Novel Hybrid Model for Tamil Handwritten Character Segmentation"*, In: International Journal of Scientific & Engineering Research, vol. 5, Issue 11, pp. 271-275, 2014.

13 M. Kumar, R. K. Sharma, M. K. Jindal, *"Segmentation of Lines and Words in Handwritten Gurumukhi Script Documents"*, In: Indian Institute of Information Technology, pp.25-28, 2010.

14 A. Rosenfeld, J. Pfaltz, *"Sequential operations in digital picture processing"*, In: Journal of ACM, 13(4), 1966.

15 R. C. Gonzalez, R. E. Woods, *"Digital Image Processing"*, vol. I. Prentice-Hall, India (1992).

16 R. Sarkar, N. Das, S. Basu, M. Kundu, M. Nasipuri, D. K. Basu, *"CMATERdb1: a database of unconstrained handwritten Bangla and Bangla–English mixed script document image"*, In: International Journal of Document Analysis and Recognition, vol. 15, pp. 71-83, 2012.

S.Sreenath Kashyap [1], Vedvyas Dwivedi [2] and Y.P.Kosta [3]

Electromagnetically Coupled Microstrip Patch Antennas with Defective Ground Structure for High Frequency Sensing Applications

Abstract: In this paper the design and analysis of the stacked patch antenna with Defective Ground Structures on substrate namely Polymide at terahertz frequency is presented. W shaped DGS structure is introduced on the ground plane of the stacked patch antenna. Electrical parameters like return loss, gain, and bandwidth are compared between stacked patch antenna - with and without DGS at resonant frequency. Our study reports significant improvement in the electrical performance of proposed antenna structure utilizing the embedded DGS in terms of gain, reflection and Bandwidth – compared to the conventional. CST microwave studio, an electromagnetic software simulation tool has been utilized to validate the electrical performance of the proposed antenna structure. Promising usage of the proposed antenna is in terahertz sensing and strategic short range communications.

Keywords: Defective Ground Structre (DGS), Terahertz Antennas, Stacked Patch Antennas, Terahertz Sensing, Terahertz Short-range Communication.

1 Introduction

Advances in wireless communication play a pivotal role in the advancement of science and technology. Decades of progress in Microstrip radiating structures is deeply appreciable in the field of wireless Communications [1]. The quest for design of microstrip based radiating structures with high bandwidth and gain enhanced operations has been reported over the past few decades [2]. The key features of microstrip patch antenna are low profile, small volume, resistant to shocks (vibrations), and ofcourse low costs – making it attractive for a

1 Marwadi Education Foundation Group of Institutions, Rajkot
kashyap.foru3@gmail.com
2 C.U. Shah University, Wadhwan City, Gujarat
3 Marwadi Education Foundation Group of Institutions, Rajkot

multiplicity of applications[3]. In terahertz antenna applications path loss is a prime factor, followed by larger bandwidth and gains – typical of terahertz's design. In the last decade, Terahertz regime has gained substantial the attention of various researching community across the globe [4][5]. The designer can bring in simultaneous improvement of Gain and bandwidth by using tuning (optimizing) techniques, for example: by varying height of substrate, or reducing the ε_r of substrate, or even utilizing different permittivity substrates depending on the application requirements [6]. At Terahertz, the permittivity of substrate has is critical in radiating structures and impact the overall performance parameters[7]. The concept of element arrays is used at Terhertz frequencies to realize multiple resonance's and bring enhancement in gain. Another best method is by placing multiple patches – stacking technique [8][9]. In this technique, a second parasitic patch is placed and strategically positioned at a height above of the active (first one) patch-element – thus the name, stacked patch antenna [10]. The DGS technique is one of the successful techniques which can be used in enhancing the performance parameters of the Antenna.

In this paper, under section-1, the analysis of the effect of stacked patch antenna with and without DGS on polymide substrate is carried out along with the variation of distance between the patches with respect to the thickness of dielectric substrate and also rotating the upper patch to 175 degrees wrt previous-below. In section-2, the geometry of the stacked patch model along with the design parameters of the proposed model on substrates namely Polymide is attended and explained. The DGS structure along with is mechanical attributes is mathematically expressed – with specs. Finally under section-3, the simulation results, using CST-Studio, for the proposed model configuration (refer figure -1) along with the comparison in terms of antenna parameters, such as: return loss(VSWR), gain, and bandwidth are specified and captured in a graph.

1.1 Antenna Configuration

In this design of the proposed antenna configuration the analysis is carried out on Polymide as a substrate whose dielectric permittivity is 3.5 and thickness 't' for stacked patch model. The lower patch and upper patch of thickness 1 μm is printed on a ground plane. The upper patch is placed at a distance 'd' above the lower patch. Instead of direct coupling, EM coupling is used for electrically thick antennas as the lower patch is directly connected to the feed line. The figure 1 shows the basic geometry of stacked patch model.

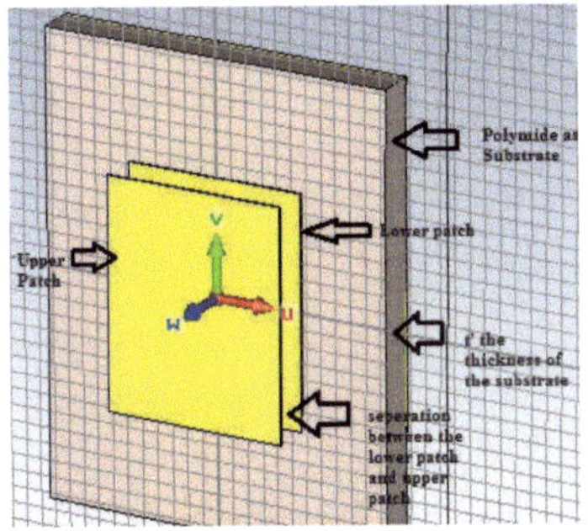

Figure 1. Stacked Patch Antenna

The width of the patch can be calculated using the equation (1).

$$W = \frac{C}{2f_0}\sqrt{\frac{\varepsilon_r + 1}{2}} \qquad (1)$$

Due to the fringing effect, the effective dielectric permittivity is to be considered. The effective dielectric permittivity can be calculated using equation (2)

$$\varepsilon_{eff} = \frac{\varepsilon_r + 1}{2} + \frac{\varepsilon_r - 1}{2}\left[\frac{1}{\sqrt{1 + 12(h/W)}}\right] \qquad (2)$$

$$L = \frac{C}{2f_0\sqrt{\varepsilon_{eff}}} - 0.824h\left(\frac{(\varepsilon_{eff} + 0.3)(\frac{W}{h} + 0.264)}{(\varepsilon_{eff} - 0.258)(\frac{W}{h} + 0.8)}\right) \qquad (3)$$

The length of the patch is 121.3 µm, width of the patch 166.6 µm are calculated using the equations 1and 3.The investigation on the proposed antenna is carried out by introducing the defective ground structure technique which enhances the performance parameters of the antenna. A "W" shaped slot is introduced on the ground plane of the proposed structure. The DGS structure is as shown in figure 2 below.

The mechanical attributes of the W shape are L = 56 µm, H = 6 µm, L1 = 56 µm, L2= 10 µm, L3=10 µm, W=9 µm, d =6 µm.

1.2 Simulation Results

The design and analysis is carried out in CST Microwave studio which uses the finet integral technique.

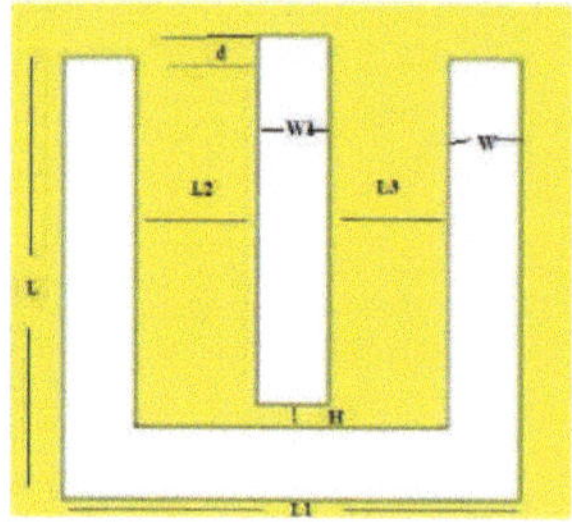

Figure 2. DGS Structure

Stacked Patch antenna without defective ground structure

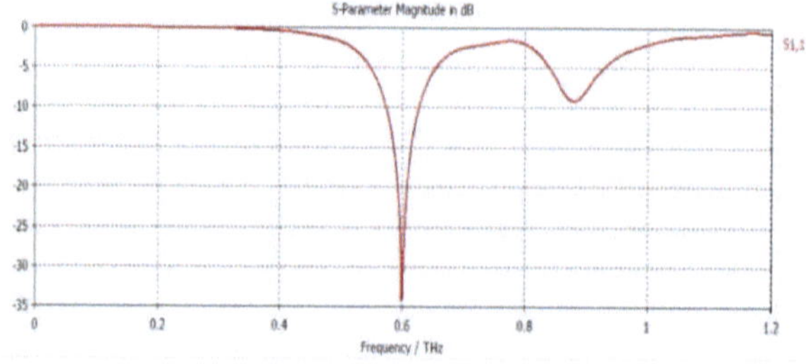

Figure 3. S11 plot of the proposed structure without DGS

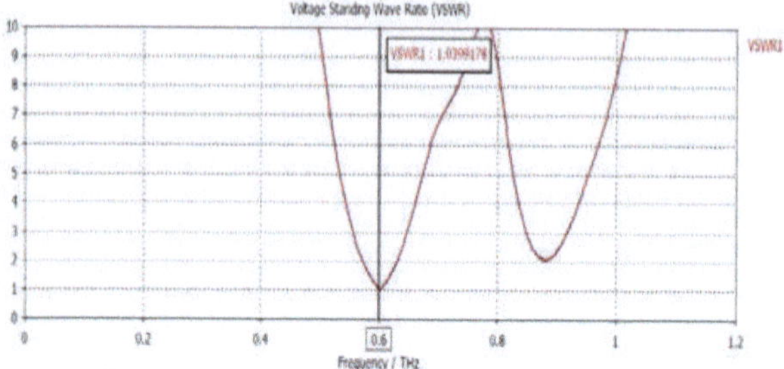

Figure 4. VSWR plot of the proposed Structure without DGS

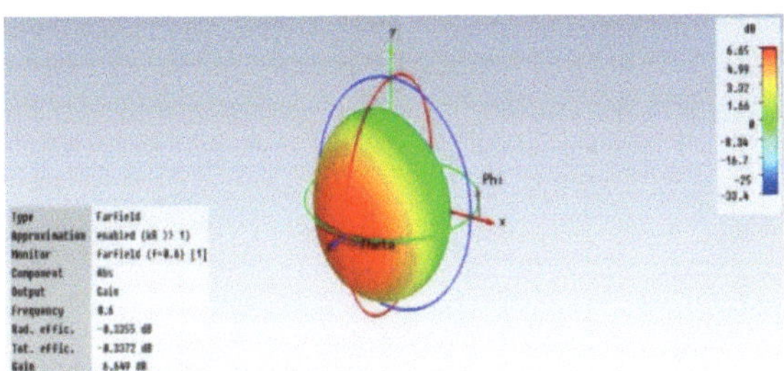

Figure 5. Gain realization plot of the proposed Structure without DGS

Stacked Patch antenna with defective ground structure

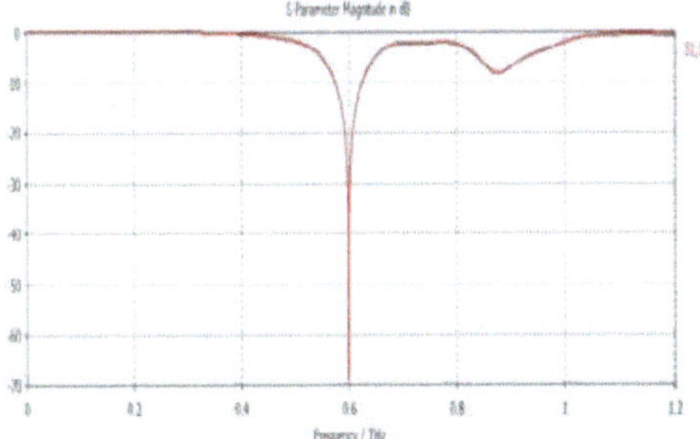

Figure 6. S11 plot of the proposed Structure with DGS

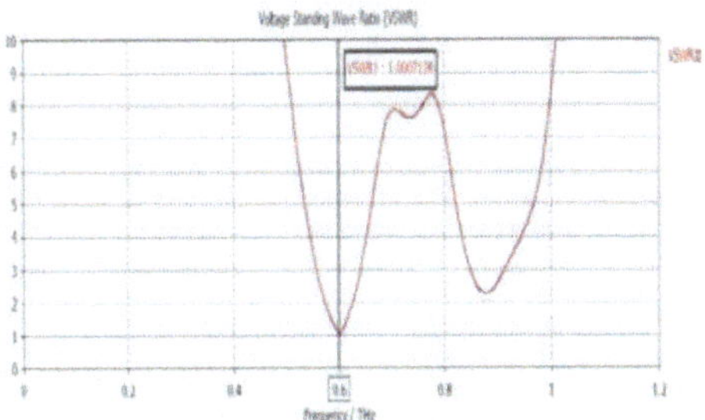

Figure 7. VSWR plot of the proposed Structure with DGS

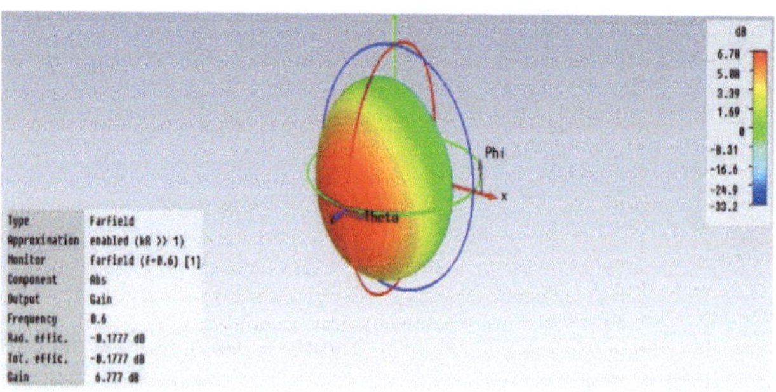

Figure 8. Gain realization of the proposed structure with DGS

The simulated results show the electrical parameters like return loss, gain
and bandwidth of the proposed models. These electrical performance
parameters are compared between stacked patch model with DGS and without
DGS.The defective ground structure model of antenna will have a large effect on
the performance parameters of the Antenna. It is observed that the model
designed using Polyimide as substrate with "W" shaped defective ground
structure is effective in enhancing the gain and with slight degradation in
bandwidth compared to without DGS. At the resonant frequency of 0.6 THz the
antenna parameters have been compared and tabled below.

Table 1. comparison table of proposed model on two different substrates for different cases

Case	Freq (THz)	S_{11} (dB)	VSWR	Gain (dB)	Bandwidth (%)
Without DGS	0.6	-34.16	1.03	6.64	9.1
With DGS	0.6	-68.96	1.00	6.77	8.5

This research study validates, through CST numerical-analysis tools for simulation that it is possible to realize antenna structures using the DGS configuration and have improvement in gain of the antenna compared to its non DGS

counterpart. Further, the overall performance of the DGS antenna can be attributed to the following;

1. Signature of the DGS structure in the ground plane
2. Finite size of the GP
3. Positing of the parasitically coupled elements
4. Angle of the positioning elements
5. Fringe-loss associated with the elements
6. Boundary conditions in terms of E_field distribution of the signature DGS structure
7. Note: it was also noticed that the radiation pattern witnessed more intense backlobes compared to conventional (although RP has not been shown in this paper).

Therefore DGS can be integrated on the ground plane in order to improve the performance parameters of the Antenna and such responses with good resonances open avenues for sensing and communication applications in Terahertz regime without any additional circuits.

Conclusion

The proposed antenna configuration demonstrates the improvement in the performance in terms of electrical and electromagnetic parameters such as return loss, Gain, Bandwidth at terahertz frequency. The effect of introducing the DGS into the ground plane of antenna model is investigated using the CST microwave studio numerical tools for electromagnetic analysis. The basic principle to model, devise, design and realize this THz antenna is its electromagnetic nature of electromagnetic aperture coupling with DGS that enhances the value of reflection, gain and bandwidth when compared with antenna without DGS. These antennas can be used for high frequency sensing and communication applications resulting into drastic reduction in antenna physical volume, weight and conductive losses. This antenna will help protecting the earth and environment from extractions from earth's crust and wastage of fuel in launching heavy bulky antennas especially used on space vehicles.

Acknowledgment

The authors wish to thank Marwadi Education Foundation and C.U.Shah
University for encouragement and support in carrying out this work.

References

1   K.R.Jha, G. Singh," Terahertz planar antennas for future wireless communications:A
Technical Review," infrared physics & Technology 60-2013.
2   C.A.balanis " Antenna Thoery Analysis and Design," Wiley student edition ,2005.
3   S .Sreenath kashyap, Vedvyas Dwivedi, " Swastika shaped microstrip patch antenna for
Terahertz applications" IEEE international conference Proceedings of ET2EN 2014.
4   Z.N.Chen, et.al " A stacked suspended plate antenna , " Microwave And optical technology
letters, vol 37, no 5, pp 337-339, 2003.
5   M. Young, The Technical Writer's Handbook. Mill Valley, CA University Science, 1989.
6   W.F.Richards, et.al, " An improved theory for microstrip antennas and applications, "IEEE
trans. Antennas propag, vol. Ap-29, no 1, pp 38-46, 1981.
7   S.Hussain et.al ," A Dual planar metamaterial based on hybrid Structure in Terahertz
regime," Progress in Electromagnetic symposium," Russia, Aug 19-23,2012.
8   Tao Chen, Suyan li, et.al, " Metamaterials applications in sensing," Sensors 2012, pp 2742-
2765.
9   S. Okhubo, Y. Onuamo et.al, " THz Wave propagation on strip devices , properties and
applications, Radio Engineering 17(2) , PP. 48-55, 2008.
10   F.Seigal, shimaburkuro, " Low loss terahertz ribbon waveguides," Appl. Opt, 44(28), pp
5937-5964, 2005.
11   Merih palandoken, Mustafa H.B. Ugar, "Compact Metamaterial –Inspired Band pass filter"
Microwave and Optical Technology Letters vol. 56, No. 12, December 2014.

Author Index

A

Aasheesh Shukla   374
Achala Deshmukh   135
Adeetya Sawant   1
Adithya H K Upadhya   204
Allam Venkata Surendra Babu   144
Amrita Roy Chowdhury   269
Annushree Bablani   415
Apurvakeerthi M   171
Arulmozhiyal R   344
Aruna Prabha B S   306
Arya G S   245
Avinash K   204

C

Chandrasekaran K   204, 330
Chitrakala S   121

D

Dhanya Sethumadhavan   462

G

Gangadhar K   157
Gaurav kumar sharma   389
Ghodekar J G   360

I

Inadyuti Dutt   280

J

Jayachandiran J   195
Jithendra K B   217, 318

K

Kashi Nath Dey   254
Katti P K   360
Kosta Y P   485
Kumaran P   27, 121

L

Leena Giri G   109

M

Madhu Sudhanan R   399
Mani Mishra   374
Manjula S H   109
Manoj V Thomas   330
Manu Sood   228
Meena P   306
Mita Nasipuri   449, 473
Mohanaradhya   73
Monica B V   99
Murali M   344

N

Naresh Kumar Yadav   57
Nath R   11
Naveen Kumar S   99
Nedumaran D   195
Neeharika K   48
Neelima N   157
Nishtha   228

O

Oormila R Varma   245

P

Parvataneni Premchand   429
Patnaik L M   109
Pawan Kumar Singh   449, 473
Piyush Shegokar   330
Poongodi P   399
Pradeepa S   144
Prakriti Trivedi   415
Pranavanand S   293
Praveen Gowda I V   109

Praveen Kumar Singh   254
Preeja Babu   462
Prema N   399
Prerna shukla   389

R
Raghuram A   293
Rajasekaran S   27
Rajkumar S   407
Ram Sarkar   449, 473
Ramesh K   186
Ramkumar Prabhu M   407
Ranga Rao J   171
Ratna Sahithi K   157
Ravi Kishan S   48
Ravi U S   99
Ravikumar H M   306
Rengarajan A   27
Rituparna Saha   269
Robert Ssali Balagadde   429
Rutooja D Gholap   84

S
Sagnik Pal Chowdhury   473
Samarjeet Borah   280
Sanjeevkumar K M   186
Sarwade A N   360
Saylee Gharge   84
Shahana T K   217, 318
Sharma B B   11
Shruti S Sharma   135
Shubham Sinha   473
Siji Rani S   245
Siva Subba Rao Patange   144
Sooryalakshmi S   245
Sreenath Kashyap S   485
Sreeparna Banerjee   269
Sri Krishna A   157
Sridevi H R   306
Sudha Gupta   1

Sujay Saha   254
Sumithra Devi K A   73
Supratim Das   449

T
Thirumalini P   344

V
Vani Hariharan   245
Vedvyas Dwivedi   485
Venugopal K R   109
Vinay kumar Deolia   374
Vinay Kumar S B   99
Vivek Sharma   11